Have you been to our website?

For code downloads, print and e-book bundles, extensive samples from all books, special deals, and our blog, please visit us at:

www.rheinwerk-computing.com

Rheinwerk Computing

The Rheinwerk Computing series offers new and established professionals comprehensive guidance to enrich their skillsets and enhance their career prospects. Our publications are written by the leading experts in their fields. Each book is detailed and hands-on to help readers develop essential, practical skills that they can apply to their daily work.

Explore more of the Rheinwerk Computing library!

Bernd Öggl, Michael Kofler
Docker: Practical Guide for Developers and DevOps Teams
2023, 491 pages, paperback and e-book
www.rheinwerk-computing.com/5650

Bernd Öggl, Michael Kofler
Git: Project Management for Developers and DevOps Teams
2023, 407 pages, paperback and e-book
www.rheinwerk-computing.com/5555

Michael Kofler
Linux: The Comprehensive Guide
2024, 1178 pages, paperback and e-book
www.rheinwerk-computing.com/5779

Michael Kofler
Scripting: Automation with Bash, PowerShell, and Python
2024, 470 pages, paperback and e-book
www.rheinwerk-computing.com/5851

Johannes Ernesti, Peter Kaiser
Python 3: The Comprehensive Guide
2022, 1036 pages, paperback and e-book
www.rheinwerk-computing.com/5566

www.rheinwerk-computing.com

Kevin Welter

Kubernetes

Practical Guide for Developers and DevOps Teams

Editor Rachel Gibson
Acquisitions Editor Hareem Shafi
German Edition Editors Dr. Christoph Meister
Translation Winema Language Services, Inc.
Copyeditor Melinda Rankin
Cover Design Graham Geary
Photo Credit Shutterstock: 1716623578/© hxdbzxy
Layout Design Vera Brauner
Production Kelly O'Callaghan
Typesetting SatzPro, Germany
Printed and bound in the United States of America, on paper from sustainable sources

ISBN 978-1-4932-2646-7

© 2024 by Rheinwerk Publishing, Inc., Boston (MA)
1st edition 2024
1st German edition published 2024 by Rheinwerk Verlag, Bonn, Germany

Library of Congress Cataloging-in-Publication Control Data
Names: Welter, Kevin, author.
Title: Kubernetes : practical guide for developers and DevOps teams / Kevin Welter.
Description: 1st edition. | Bonn ; Boston : Rheinwerk Publishing, 2024. | Includes index.
Identifiers: LCCN 2024034038 | ISBN 9781493226467 (hardcover) | ISBN 9781493226474 (ebook)
Subjects: LCSH: Virtual computer systems. | Cloud computing. | Application software--Development--Computer programs. | Kubernetes.
Classification: LCC QA76.9.V5 W45 2024 | DDC 005.4/3--dc23/eng/20240802
LC record available at https://lccn.loc.gov/2024034038

All rights reserved. Neither this publication nor any part of it may be copied or reproduced in any form or by any means or translated into another language, without the prior consent of Rheinwerk Publishing, 2 Heritage Drive, Suite 305, Quincy, MA 02171.

Rheinwerk Publishing makes no warranties or representations with respect to the content hereof and specifically disclaims any implied warranties of merchantability or fitness for any particular purpose. Rheinwerk Publishing assumes no responsibility for any errors that may appear in this publication.

"Rheinwerk Publishing", "Rheinwerk Computing", and the Rheinwerk Publishing and Rheinwerk Computing logos are registered trademarks of Rheinwerk Verlag GmbH, Bonn, Germany.

All products mentioned in this book are registered or unregistered trademarks of their respective companies.

Contents at a Glance

1	Introduction to Kubernetes	21
2	Basic Objects and Concepts in Kubernetes	95
3	Everything as Code: Tools and Principles for Kubernetes Operations	187
4	Advanced Objects and Concepts in Kubernetes	235
5	Stateful Applications and Storage	265
6	Kubernetes Governance and Security: Prepare for Production	299
7	Developing Applications for Kubernetes: Ready for Production	321
8	Orchestrating Kubernetes Using Helm	357

Contents

Preface 13

1 Introduction to Kubernetes   21

1.1 Basic Principles and Concepts: Why Use Container Clusters at All? 21
- 1.1.1 Why Use Containers at All? 23
- 1.1.2 Why You Need a Container Management Tool 27
- 1.1.3 Of Pets and Cattle 28
- 1.1.4 Stateless and Stateful Applications 29
- 1.1.5 Separation of Concerns 31

1.2 Kubernetes, the Tool of Choice 34
- 1.2.1 Why Do Companies Want to Use Kubernetes? 34
- 1.2.2 The Promise of Kubernetes 37
- 1.2.3 Major Features 41
- 1.2.4 For Which Companies Is Kubernetes Useful? 42
- 1.2.5 Which Companies Should Not Use Kubernetes? 44

1.3 Architecture and Components 45
- 1.3.1 Master Nodes 46
- 1.3.2 Worker Nodes 49
- 1.3.3 API Call Flow 51

1.4 A Kubernetes Cluster on Your Computer 53
- 1.4.1 Minikube on macOS 54
- 1.4.2 Minikube on Linux 55
- 1.4.3 Minikube on Windows 57
- 1.4.4 Launching Minikube 58
- 1.4.5 Controlling Minikube 58
- 1.4.6 Possible Errors when Starting Minikube 59
- 1.4.7 Container Registry of Minikube 59

1.5 Interaction with Kubernetes via the Command Line and Dashboard 61
- 1.5.1 Minikube Comes With kubectl 61
- 1.5.2 Installing kubectl 61
- 1.5.3 Accessing the Cluster Using Kubeconfig 65
- 1.5.4 Namespaces 67
- 1.5.5 kubectl Commands 68
- 1.5.6 Switching Clusters and Namespaces Easily 77
- 1.5.7 The Kubernetes Dashboard 78

1.6	**Lens: The IDE for Kubernetes**	81
	1.6.1 Overview of Lens	82
	1.6.2 Advantages over the Kubernetes Dashboard	83
	1.6.3 The Lens Reference	86
1.7	**The Kubernetes Cluster from Raspberry Pis**	89
	1.7.1 Choosing the Right Raspberry Pis	90
	1.7.2 Installation of Kubernetes	92
	1.7.3 Using the Kubeconfig File of the Pi Cluster	93

2 Basic Objects and Concepts in Kubernetes 95

2.1	**Pod and Container Management**	98
	2.1.1 Container Engines	101
	2.1.2 Your First Own Pod	104
	2.1.3 Multiple Containers within a Pod	106
	2.1.4 Communication between Containers	109
	2.1.5 Init Container	110
	2.1.6 Pod Phases and Container Statuses	113
	2.1.7 The Restart Policy of Pods	114
	2.1.8 When the Pod Comes to an End	115
2.2	**Annotations and Labels**	118
	2.2.1 Using Labels and Selectors	119
	2.2.2 Field Selectors	121
	2.2.3 NodeSelector	122
	2.2.4 Node Affinity and Antiaffinity	124
	2.2.5 Pod Affinity and Antiaffinity	128
	2.2.6 Taints and Tolerations	133
	2.2.7 Annotations	136
2.3	**Deployments and ReplicaSets**	138
	2.3.1 The Role of ReplicaSets	139
	2.3.2 Creating Deployments	142
	2.3.3 Rolling Updates via the Deployment Object	144
	2.3.4 Rollback via Deployment	150
2.4	**ConfigMaps and Secrets**	152
	2.4.1 What Are ConfigMaps?	154
	2.4.2 What Are Secrets?	162
2.5	**Establishing a Communication with Services and an Ingress**	171
	2.5.1 Communication between Pods	173

	2.5.2	Communication via a Service	174
	2.5.3	Communication via Ingress	180

3 Everything as Code: Tools and Principles for Kubernetes Operations — 187

3.1	Declarative Configurations		188
3.2	YAML: The Language for Kubernetes		192
	3.2.1	Basics of YAML Syntax	192
	3.2.2	Data Types in YAML	194
	3.2.3	Anchors and Aliases	196
	3.2.4	Single-Line YAML Notation in Documentation	197
	3.2.5	Weaknesses of YAML	197
	3.2.6	Tips for Practical Use	199
3.3	Version Management of Kubernetes Manifests		200
	3.3.1	Using Git	201
	3.3.2	Managing Numerous Kubernetes Manifests	203
	3.3.3	Branching Strategies	207
	3.3.4	Division of the Repositories	211
3.4	Continuous Integration and Continuous Delivery		213
	3.4.1	Pipeline Steps for Kubernetes	213
	3.4.2	Pipeline Architectures	218
	3.4.3	GitOps	223
3.5	Templating Using Kustomize		225
	3.5.1	Basic Principles of Kustomize	226
	3.5.2	Resource Generator	231
	3.5.3	More Kustomize Built-Ins	233
	3.5.4	Conclusion on Kustomize	234

4 Advanced Objects and Concepts in Kubernetes — 235

4.1	DaemonSets		236
4.2	Jobs in Kubernetes		239
	4.2.1	Real-Life Kubernetes Jobs	240
	4.2.2	Queue Worker with RabbitMQ	242
	4.2.3	Kubernetes CronJobs	246

4.3	Custom Resources and Custom Resource Definitions	248
	4.3.1 Example: A Monitoring CR	249
	4.3.2 Validation in CRD	252
	4.3.3 Operators	255
4.4	Downward API	258
4.5	Pod Priority and Preemption	261
4.6	Versioning Objects in Kubernetes	263

5 Stateful Applications and Storage   265

5.1	Stateful Applications in Kubernetes through StatefulSets	266
	5.1.1 Pod Management Policy	269
	5.1.2 Strategies for Updates	270
	5.1.3 Retention Policy for Persistent Volume Claims	272
5.2	Persistent Volumes and Persistent Volume Claims	273
	5.2.1 Storage Types for PVs	278
	5.2.2 CSI Drivers for External Storage Media	281
	5.2.3 Storage Classes and Dynamic PVs	283
	5.2.4 PostgreSQL as StatefulSet with Persistent Volume	286
5.3	Ephemeral Volumes	289
5.4	Other Features of Volumes	291
	5.4.1 Volume Snapshots	292
	5.4.2 Projected Volumes	295

6 Kubernetes Governance and Security: Prepare for Production   299

6.1	Pod Security	301
6.2	Pod Security Admission	304
6.3	Admission Controller	306
6.4	Kubernetes Policies	308
6.5	Policy Objects	311

6.6	**Role-Based Access Control in Kubernetes**	313
	6.6.1 Subjects: Users, Groups, and Service Accounts	315
	6.6.2 Roles and Role Bindings	317
	6.6.3 Conclusion	319

7 Developing Applications for Kubernetes: Ready for Production — 321

7.1	**Managing Pod Resources**	322
7.2	**Readiness, Liveness, and Startup Probes**	325
	7.2.1 How to Define Probes	328
	7.2.2 Testing Probes Using an Example	330
7.3	**Scaling and Load Balancing**	335
	7.3.1 Horizontal Pod Autoscaling	335
	7.3.2 Vertical Pod Autoscaling	339
	7.3.3 Cluster Autoscaler	341
7.4	**Monitoring**	342
	7.4.1 Introduction: Prometheus, Grafana, and Alertmanager	343
	7.4.2 Monitoring on the Pi Cluster	345

8 Orchestrating Kubernetes Using Helm — 357

8.1	**Helm: The Kubernetes Package Manager**	358
	8.1.1 Creating a First Helm Chart	361
	8.1.2 Deploying a Helm Chart via the Command Line Interface	361
	8.1.3 Setting Up and Managing a Helm Repository	363
	8.1.4 Deploying a Helm Chart via Lens	365
	8.1.5 Updating and Deleting Helm Releases	366
	8.1.6 Downloading Helm Charts from a Repository	368
8.2	**Reading and Developing Helm Charts**	368
	8.2.1 The Templating Engine and the Language of the Charts	369
	8.2.2 Configuring Charts with Values	374
	8.2.3 Conditions in Helm Templates	378
	8.2.4 Other Operations and Control Structures	380
	8.2.5 Helm Diff for Checking Changes	383

8.3	**Developing Custom Charts**	385
	8.3.1 The Framework of Your Helm Chart	386
	8.3.2 Packaging Charts and Storing Them in the Repository	387
	8.3.3 Managing Dependencies in Helm Charts	390
8.4	**Conclusion**	394

The Author	395
Index	397

Preface

Be water, my friend.
—Bruce Lee

Perhaps you know the interview with Bruce Lee from which this quote comes. The idea at the heart of his statement is the adaptability of water. What Bruce Lee meant was that you must adapt to your opponent in a fight. Adapt when necessary. Be open in order to be able to react appropriately to the environment and changing circumstances. This sentence has been with me for quite a while—not only because I do martial arts, but above all because this metaphor is also very appropriate in IT.

When I started my training as an IT specialist in system integration in 2011, I had no idea what kind of world would open up to me. I learned the IT craft from scratch. I can still remember my first projects very well: We installed a network for a law firm. I drilled the holes myself, pulled the cables, and crimped the network connectors. I also configured and installed a new server for a medium-sized company, fixed the cables, and set it up in a server room on site.

If you need a server today, you simply need to click **Launch** in your cloud provider's frontend. A virtual machine (VM) is then activated for you by magic in some high-security data center. The world is constantly changing, and so is technology. Abstraction makes it increasingly easier to use, but this does not bring only advantages.

I first came into contact with the cloud, DevOps, Docker, and Kubernetes after graduating in 2017, and the topic has stayed with me ever since. I still remember the days during my studies when I asked myself: "How does my software actually get to the customer?"

I learned programming during my apprenticeship and studies. First it was Java, then C, then C++. I also learned what software engineering is, how to create unified modeling language (UML) diagrams, and how to develop in a machine-oriented way in assembler. But there was one thing I always missed: How does the software ultimately reach the customer? The operation itself had never really been part of my training. Thus, I am concerned with questions such as the following:

- How is the software built and packaged?
- How is the software delivered?
- What happens if the software doesn't work?

Most of the time, I ran my own development on my computer or in the integrated development environment (IDE). For a long time, no one was able to give me a satisfactory answer to my questions about the last piece of the puzzle.

I entered a dual-study program, and during the practical phases, I worked in departments that created software concepts. The work of my colleagues was to think about what the business requirements were and how they could be translated into software, and we wrote hundreds of pages of specifications and drew UML diagrams. The software was then developed both onshore and offshore by partner companies, which usually worked less than optimally. They were classic waterfall projects. Again, I never understood the software development process from start to finish. I always thought: "There is something missing. Somehow it doesn't fit yet." And sometimes, I thought: "Maybe I'm just too stupid for that."

Today, I know that I'm not too stupid and that I was just missing the last little piece of the jigsaw. After my studies, I came into contact with modern agile software development for the first time. One team designs, develops, and delivers the software, while another team takes care of operations. That was also the first time I came across terms such as *cloud*, *Docker*, and *Kubernetes*. I had already learned about agile software development during my studies, but all the tools used for it were new to me. My world was completely turned upside down.

I was familiar with virtual machines from my training, and I also knew that a cloud service is more than just a storage service like Dropbox or OneDrive. Today, among other things, I hold the *AWS Certified Solutions Architect—Professional* certification, and as I write about my past, I start smiling. The world of IT is so much bigger than I could have ever imagined, and I have really discovered my passion.

Since I've been using the cloud, Docker, and Kubernetes, it's felt really smooth for me. I now understand how modern software operation works and what is needed for it. For this reason, I want to share my findings with you in this book.

In 2017, I published an online course on the Udemy platform about getting started with Docker. My aim was to make getting started easier so that the participants get a feel for Docker by getting involved themselves. That's what I've been missing in my dual studies so far. I am a hands-on person and learn best when I do something with my own hands.

This book is also written in such a way that you get a quick introduction to the topic of Kubernetes. You will set up a cluster yourself and deploy your first services. I will take you on a journey and introduce you to the topic in a structured way because I love to keep things simple. It will get complicated all by itself, and pretty soon, so you will learn everything step by step that is important to make your software fit Kubernetes.

> **Acknowledgments**
>
> To my son, Levi Ace: You show me every day what is truly important in life.
>
> To my wife, Nicole: Thank you for always having my back, even when I'm writing until late at night.
>
> To my best friend and business partner, Fabian: Thank you for our journey together over a decade.

Structure

Let me briefly explain what you can expect on the following pages. You are already in the middle of the Preface. I want to pick up where you are right now and introduce what awaits you.

Chapter 1 and **Chapter 2** have been designed as a tutorial. Each section builds on the preceding one. You will get to know the basic principles and concepts and then get down to work very quickly. After these chapters, you will be prepared for Kubernetes and can then delve deeper into individual topics. From **Chapter 3** onward, we will take a closer look at individual aspects. You can read those chapters in the order in which you need them.

> **Note**
>
> Some sections in Chapter 1 and Chapter 2 are very well suited as reference sections. If you have the feeling in a particular section that this is not the right time for it, then you can just skim through it. This will feed your subconscious and you can come back to it when you need to read it in more detail.

In **Chapter 3**, I will take you on a tour of infrastructure as code (IaC). You will learn about YAML and the difference between declarative and imperative work.

> **Note**
>
> As you will already be working with YAML and IaC in Chapter 2, you are welcome to skip to Chapter 3 for a small digression before continuing with Chapter 2.

In **Chapter 4**, we will delve into more advanced concepts and objects.

Chapter 5 is dedicated to the topic of storage and stateful applications. What do you do with applications that have a state, much like databases have? What types of storage are available in Kubernetes, and how can you best manage your data?

Security and governance is a major topic in IT. In **Chapter 6**, I will introduce you to topics such as user and rights management, pod security, and Kubernetes policies. You will get to know the basic principles to prepare your application for production.

In **Chapter 7**, you will learn everything you need to know to make your application "ready for production," such as resource management, health checks, and scaling for your applications.

Finally, in **Chapter 8**, you will get to know Helm, the Kubernetes package manager. Note that you will have already used Helm in earlier chapters to deploy finished applications in examples. Helm will make your life as a developer very easy and help you to make your application fit for multiple environments.

What You Should Already Know Now

Kubernetes is software that builds on knowledge of other topics, such as the topic of containers, which themselves can fill entire books. For this reason, there are some prerequisites that you're expected to meet in this book so that you can be introduced to the topic of Kubernetes quickly without us getting bogged down in the details.

The Kubernetes tool is a container management system, which is why some basic knowledge of containers is required. You should be able to answer questions like the following:

- What is a container?
- How is a container structured?
- How can I build container images and bring my software into a container?
- How can I start and stop containers?
- How does the container tool work on my computer?

I use Docker Desktop as a basis, build containers using Dockerfiles, and will set up a test cluster using Minikube as a container with you in Chapter 1, Section 1.4. Depending on your operating system, you can of course also use other tools, such as Podman. You are not dependent on Minikube either and can use other test clusters if you are familiar with them.

> **Note**
>
> If you use a company computer, you may need a license for Docker Desktop. If you are unsure, it is best to ask or to use your private computer.

The containers used in this book are all based on Linux images. It is therefore an advantage if you are familiar with the basics of Linux. You should also be able to use the command line through tools such as Bash or another shell. If you've ever written Bash scripts before, then what we use in this book will be a breeze for you. If you run the examples on a Windows operating system, you should be able to use PowerShell. However, I will also provide you with the most important commands in that respect.

We will use command line interface (CLI) tools such as kubectl and Minikube. These are programs that are executed on the command line to operate Kubernetes, for example. Here I will guide you step by step, but you will find your way around more quickly if you have already used CLI tools previously.

In general, however, this book is suitable for beginners. You will get to know Kubernetes from the ground up, and I will try to pick you up as best I can from where you are right now. This means that even if you have little experience with the tools mentioned so far, you will be able to work through this book. In some places, it may be advantageous for you to put the book away and grab additional learning content on Docker and the like.

What You Will Learn

As mentioned earlier, this book is aimed at developers and DevOps engineers who want to get to grips with Kubernetes—whether you've only recently heard of Kubernetes or have been using it for some time. After reading this book, you will have the tools you need to develop and run your applications for Kubernetes. You will be able to build resilient, scalable, and reliable infrastructures. Your applications will be ready for production environments thanks to self-healing and load balancing.

You will not or will only marginally learn how to install or administrate a Kubernetes cluster in this book. However, what you will learn is how to run applications in a cluster and what you need to bear in mind as a developer. In addition, you will learn how to operate and control a Kubernetes cluster. For this purpose, you will install a test cluster on your computer based on Minikube. This will help you to try everything out in a test environment.

Important to Know

In the advanced chapters, you will delve deeper into the peculiarities of Kubernetes. I use simple applications as an example. If you try to run through the examples directly with your own applications, the learning effect is significantly higher, and you can then implement what you have learned much better in your daily work.

Perhaps you know this too: Imagine you want to go on a vacation to Italy and use a language app at home to learn the most important terms such as *hello*, *goodbye*, and *apple*. Even short phrases like "A coffee, please" and "I'd like to pay."

Then the time has come. You have arrived at your vacation destination and enter the first café. The waiter asks you what you would like to have and suddenly everything you have learned is gone. Not a word escapes your lips.

Here's what memory research has found out: The recall of a new skill is most successful when the circumstances are as similar as possible to those under which the neuronal connections took place. So when we study at home at our desks using an app, we find it easiest to retrieve the information by using that very same app. In a new situation, such as an Italian café, the circumstances are different, and we can no longer recall the information.

For this reason, it is important that you try out each chapter's content either directly or additionally with your own projects. This will make everything more interconnected and you will be able to apply the content much better. Have confidence in the process of this book. In the end, everything will fall into place and you will be able to use Kubernetes successfully in your environment. We still have a few steps to go before then, but I will accompany you.

I also want to introduce you to a model developed by the Canadian psychologist Albert Bandura (see Figure 1).

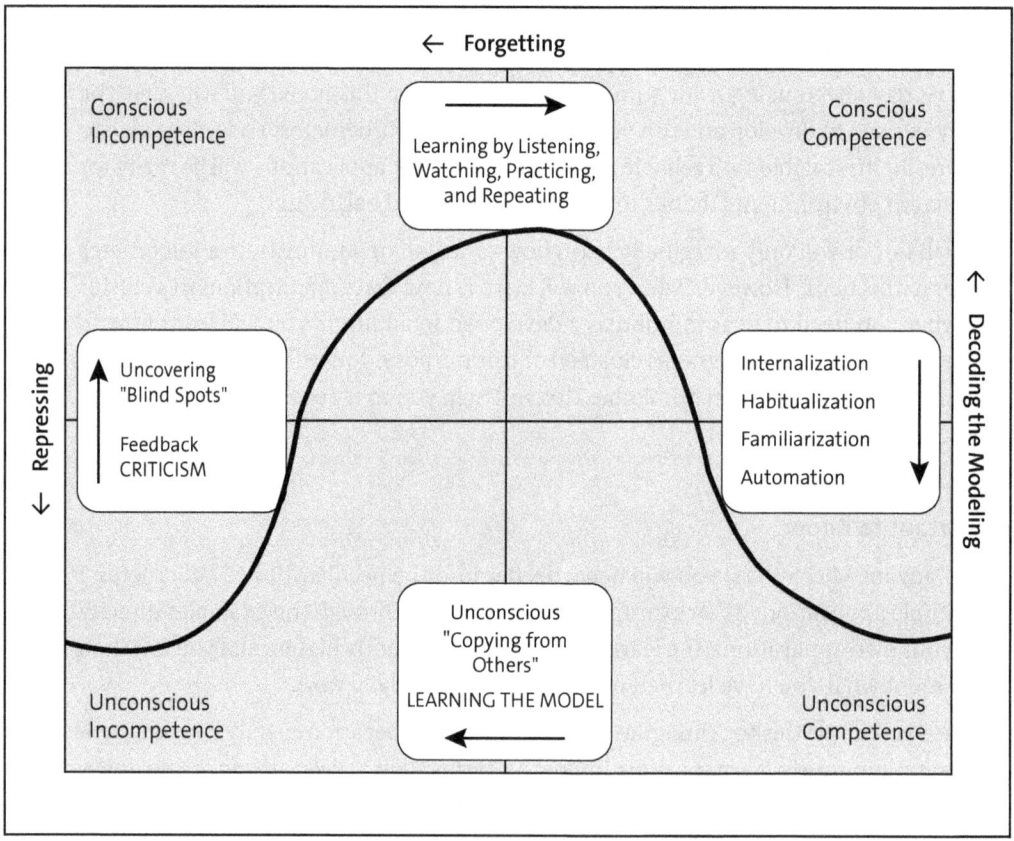

Figure 1 Four Stages of Learning According to Albert Bandura

You may even already know it, because it is a well-known model of how learning works for us humans. Bandura divides human learning into four stages:

1. Unconscious incompetence
2. Conscious incompetence
3. Conscious competence
4. Unconscious competence

Think about your driver's license. Can you remember what it was like for you when you first got behind the steering wheel? Or stalled the car at traffic lights for the first time? You suddenly realize that you are missing a skill. You want to drive but you also realize that it won't be an easy path because the only way from conscious incompetence to conscious competence is hard work. You learn and practice over and over again.

After a while you will be able to drive. But there's always that little voice in your head saying: "Shift gears now" or "Look over your shoulder and signal." You must drive with full awareness and cognitive effort. Only after many hours of driving does the activity gradually become easier and you develop automatism. Today I can drive the car while listening to children's music for my son's sake and singing along loudly. Driving itself has become unconscious and easy.

By purchasing this book, you have already left the first phase behind you. You are aware that you have to learn. The transition from phase 2 to phase 3 is the hardest and is often associated with a lot of frustration. The book will make the transition as easy as possible so that you can enjoy implementing it and become a Kubernetes expert with ease. I wish you lots of success and fun reading, learning, and playing.

Book Resources

Sample listings are available for you to download from the website for this book. Go to *www.rheinwerk-computing.com/5964*. Scroll down to the **Product Supplements** box. You will see the downloadable files along with a brief description of the file content. Click the **Download** button to start the download. Depending on the size of the file (and the speed of your internet connection), it may take some time for the download to complete.

Chapter 1
Introduction to Kubernetes

Kubernetes, also known as K8s, is an open-source system for automating deployment, scaling, and management of containerized applications.
—*kubernetes.io*

Developing containers and running them on your laptop computer is easily doable with a little know-how. However, operating hundreds of containers across multiple host systems, scaling them as required, and not risking any downtime is a lot more complicated.

Kubernetes (K8s) was developed to address and solve these types of issues. The name has its origins in Greek and means *helmsman*. This is also where the Kubernetes logo comes from. Developed by Google and continued as an open-source project, it is now an integral part of many companies.

Before I guide you through installing your first Kubernetes cluster on your computer, let's dive into the basics of Kubernetes.

> **Good to Know**
>
> The abbreviation K8s comes from replacing the eight letters of *ubernete* with the number 8.

> **Note**
>
> The book is based on Kubernetes version v1.27. If your company uses clusters of an older version, then some features are probably not available. For new features, I will point this out separately in the corresponding chapter.
>
> If you are unsure whether a feature can be used in your company, you can check the Kubernetes documentation at *https://kubernetes.io/* or ask your administrator.

1.1 Basic Principles and Concepts: Why Use Container Clusters at All?

To better understand Kubernetes, I want to take you back to the past and the system's origins. Kubernetes saw the light of day on June 7, 2014, at least in the public world, as the first commit was published on GitHub on that day. However, the idea for a container management platform was not new. It originated back in the 2000s at Google as even then developers had to operate several hundred thousand containers there.

With so many containers, Google needed a system that would simplify the administration and operation of a large number of servers. But at that time there was not yet a large market for it, and the developers at Google built their own solution. That was when Borg was born. As Google states in their abstract, "Borg simplifies life for its users by offering a declarative job specification language, name service integration, real-time job monitoring, and tools to analyze and simulate system behavior" (see *http://s-prs.co/v596463*).

Google used Borg to tackle a variety of challenges related to managing large clusters of machines. Problems that were solved by Borg included, for example:

- **Resource management**
  Borg automated the scheduling, starting, stopping, restarting, and monitoring of containers. This allowed developers to focus on their development work instead of managing resources.
- **Efficiency and capacity utilization**
  Using techniques such as overcommitment, Borg enabled a high utilization of the available resources. This saved Google from high data center costs.
- **Error handling**
  Borg offered runtime functions and scheduling rules that reduced the time needed for troubleshooting.

Good to Know

The introduction of Borg was a decisive step for Google to manage its infrastructure efficiently. Where they used to monitor and manage the servers themselves, this could be taken over by Borg.

In the course of this book, you will also see that Kubernetes automatically moves all your applications to a functioning server in the event of a server hardware failure. This saves you time and reduces downtime simultaneously.

Even today, we still expect exactly these benefits from a management system. But the world has moved on since then, more and more companies have opted for containers, and Borg has also needed to evolve.

Kubernetes was to be a new development for the existing container management tool. Years of experience with Borg were to flow into a new design. Parts that worked were adopted and other parts were optimized. What is probably the biggest difference from Borg is the new license model. The developers at Google opted for an open-source model and donated Kubernetes version 1.0 to the Cloud Native Computing Foundation. This makes Kubernetes an open and independent system, which is perhaps why it is currently so popular.

> **Good to Know**
>
> The Cloud Native Computing Foundation is part of the Linux Foundation, which introduces itself on its website as follows: "The Linux Foundation provides a neutral, trusted hub for developers and organizations to code, manage, and scale open technology projects and ecosystems."
>
> In my view, a foundation as a company for an open-source technology increases the trust and independence of Kubernetes.

You can find out more about the history of Borg and the origins of Kubernetes at the following two links:

- *http://s-prs.co/v596401*
- *http://s-prs.co/v596402*

1.1.1 Why Use Containers at All?

Perhaps you have already developed and operated containers yourself. Docker is currently the best-known representative of containers, and it is usually used as a synonym for container. Just as Kleenex is the paper tissue, Docker is the container. Docker did not invent the concept of containers, but it has done a great deal to make it so widespread today. This is understandable, because containers

- are lightweight,
- are easy to use, and
- run on virtually any server that has a container runtime.

In addition, container images are easy to transport and contain everything your application needs. You no longer face a common problem: "But it doesn't run on my computer!"

When we compare containers with virtual machines, the biggest advantage is obvious: you do not need to install a complete operating system with a container.

You may already be familiar with the evolution of virtualization, as shown in Figure 1.1.

You can see how the deployment of applications has evolved over time from bare metal to virtualization. (This is not to say that containers are replacing virtual machines, but they outstrip them in many application scenarios.)

But why has it developed like this? Let's consider a very simplified example.

Think of a data center. There are racks there that can contain multiple bare metal servers. A rack has a maximum capacity of servers that it can hold, and the data center has a maximum capacity of racks it can hold. If you now think of a regional web store that runs on one of the servers, it is busier in the evening than in the middle of the night. The server therefore has nothing to do during the night and heats up the data center unnecessarily. If you only have servers like this, you will have very poor capacity utilization throughout the data center and therefore high costs.

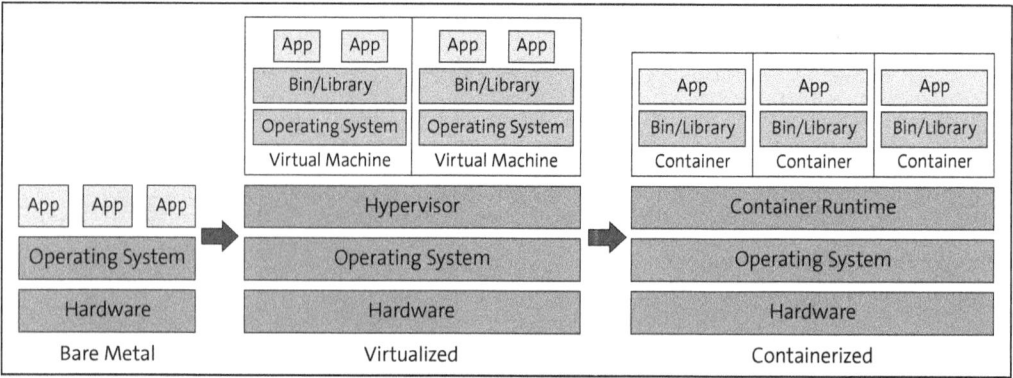

Figure 1.1 Evolution of Virtualization

In addition, you have to design the server for the peak load so that every customer can buy your products even at peak times. This means you generally have an oversized server. Another point is the dependency on the operating system and the underlying hardware. You have a single operating system with the drivers for the server's hardware. You cannot simply make a clone of it and install it on another server, which in turn makes backup and recovery more difficult.

> **Note**
> Of course, you can also install multiple applications on one server. If, for example, another application runs batch jobs and performs billing at times when nothing is happening in the web store, you also increase the workload, but virtualization brings even more to the table.

How can you increase the utilization of a server and overcome the difficulties of the bare metal server? If you enable the server to run virtual machines, you can run multiple virtual servers on a bare metal instance. With more instances, you create better utilization and even spread the costs of the server across multiple virtual servers. But it's not just capacity utilization that will improve:

- You are more independent of the actual hardware and can also run your virtual machine (VM) on other servers without much effort.
- You can set up backup and recovery processes very easily using VMs.
- You can use golden images to set standards that are easy to use.

Thus, virtual machines are an optimization of the bare metal server, each with its own operating system. They behave like real servers: they have to boot up everything at startup and still have the overhead of an ordinary server. But then the following question arises: Could it be even simpler and smaller? The answer is found in containerization.

Good to Know

Companies that operate their Kubernetes clusters in the cloud usually even build their clusters on virtual machines. This makes sense, as bare metal instances on Amazon Web Services (AWS) only start at 48 CPU cores and 384 GB of RAM. You could easily run all containers from smaller clusters on a single instance, but that would be fatal in the event of a hardware error.

For redundancy and scalability, it is therefore better to have smaller instances, but more of them.

Let's take the web store and pack it and everything we need for operation into a container. For this purpose, we separate the application from the virtual machine and can use the web store independently of it. All you need for a container is a runtime that is installed either directly on the bare metal instance or in a virtual machine. This allows you to benefit from the advantages of containers:

- Containers use significantly fewer system resources than virtual machines, as they do not require a complete operating system.
- Thanks to containerization, applications can be used without much effort across different operating systems and hardware environments.
- By using containers, applications can be rolled out, updated, and scaled more quickly.
- Containers also speed up the development process, and the portability of the images means they can be run on any developer computer.

Containers therefore have a number of advantages over virtual machines, but they do not replace virtual machines or bare metal servers. All of these have their right to exist and a corresponding use case. However, this example shows why applications are nowadays almost exclusively developed in containers.

Good to Know

Compared to virtual machines, a container is even better for the utilization of your servers. The smaller the unit, the easier it is to find a gap on a server.

Take a jar full of marbles, for example, as shown in Figure 1.2. There is still enough air between the marbles to fill in small beads, and then there is still enough air between the beads to fill in fine sand.

1 Introduction to Kubernetes

Figure 1.2 Jar Containing Marbles and Beads

If we look at modern applications, it is much easier to handle them in containers. Even the startup is significantly faster, and that again changes the way scaling works. In the past, the physical server received a CPU or memory upgrade so that the monolithic application had more power.

Today, all you need to do is start another container with the same application, and the load is distributed to the new container within a few minutes.

If there is no more space on the server, a new server in the cloud starts up as if by magic and the container is deployed there. The trend is moving from vertical scaling to horizontal scaling—but this is not as simple as it may sometimes sound: there is a lot of know-how and work that goes into such a cluster setup.

Good to Know

The good thing about horizontal scaling is that you no longer have to rely on a large server as a single point of failure. If the load is distributed across many smaller servers, you can compensate for a failure much more easily.

The technologies and software architecture currently work hand in hand. New applications are usually only developed in a microservice architecture, and communication must ideally be asynchronous and event-based. Companies want to outsource their workloads to clouds and only pay for what they really need thanks to automatic and requirements-based scaling.

1.1.2 Why You Need a Container Management Tool

The use of many microservices and horizontal scaling raises new challenges. Suddenly there are hundreds or thousands of containers that have to be operated and monitored simultaneously.

I have fond memories of when I first came into contact with containers. In 2017, I worked in a company that was undergoing a major transformation. It had just made the decision that all software, whether legacy applications or new developments, should be migrated to Amazon's cloud. As part of this, a program was set up to redevelop an old distribution platform, and all new microservices were to be containerized and run on Kubernetes.

That was just two years after the release of Kubernetes version 1.0. The new container world works completely differently from the applications that were developed decades ago. In the past, when a web application was developed, it usually had a monolithic design and ran as a virtual machine on a server in the data center. These applications were also often designed for a specific number of users. Scaling according to demand was not easily possible. This was also the case in that company.

If the old sales platform received more traffic than expected—for example, because there was a Christmas campaign—then the application had to be assigned more CPU and memory. It could therefore only be scaled vertically. In most cases, this was accompanied by weeks of preparation and planning, with employees only concerned with capacity management. The rest of the time, this application ran at a 30% load and unnecessarily heated up the data center.

The architecture of new applications is moving away from monolithic designs and toward microservice architectures: smaller, independent services that communicate asynchronously and are therefore loosely coupled. In the example of a web application, a web server such as Nginx can simply be scaled as required. Whether the user's request is answered by one or another Nginx instance does not matter to either the web server or the user. The main thing is that the answer is the same. Loose coupling allows us to scale precisely that part of the system that is currently experiencing a capacity bottleneck.

So it is not the case that we have to pack the monolithic applications into containers and deploy them on Kubernetes in order to arrive in the brave new world. It is instead based on a completely new concept and requires a new and modern microservice architecture, which in turn requires a rethink within a company. The microservice architecture means that there are suddenly not just a few monolithic applications, but several small applications. The number of these in larger projects quickly reaches double figures, and in companies it can easily be in the hundreds or thousands. This also changes the challenges in a company.

Smaller services often provide the opportunity to deliver updates by way of continuous integration and continuous delivery (CI/CD). This has the following advantages:

- Developers can deploy more easily and quickly.
- More responsibility lies with the developers.
- The burden on operations is reduced.
- The focus shifts more toward a higher quality.

But there are also disadvantages:

- Every change harbors the risk of errors.
- Dependencies on other components can be forgotten.
- Release processes are ignored.

The degree of automation must also be significantly increased so that tickets are not opened every minute in IT operations causing the phone to not cease ringing.

Another aspect is the change in processes and sometimes the entire organization. It is usually not just the technology that changes, but cultural changes through the principles of DevOps or process optimizations through Lean and ITIL go hand in hand with the technology. As a developer, you are caught in the middle of all the changes and are expected to quickly develop outstanding software that is stable in operation and makes the end customer happy.

Amid all the chaos, Kubernetes comes into play. The container management tool allows you to monitor your containers automatically and scale, restart, or terminate them as required. As a platform, it fits very well into modern processes and gives you as a developer more personal responsibility. Kubernetes also ensures that users do not experience any downtime during releases, and we will take a closer look at why Kubernetes simplifies the operation of microservices.

1.1.3 Of Pets and Cattle

You already know that there needs to be a rethink within companies, and you may already be familiar with the classic comparison of pets and cattle. Pets are animals you have at home, while cattle are the farmer's livestock. In the world of monolithic applications, servers are usually treated like pets. A pet is fed, cared for, and loved. It has a name and belongs to the family, and it cannot simply be replaced by another animal.

The servers were also difficult to replace in the infrastructure of the monolithic applications. If the infrastructure runs out of support, migrations are necessary, which are associated with major risks. The goal is therefore to prevent the server from going bad, and a new server is out of the question. Spare parts, such as hard disks and power supply units, are in the safe waiting to be used so that the applications can continue to run.

If we look at a farmer's cattle as a counterexample, we can observe a completely different kind of love. A farmer also looks after their cattle, but the cows have numbers on their ears rather than names to identify them. They have a clear task, to provide milk and meat. After a certain time, when an animal becomes too old or ill, it is simply replaced by another one.

> **Note**
>
> The standard example with pets and cattle is perhaps a little macabre. A nice alternative is the comparison between wildflowers and bonsai.
>
> A bonsai requires constant care and attention. You need to monitor its shape, size, soil type, and supply of fertilizer on a regular basis. Moving a bonsai to a new environment or changing its care conditions can have a significant impact on its well-being and growth.
>
> Wildflowers, on the other hand, are robust by nature. They grow where conditions permit, without making specific demands on the location or the environment. If an area is no longer suitable, you can simply sow them in a new location without having to consider the previous position or special conditions.

Since the advent of cloud computing, the handling of infrastructure has been changing more and more in the direction of cattle. If a server no longer works, a new one is set up to take over the task. The setup and migration of the applications is automated.

Kubernetes can take over these tasks so that you don't need to worry about the infrastructure. The advantages are obvious: infrastructure issues are resolved automatically, and the infrastructure can be scaled as per your requirements. This saves companies a lot of money because they only pay for as much computing power as they actually need, and automation prevents the odd on-call assignment.

But it is not only the servers that are treated as cattle. Even the applications that run in containers are no longer pets. Have you heard of Chaos Monkey yet? This is a tool developed by Netflix to check the stability of production systems. Imagine that a monkey has broken into your data center. It accidentally bites through cables and hits the servers with a hammer. Would your application survive this?

You don't need to introduce a chaos monkey in your company right now, but the idea behind it is a good one. Just ask yourself a question: Would users notice if a component or container failed? If the answer is yes, then there is definitely room for improvement.

1.1.4 Stateless and Stateful Applications

To develop an application for Kubernetes, one key question is important: Is your application stateless or stateful? Does your application have to remember a state? But what is the difference between these two concepts?

Imagine you are in Italy sitting in a cafe in a small town near Venice. You have a direct view of the Adriatic and order a coffee. The waiter brings you your coffee, and milk and sugar are provided, but you drink your coffee black.

This is a good example of *stateless*. The cafe itself does not store any information about you as a customer and does not know your previous orders. When you order something, you get the same items as any other customer making that order. Each order is treated in isolation, without any previous history being taken into account. It is therefore stateless as it does not store any permanent state or information.

Let's now take a look at the *stateful* concept. When I take my buddy Fabian to our favorite cafe, he just nods to the waiter and we both get a black coffee. No milk, no sugar. The waiter simply knows us.

It's like being a member of a gym and saving your personal data and workout progress in your member account. You go to the leg press, insert your card into the machine, and it suggests the right weight for your workout progress. Here, the status is saved and continuously updated to provide a personalized experience. So the gym or your favorite coffee shop is stateful because it stores and uses information about you to improve your workout routine or bring you your favorite drink directly. The saved state brings convenience, but also more responsibility. You need to think about how the status is saved. For example, what do you do if the waiter who knows you is out sick?

States bring challenges with them:

- Data must be kept consistent across all instances. (Every waiter must know us.)
- Horizontal scaling is more difficult. (A new waiter must be trained first.)
- You need to think about backup and recovery. (How does the data get restored if the waiter is absent?)

When we transfer all this to the world of IT, you can compare a simple website with an online store. If you imagine the website of a small carpenter's workshop from a neighboring village, you will find pictures of projects, information on how to reach the workshop, and perhaps a contact form. This is all data that is displayed to every user when they access the website. No data needs to be kept or stored, and even the contact form simply sends an email to the managing director. The website is stateless.

The online store of a large furniture store, on the other hand, provides features that require a state. Think of the shopping cart, for example. It contains all the products you want to buy. The order history and invoices also represent data about you that must be stored. Thanks to the data, the online store can also make suggestions to you, such as "Customers who bought a table also bought a chair" or categories tailored to you.

In more simple terms, *stateless* means that each interaction is independent and contains no information about previous interactions. *Stateful*, on the other hand, means that information about past interactions is stored and used to provide a continuous experience. However, you will notice from the examples that there are states in most applications that need to be saved.

Usually, we cannot control the fact that states have to be saved. What we can control, however, is the way we design our applications. As many of them as possible should be stateless because they are then much easier to handle.

Stateless applications

- are easier to scale and provide higher performance,
- are easier to deploy, and
- are better to manage and debug.

> **Good to Know**
>
> An application must be handled differently in Kubernetes depending on whether or not it stores data.
>
> A database is stateful, and if it is operated in Kubernetes, Kubernetes cannot simply terminate or rebuild it. For this type of use case, there is a separate object that has precisely these properties for running stateful applications. You will get to know this object in Chapter 5, Section 5.1.

1.1.5 Separation of Concerns

Separation of concerns is a design principle in software development. It aims to divide complex systems into several components or modules. Each module has a clearly defined and limited responsibility or task. This division significantly improves the maintenance, further development, and comprehensibility of the software.

A classic example is the three-tier architecture, as shown in Figure 1.3, which you may already be familiar with. Let's assume you are developing an online store.

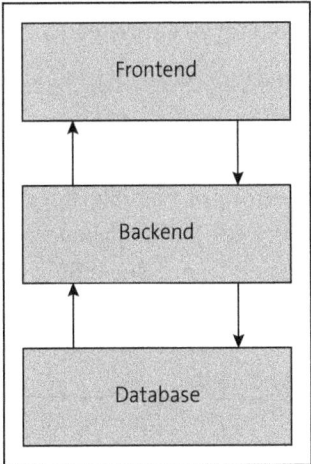

Figure 1.3 Simple Three-Tier Architecture

You could, for example, separate the frontend, backend, and database. The *frontend* is responsible for providing the HTML, CSS, and JavaScript files. In the end, that's what the customer sees. They can add products to their shopping cart, check them out at the end, and buy them.

The *backend* is the application that runs in the background and processes requests from the frontend. Once the customer has entered their credit card details, the backend can take care of the actual billing process.

Finally, persistence is required, which can be mapped with a *database*, which will be used by the backend to store or retrieve data. For example, the customer's order history is stored there. There are API interfaces between the individual components that enable communication between them.

Good to Know

The three-tier architecture is one possibility, but there are others. One of my customers has multiple software products in use. Some of these are legacy applications in the data center and others are software-as-a-service (SaaS) applications in the cloud. In order to exchange data between the systems, we have developed connectors, with a separate connector for each data path, based on the extract, transform, and load (ETL) principle.

This separation allows developers working on the user interface to do so without having to worry about how the business logic or database access works.

The API defines types of communication with the backend and which functions it offers. At the same time, backend developers can work on the business logic without having to think about the frontend. This separation makes it much easier to maintain and expand the application as there are different programming languages for the respective components and usually also different developers or even entire teams.

The layout then often looks as shown in Figure 1.4, and the developers can focus on specific modules.

In the world of containers, attempts are often made to divide the components into atomic units in order to make the separation of concerns as strong as possible. It is not uncommon for the backend to be split into multiple microservices. You then have separate microservices for the shopping cart, billing, ordering, and so on. Depending on the system, an even more granular categorization may make sense. The smaller the system, the greater the effort required to maintain it. The overhead then increases, which is why it is important to check where the cut needs to be made.

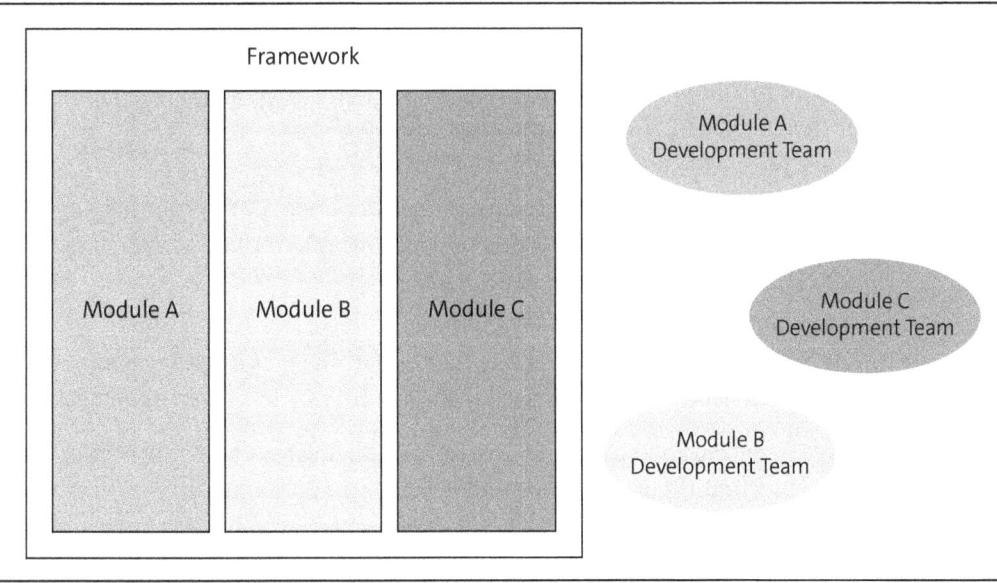

Figure 1.4 Responsibility of Developers

> **Good to Know**
>
> The term *atomic* is derived from the Greek word *atomos* and means *indivisible*. When I talk about *atomic components* (which you provide in containers), this always means that they cannot be further divided into smaller units. They are part of a larger application, but the separation of tasks and responsibilities helps in many areas.
>
> Just as you as a developer want to keep individual functions in your code small, you also want to keep the components small. They should have a task and nothing more.

In development projects, often not everything can always be considered in advance. You should therefore not be afraid to separate or combine components later on in the project. It is also always good to look for the sweet spot so that you don't fall into one extreme (too small...) or the other (... or too large components). What advantages can you achieve through good partitioning? Here are some examples:

- **Granular scalability**
  With atomic containers, you can scale exactly the parts of your application that are needed. If, for example, the number of web servers needs to be increased, you can scale it without having to scale the database directly.

- **Independent updates**
  Atomic containers make it possible to update parts of your application independently of each other. This minimizes downtime and simplifies deployment.

- **Improved resource utilization**
  As each component is isolated in its own container, you can distribute resources as you wish. In combination with simple scaling, your operations are significantly more flexible, saving resources and, thus, money.
- **Increased security and stability**
  Separation increases security, because if one container is compromised, the other containers are not necessarily at risk. In addition, an error in a container does not directly lead to the failure of the entire system.
- **Easy maintenance and further development**
  The separation into atomic components also facilitates further development and maintenance, which means you need to do significantly less reverse engineering.

If you have no separation, then that is just as bad as being too compartmentalized. The challenge is to find the right arrangement for your application. If you start with a hypothesis, you will find out over time whether you need to subdivide further or combine components again.

Just have the confidence to get started.

1.2 Kubernetes, the Tool of Choice

You have now learned what a container management tool is needed for and why and have become familiar with the most important overarching concepts. You also know how Kubernetes came about. But why is Kubernetes the tool of choice?

I want to start by looking at the reasons that companies want to use Kubernetes in the first place, because let's be honest: a company does not optimize its IT in order to use the latest software tools without a business purpose behind it.

I will then briefly take you through the arguments with which Kubernetes is entering the market and give you an insight into my experience of this in the real world. Does Kubernetes really deliver what it promises?

To conclude this section, I will clearly emphasize for which companies Kubernetes is useful and for which companies it is not. Kubernetes is certainly not the panacea for every IT problem.

1.2.1 Why Do Companies Want to Use Kubernetes?

The first question I ask my customers when they want to introduce Kubernetes is: "What goal do you want to achieve with it?" The answers can be varied. In my experience, they can always be broken down into the following three aspects:

- Faster time-to-market
- Saving costs through optimized processes
- Opening up new markets through new software

But in the end, it always comes down to clear, measurable facts, usually about making more money or spending less money. What does it look like in your company? Has your company been using Kubernetes for some time, or are you one of the first in the development team to use Kubernetes? Do you know the goals behind it?

You have likely experienced this yourself. You develop a prototype, and a decision is made from one day to the next: the project is canceled. Only if you as a developer know the company's goals in advance can you make sure that the project will become a success during development. You make a major contribution to the success of your company and can also prioritize your tasks much better thanks to the clarity of your goals.

The decision to use a platform such as Kubernetes usually has a greater impact on a company than a prototype. To avoid one project relying on Kubernetes, another on Amazon ECS, and a third one on Docker Swarm, it is necessary to set a standard.

What motivates your company to invest time and money to introduce Kubernetes?

To help you better understand the decisions in your company, I would like to briefly digress on the purpose of a company and the resulting value chain.

> **Good to Know**
>
> For us as computer scientists, the technological advantages are usually the most important elements—for example:
>
> - Automated rollouts/rollbacks
> - Service discovery/load balancing
> - Horizontal scaling
> - Memory orchestration
> - Self-healing
>
> Unfortunately, technical excellence is often not (only) important for decision-makers. It is our task to translate these technological advantages. In the end, they can always be broken down to the following: we save time and money using Kubernetes or can develop and deliver software faster.

Every company has a purpose and must create added value for society in order to survive. Try to imagine a village 5,000 years ago. Everyone in this village had a job. The miller ground the farmer's grain so that the flour could be processed into bread by the baker. The butcher processed the hunters' game. Everyone contributed to the village community to ensure its survival. In the worst case, anyone who was unable to contribute to the community was cast out.

Even back then, small chains of value creation emerged where several people worked on products to process them and turn them into something "more valuable." For example, the chef could cook a dish from the butcher's meat. Figure 1.5 illustrates what such value chains looked like. Each individual contributed something to the village community and received something in return.

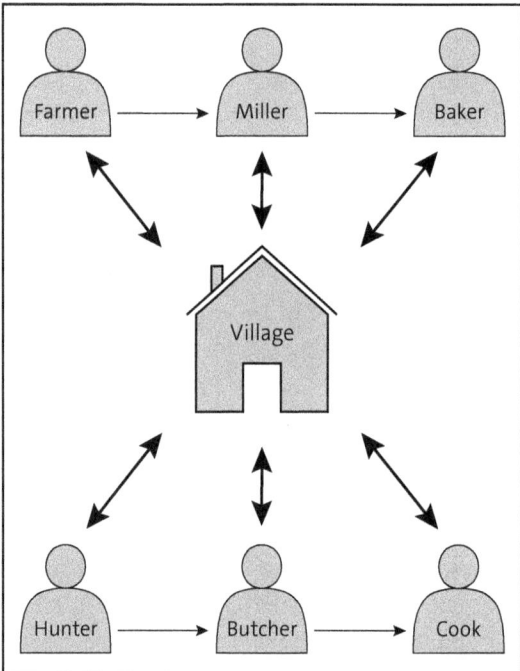

Figure 1.5 Added Value in Village Community

Today, value chains are much more complex than they were back then, but every company still has to contribute something to society in order to earn money. Money is the currency used to express how much value it has for society or for an individual. Money has enabled value chains to become larger and companies to grow. Unfortunately, money also contributes to the chains being forgotten. But our society can only function in this way today because every company works like a cog in a clockwork mechanism. We notice this above all when a cogwheel no longer runs smoothly.

> **Good to Know**
>
> The introduction of the monetary system made a growing society possible in the first place. Bartering was still possible in a village community, but imagine what it would be like if you spent months developing an application for a car company on the barter system. You would get a car in exchange—but you wouldn't have any gas, food, or drink, and you wouldn't even be able to pay the rent or mortgage on your house.

There is also a value chain in your company, and you are part of it. The question is: Do you know what part you play in this? Because the better you understand it, the better you can pay into it. So if you're part of the value chain, then Kubernetes is a tool to optimize the value chain, and we're going to look at exactly how Kubernetes can do that.

> **Good to Know**
>
> Any software such as Kubernetes that you use in your company should be measured by how well it contributes to the optimization of value creation. This is the only way to determine whether it is a sensible business decision.

1.2.2 The Promise of Kubernetes

Let's first take a look at what Kubernetes promises. I want to introduce you to the three core competencies that Kubernetes lists on its website. These are probably on every management slide, and they provide a picture at a high strategic level to convince a company's decision-makers.

But I think they are also important for you to understand how Kubernetes is positioned in the market and what questions you may face when you say, "I'm developing applications on Kubernetes."

Planet Scale

As you learned in Section 1.1, Google developed the core concepts of Kubernetes on the basis of Borg. Kubernetes is therefore based on the same principles Google has used to run billions of containers per week. For you, the *Planet Scale* competency means that Kubernetes can grow with your requirements. So it doesn't matter whether you operate hundreds, thousands, or millions of containers.

But is it really that simple?

In this book, you will not learn what a good cluster should look like or what the perfect cluster setup for your company looks like; however, I want to give you a few insights into what clusters look like in German companies.

In theory, Planet Scale is probably possible, but in real life I have rarely seen companies that have a huge cluster for everything, as was the case with Borg and Google. Instead, there is always an organizational separation at a certain point, be it between different company subsidiaries or between different subject areas. The clusters can be divided into two types:

- Large clusters, which I refer to as *cluster as a service*
- Individual clusters that are provided for large applications with multiple microservices

I call the large clusters *clusters as a service* because they usually provide a home for many different small applications that are created in a company but cannot be directly assigned to a large project. A small project team can thus quickly and easily get a container live and does not have to worry so much about cluster operation.

This approach already goes in the direction of how Kubernetes is actually intended: a cluster for all of the company's containers. That's no problem with Planet Scale either, but one cluster for many brings operational and configuration challenges that cause companies to structure clusters to fit the hierarchy. Here are some of the challenges that I have already encountered with customers:

- **Roles and rights**
  These must be clearly thought out so that teams cannot influence each other.
- **Resources**
  Allocation is more difficult, and containers from different teams could fight over resources.
- **Billing**
  Who produces which costs? Breaking down server costs into containers is much more complicated.

> **Good to Know**
>
> The fact that an IT system adapts to the structure of the company and its hierarchy is also known as *Conway's law*.

This makes it much easier to separate individual clusters. If there is a separate cluster for each project, there will be no complicated separations that need to be observed and maintained and monitored by an operations team.

However, the separation into individual clusters also has its disadvantages. What is often forgotten is that the basic setup of a cluster can cost a lot of money: the cluster requires management services that need computing power, and a setup of two servers is far from sufficient to be highly available. The budget for this may be a little looser in corporations for security reasons, but the situation is somewhat different in medium-sized or small companies.

Nor can the appropriate budget be made available for every prototype or for small applications. A prototype in particular does not yet have a reliable business case, and it must be quick, simple, and cost-effective to operate. A dedicated Kubernetes cluster is simply oversized here.

> **Note**
>
> Take a look at the clusters in your company. How are they designed? Do you think it's right?

Here's the bottom line on Planet Scale: A cluster structure must be well thought out and adapted to the company; there are many facets that need to be considered. But let's be honest: for very few companies is Planet Scale relevant at all.

I would like to take a closer look at one promise in this context—namely, that you can scale clusters as you like without having to increase the size of the operating team.

From my experience, I can state quite clearly that we are still a long way from that end as each scaling stage brings further challenges. For example:

- A cluster setup for 5,000 containers is different from a setup for 50,000 containers.
- A larger number of applications also means more responsibility and higher demands on a cluster.
- For every additional virtual machine in the cluster, the probability that a machine will fail also increases.

I have yet to experience a company in which principles of no operations (NoOps) really take full effect. A high degree of automation is constantly being used in an attempt to reduce operations to just a few supporters. But a lot of steps are needed to get there, and I don't think it's realistic that the vast majority of companies will ever achieve this form of automation.

In this book, we will go into a few more points that are relevant for you on the way to NoOps. You will learn how to prepare your application for operation in a cluster.

Never Outgrow

How long have you been a software developer? Can you still remember the days when the runtime of a function specified in Big O notation was important for performance?

I can still remember when it was said: "Bubblesort has a runtime of $O(n^2)$." To be honest, that was the last time I gave much thought to resources or runtimes. Today, optimized libraries and cheaper hardware save us a lot.

The *Never Outgrow* competency of Kubernetes follows exactly the same line. What could be worse than waiting three months for a test environment until the hardware is set up in the data center? You may smile at the thought, but unfortunately this is still the standard in many companies.

In Kubernetes, you can create a new namespace at the click of a mouse, and your applications can be deployed for a new test in no time at all. You can delete them just as quickly when the test is finished.

Kubernetes promises

- that a cluster can grow with you and your requirements,
- that you can retain flexibility in scaling, and
- that everything is billed according to the pay-as-you-go principle.

Admittedly, developers have never had so much freedom before. I ran several Kubernetes clusters in an operations team for several years. We had built the clusters on AWS infrastructure and automated them to a fairly high degree. For example, if more resources were needed because all servers were at capacity, a new server was automatically scaled to accommodate new containers. That made our work much easier. Sometimes we were told things like: "We would like to do a load test and need an environment for it." The best feeling was always being able to say: "You have it in your own hands and can do it without us."

I am convinced that more responsibility for developers leads directly to better products. It increases the speed of development because you don't have to wait a couple of weeks for someone from the ops team to finally have the time to provide new hardware; that's especially frustrating when you as a developer could do it yourself.

> **Note**
> More responsibility for developers doesn't happen overnight, and the path to it is sometimes a little unfamiliar. But the work is worth it. I have had mostly positive experiences with it so far.

Here's the bottom line on Never Outgrow: If the cluster is set up well, it simply grows with your applications. You are more flexible and can focus on what is important: the development of your product.

But one thing is important to me: an increasingly higher degree of abstraction and a growing number of cheap resources mean that we as developers are thinking less and less about runtimes and efficient programs. It is therefore important to me that we do not forget our craft and do not throw more resources at every problem. I just had to optimize the runtime of an application a few weeks ago and am glad to have the tools to do so. Never outgrow is a nice feature, but the rest should not fall by the wayside.

Run K8s Anywhere

Are you already in the cloud or are you still working on premise?

You know what? In my opinion, this question is actually unimportant—or at least, there is no general right or wrong. It always depends on the specific area of application.

The nice thing is that Kubernetes doesn't care which infrastructure it runs on, per the Run K8s Anywhere competency. You can also set up a cluster on premise in such a way that you can fully utilize the features of Kubernetes.

Kubernetes promises

- to run in the cloud, on premise, or in a hybrid environment;
- to easily move workloads as per your requirements; and
- to give you all the freedom you need due to being open source.

And from my point of view, it lives up to it. The combination of these points makes Kubernetes an interesting abstraction of your applications from the actual hardware. In this book, you will also learn how workloads can be moved from one node to another and how you can influence where containers run.

A lot has happened with Kubernetes in recent years, and open standards have been developed that allow you to connect different storage systems according to the same schema, for example. This makes it really easy to operate Kubernetes in your own data center.

What I regard as extremely positive is the development of managed Kubernetes services. Due to their great popularity, the major cloud providers also offer managed clusters. This makes it particularly interesting for small companies that do not want to worry about operating a cluster.

Here's the bottom line on Run K8s Anywhere: Yes, you can run Kubernetes on many platforms and have a lot of freedom. However, setting up a cluster involves a great deal of effort, so you don't have the option of quickly switching from an on-premise cluster to a cloud cluster. It takes something to set up the infrastructure components properly. Many questions need to be answered and aligned with individual objectives.

The big advantage for you as a developer is that if the cluster is in place, your application does not care where the cluster is located. You simply have an additional abstraction layer, which is very useful for developers.

1.2.3 Major Features

You have now learned about the three most important core promises of Kubernetes, but you can certainly imagine that a cluster adds complexity to the company in addition to costs. So it makes sense to ask the following question: "What's in it for me?"

Let's also take a look at the features that are boldly presented on the website. What we expect from Kubernetes is a simplification in the operation of containers. But how exactly does Kubernetes want to achieve this? It implements the following:

- **Automated rollouts and rollbacks**
  Automated rollouts have become the standard in the world of the cloud, but they are not being implemented enough in most companies. Kubernetes offers the possibility of rolling out rollouts automatically and without downtime. You can even monitor the application and roll it back automatically in the event of errors. I haven't seen many automated rollbacks with my customers so far, but even manual rollbacks are practically possible at the touch of a button.
- **Service discovery and load balancing**
  You don't want to worry about exactly where your application is running and how the traffic is routed there. Kubernetes takes care of this for you and knows at all times where a container is and how the load can be routed to it. This also opens up the world of autoscaling.

- **Horizontal scaling**
  If your application receives more requests than it can handle, it should be scaled as automatically as possible. In Kubernetes, this works almost automatically by starting up a new container, a process that is referred to as *horizontal scaling*. We will look at exactly how this works in Chapter 7, Section 7.3. In my opinion, this feature is one of the biggest advantages of Kubernetes.

- **Storage orchestration**
  You don't want to have to deal with storage personally either when the hard disk is full or a new disk needs to be connected. Through abstraction, Kubernetes offers you a standardized API to use storage for your application. In Chapter 5, you will learn how easy it is.

- **Self-healing**
  Have you ever been on call at night? How nice would it be if your application could heal itself—or at least keep running until normal operations can take care of the problem? Kubernetes offers technical self-healing and takes care of containers when they are no longer running.

  But let me be honest: Kubernetes cannot do more on its own than switch on and off. You have already gained a lot with this, but for full self-healing, a little more brainpower is required from the developers.

- **Secret management and config management**
  Every application needs configs and secrets. When planning and programming your application, you must decide how best to deal with this. Sensitive data such as passwords or private keys in particular must be treated with special care. Kubernetes offers a simple way to manage this data. We will look at this in detail in Chapter 2, Section 2.4.

- **Batch execution**
  Do you have regular jobs that need to be processed? In the classic world, these are executed by cron jobs on virtual machines. Kubernetes provides an option for processing such tasks, which you will learn about in Chapter 4, Section 4.1. And it comes with all the advantages of Kubernetes.

From my experience, I can tell you that all these features in one tool are very useful. As with self-healing, not everything is always as promised, but over time working with Kubernetes becomes very pleasant. In the course of this book, you will learn about each of these topics in detail and be able to form your own opinion.

1.2.4 For Which Companies Is Kubernetes Useful?

Imagine a family that is about to decide on a new car. The father travels a lot for work and needs a car that he can drive hundreds of miles on the freeway in one go. It should contain all possible assistance systems and make driving as pleasant as possible. At the same time, he wants to have fun driving and be able to hit the gas. The mother, on the other hand, wants a comfortable and spacious car so that she can easily load her three children and the dog. It has to be safe, and an electric car would suit her best for short journeys.

The family goes on vacation twice a year and goes camping once. To fit everything in the car, it must have a large trunk and preferably a hitch.

It's extremely difficult to choose a car that suits every situation. If it is a large SUV, then it is good for the family, but not economical for business trips. If it's a normal sedan, then it will be difficult to go on a camping vacation.

If buying a car is so difficult, how complicated is it to commit to a container platform? And for which companies is it even suitable?

If we look at the core competencies and features of Kubernetes, the question arises as to which companies it makes sense to use Kubernetes for. There is no one-size-fits-all answer, but there are indications of when Kubernetes is helpful.

A company develops applications that consist of multiple services, microservices or components. A separate stack of applications is deployed for each customer. The more services you have at the same time, the more sense it makes to have a platform that simplifies the management of applications and makes it more efficient. Kubernetes can help with the deployment, scaling, and management of these applications and reduces complexity by automating standard operational tasks.

If the following aspects apply to your company, then you are on the right track with Kubernetes:

- **Scalability**
  Do you want your applications to cope with peak loads or strong growth? Kubernetes enables the automatic scaling of resources to meet utilization and performance requirements.

- **High availability**
  Do your applications have high availability requirements? The simple distribution of applications across multiple nodes and the automatic restart in the event of errors or failures means that Kubernetes guarantees a high level of reliability.

- **Efficient resource utilization**
  Would you like to increase the utilization of your hardware? Kubernetes makes it possible to scale resources such as CPU and memory according to actual requirements and thus to pay only for what you actually use.

- **Flexibility and portability**
  Do you develop platform-independent applications that are deployed in different environments? Kubernetes provides a standardized platform for the execution of containers. This facilitates the portability of applications between different cloud providers or local data centers.

However, Kubernetes can also be useful for smaller companies, especially if they develop complex applications or are growing rapidly. It provides flexibility, scalability, and improved efficiency in the provision and management of applications.

1.2.5 Which Companies Should Not Use Kubernetes?

Kubernetes is not suitable for each and every company, just like your doctor doesn't always recommend the same balm for every skin problem. The balm can help in many cases, but not in every case.

The introduction of new technologies brings not only opportunities, but also obligations. The following points play an important role in this regard:

- **Specialist knowledge**
  The implementation and maintenance of Kubernetes requires extensive knowledge.
- **Cost**
  The need for specialized personnel can significantly increase the cost of using Kubernetes.
- **Change**
  The use of Kubernetes leads to a change in working methods and process design in IT operations.
- **Avoiding over engineering**
  Premature or excessive technical fine-tuning can hamper the development of a company.
- **Growth**
  Excessive complexity and high operating costs can slow the growth of a startup or small business.

Let's go through two sample situations in which I would not recommend the use of Kubernetes from the outset.

First, imagine that you work in a startup and create a simple application with a three-tier architecture (frontend, backend, and database). The technology stack is straightforward; the operation of your application is uncomplicated and can be managed with minimal effort. You only have a few customers and are currently more concerned with the further development of features and customer acquisition.

In this case, a single server with containers or a cloud service for container operation can be completely sufficient.

In another case, imagine that your applications have special infrastructure requirements. Especially with legacy applications that are to be migrated to Kubernetes, it can happen that things simply don't "fit." For example, I have seen a company's application that absolutely needed a fixed IP address so that it could be enabled for access to other resources in the firewall. Although there are ways to implement such requirements in Kubernetes, you will lose all the advantages of the system, which means you can save yourself the effort.

In addition, particularly strict guidelines on compliance, security, or data protection can also mean that the use of Kubernetes does not make any sense.

We had a similar challenge with another application. The company wanted to migrate a data import job that was previously running on AWS ECS to Kubernetes. However, the application was developed in such a way that a function based on AWS Lambda checked the start condition and then started the job if necessary. The container for the import itself required around 25 GB of RAM and did not fit into the company's other microservice landscape. The challenge was to determine how this service could be meaningfully mapped using Kubernetes resources.

Sometimes, you conclude that it just doesn't make any sense. In this example, however, we migrated the job to Kubernetes with some development work and rebuilt it accordingly.

In such cases, you need to clearly weigh whether migration to Kubernetes makes sense and what benefits you want to gain from it. To put it more generally: there are no one-size-fits-all solutions, and Kubernetes is not a panacea either. Each company must examine its own requirements and assess whether the use of the system brings more advantages than disadvantages. As a developer, you also need to check whether Kubernetes is the right platform for your applications.

1.3 Architecture and Components

Let's now move on to the architecture of Kubernetes. With Kubernetes, several components work together to ultimately provide what you call by the generic term Kubernetes. In the following sections, you will see how Kubernetes works at its core. It always makes me realize that Kubernetes is not witchcraft either, but just introduces another level of abstraction. It takes work off our hands because we no longer have to worry about the underlying hardware, but under the hood there is the same technology as ever.

With Kubernetes, a distinction is made between master nodes (*masters* for short) and worker nodes (*workers* for short). A *node* is a server that is part of a Kubernetes cluster. The masters are also referred to as *control plane nodes* because they run the services that are considered the brains of Kubernetes. The workers are controlled by the masters and receive commands to start containers and provide them with everything they need: from storage and secrets to network connections.

1   Introduction to Kubernetes

Figure 1.6 shows the services divided into masters and workers and how they communicate with each other. Use this overview to see what the big picture looks like in the following sections. I will now go into each individual component and its significance.

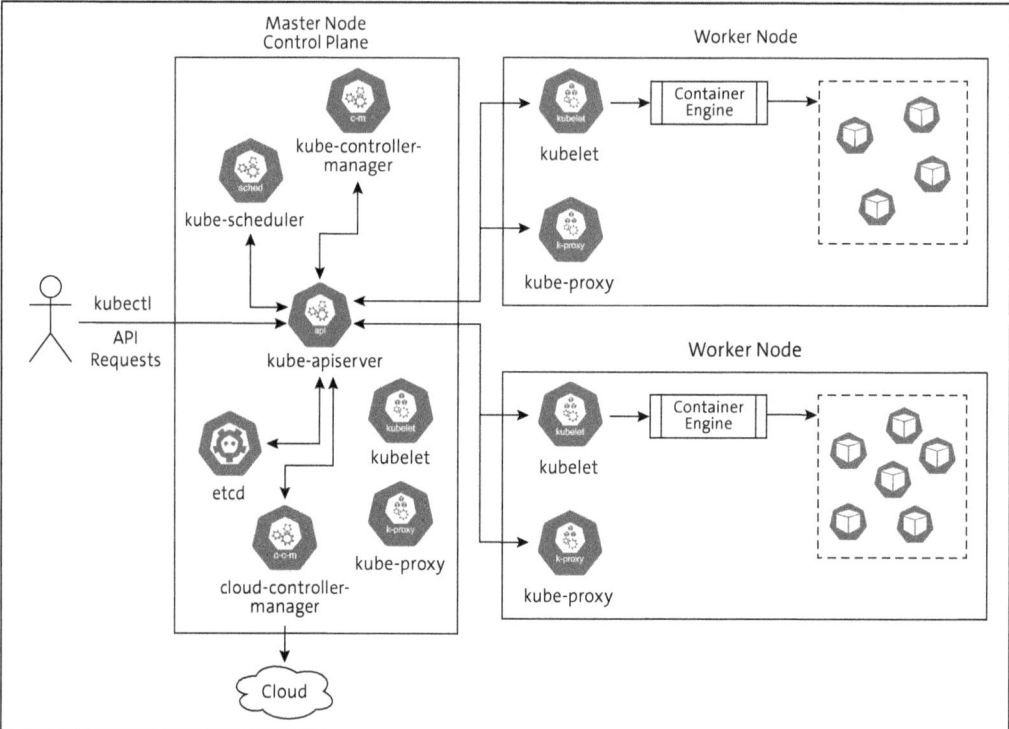

Figure 1.6 Kubernetes Architecture

1.3.1   Master Nodes

Let's start with the master nodes. In the simplest version, your Kubernetes cluster has a master or control plane, which controls the cluster, registers workers, and manages resources.

The masters are the brains of the cluster. They store configuration and status data, provide the API, and ensure that new containers are deployed. The masters monitor the cluster and its resources and decide on which worker containers will be executed. There are several services that take over the tasks. I will introduce you to these in a moment.

The Kubernetes master executes various server and manager processes for the cluster, which themselves also run in containers. As the software has matured, new components have been developed to meet specific requirements, culminating in what Kubernetes is today. Let us now take a closer look at the individual components and their function.

kube-apiserver

The *kube-apiserver* component is of central importance for the operation of the Kubernetes cluster. It is the center of communication, as you will see in Section 1.3.3. All calls, for both internal and external traffic, are processed via this component.

In addition to providing the API, it is also responsible for tasks such as the following:

- Validating requests and manifests
- Checking authorizations
- Monitoring rate limits and quotas

In addition, *kube-apiserver* is the only component that establishes a connection to the *etcd* database. The cluster would not work without it. You could no longer control anything, and nothing could change within the cluster.

The advantage of this central component is that nothing happens without *kube-apiserver* being aware of it. For example, you can implement `watch` requests to receive information when certain resources change or are newly created.

kube-scheduler

If you want to know where there is still room for a container on your cluster, it is best to ask *kube-scheduler*. This component knows your nodes and how much CPU and memory are available. It also has a plan of how many resources have already been reserved, and it knows all your rules that allow you to influence pod scheduling. There are affinities, taints, and tolerations for this, which we will look at in more detail in Chapter 2, Section 2.2.3.

kube-scheduler takes all of this into account in its algorithm to determine which node can best host additional containers. It always tries to achieve a certain balance across the cluster and uses preemptions to "displace" containers to new nodes if necessary.

kube-scheduler is always in close contact with *kube-apiserver* to receive new requests and information about nodes and containers.

The etcd Database

Looking for the brain of Kubernetes? Then *etcd* is the right place for you. As a key-value database, *etcd* is not only used by Kubernetes, but is also of interest for other distributed systems. It uses the so-called raft consensus algorithm to provide highly available data persistence with the quorum concept.

> **Quorum**
>
> The *quorum* is a concept in the theory of distributed systems that refers to the minimum number of nodes required to perform a certain operation in a distributed system or to make a decision. This ensures consistency in a cluster.

> Imagine the following scenario: You have a database in which you store your data. When you retrieve data, you receive it from the database. So far, so good—but if you now operate the database with two distributed instances to increase reliability, things get complicated. What happens if you query both databases and they each return a different result? How do you decide which of the two is right?
>
> The quorum is a way of maintaining consistency, because in this case the majority of nodes is right. To avoid a stalemate, an odd number of nodes is always used in a cluster. This means that in a cluster with three nodes, one can fail without any problems; with five nodes, it's two; and so on. In most cases, a cluster of three nodes is used in a production environment.

As long as the *etcd* database can provide its data, your Kubernetes cluster will be able to get out of any predicament. *etcd* saves all manifests of resources of the cluster and thus always maintains the desired state. For example, if a node fails, Kubernetes can roll out your containers again on a new node using your manifests.

The only interface for the *etcd* database is *kube-apiserver*. Everything that goes into or out of *etcd* can therefore only be carried out via *kube-apiserver*. This ensures that only authorized actions can manipulate the stored information.

cloud-controller-manager

The *cloud-controller-manager* component handles communication with other cloud services. This allows Kubernetes itself to remain independent, as the cloud services are integrated via *cloud-controller-manager*. For this purpose, a plug-in mechanism is used that makes virtually anything possible:

- Managing the cluster
- Deleting Kubernetes resources
- Creating infrastructure in the cloud such as load balancers when a specific Kubernetes object is created
- Deleting nodes when the infrastructure in the cloud is deleted

cloud-controller-manager therefore makes operating a cluster in the cloud more convenient. However, cluster management tools such as Rancher also use these components to gain access to the cluster and manage it.

kube-controller-manager

Perhaps one of your nodes doesn't seem to be doing so well, and the containers on it are having problems. Or maybe a container had an error and was terminated. Fortunately, you can use *kube-controller-manager*, which carries out the monitoring of various functions independently and automatically. It is comparable to a worker in the engine room.

There are many different controllers with different tasks. To simplify matters, these controllers are grouped together under their manager and provided as a single binary file. An overview of this is shown in Table 1.1.

Controller	Function
Node controller	Monitors all nodes and will actively evacuate the containers from a node if it is no longer intact.
Replication controller	Regularly checks the correct number of containers. If one is no longer functional, it takes care of starting a new one.
Endpoints controller	Takes care of the connection between services and containers. You will become familiar with this process in Chapter 2, Section 2.5.
Service account controller	Creates standard service accounts and API access tokens for newly created namespaces.

Table 1.1 Controllers under kube-controller-manager and Their Functions

At this point, let's take a closer look at the node controller. This controller recognizes immediately if a node is not working correctly and cannot be reached, for example. The node controller communicates regularly with the nodes, and each node needs a so-called `kube-node-lease` that it must renew on a regular basis. This is a heartbeat that allows the node controller to recognize that the node is still alive.

If this heartbeat does not occur within a certain time slot, the node controller becomes active and takes care of the evacuation of the containers and ensures that they are rebuilt on a functioning node. It also maintains a list of available nodes and updates it when new ones are added or old ones need to be removed. It also takes care of the onboarding of new nodes and assigns Classless Inter-Domain Routings (CIDRs) to them, for example.

The node controller is therefore an important component for keeping a Kubernetes cluster alive.

1.3.2 Worker Nodes

The kubelet and the kube proxy—as well as the container engine, which starts and keeps the containers running—are executed on all worker nodes. You will get to know the container engines in Chapter 2, Section 2.1.1. The worker is the one who carries out the work at the end. No management processes run on it, but only application containers, which makes workers interchangeable. This is precisely where the magic of Kubernetes lies, because it means that a worker can fail or be replaced and the applications will still continue to run. Let's now look at the components that run on the worker nodes.

The Kubelet

Imagine you own an apartment building with 60 residential units. As the owner, you don't want to take care of the management of the apartments yourself and so you hire a janitor. The janitor looks after the apartments and ensures that each one is in the desired condition. If a new tenant moves in, you as the owner establish clear rules for how the apartment is to be used and the janitor takes care of enforcing them.

In this metaphor, you are *kube-apiserver*, the janitor is the kubelet, and the apartments are the containers. Thus, the kubelet is the central component on all nodes, which also takes care of the registration of new nodes. To do this, it registers with *kube-apiserver*. The *kube-apiserver* component can then pass jobs to the kubelet in the form of manifests in order to deploy containers. A *manifest* describes everything that is necessary to execute the container. If a container requires access to storage, secrets, or configurations, the kubelet ensures access.

Another task of the kubelet is to monitor the containers. The status is also sent back to the *kube-apiserver*, which in turn saves it in the *etcd* database. The kubelet does not work alone, but interacts with the underlying container engine, which is ultimately responsible for executing containers.

Good to Know

The kubelet also runs on the master nodes, as these also run containers that the kubelet takes care of.

The Kube Proxy

Let's continue to use the image of your apartment building. Imagine you don't just have one house, but 10 of them combined into one building complex. There is a doorman in the entrance area of every house. Every doorman knows exactly which tenant lives in which apartment and in which building, because you as the owner always let them know when a new apartment is being moved into. Now, when the courier arrives, the doorman directs the mail to the right house, so that packages always arrive at the right apartment.

The doorman is the kube proxy that is contacted by the *kube-apiserver* when a new container gets deployed. Every kube proxy on every node knows about this at all times. This is the only way to ensure that the data packages reach the right container. The kube proxy is therefore responsible for managing the network connectivity to the containers.

1.3.3 API Call Flow

Kubernetes consists of numerous components that have to interact with each other. To give you a better idea of how the communication between the components takes place in order to deploy a container at the end, I will take you through a simplified example. I will leave out the different Kubernetes objects that you will learn about in Chapter 2 for now, as these make it much more difficult to understand the API flow.

Assume that you want to deploy a container in Kubernetes. Using the `kubectl` tool, which you will learn more about in Section 1.5, you can send a request with a manifest of your desired container to *kube-apiserver*. The manifest is written in the YAML language, which we will take a closer look at in Chapter 3, Section 3.2.

The manifest contains everything Kubernetes needs to know to set up the container. As soon as you send the request, *kube-apiserver* starts a process that basically runs as follows:

1. *kube-apiserver* accepts the request and saves the manifest in the *etcd* database.
2. *kube-controller-manager* becomes active and receives information from *kube-apiserver* that there is a new manifest.
3. *kube-controller-manager* asks *kube-apiserver* whether the container has already been deployed according to the manifest and, if so, whether the current status corresponds to the desired status.
4. *kube-apiserver* responds that the container does not yet exist.
5. *kube-controller-manager* gives *kube-apiserver* the command to create the container.
6. *kube-apiserver* contacts *kube-scheduler* to check which worker the container can be deployed to. The scheduler then responds to it.
7. *kube-apiserver* sends the necessary information from the manifest to the kubelet of the corresponding worker that is to build the container.
8. In addition, *kube-apiserver* sends network information to each kube proxy that this container is made available on the corresponding worker.
9. The kubelet on the worker will then ensure that the container is created in the container engine and receives all the necessary resources, such as secrets or volumes, that are requested in the manifest.
10. The kubelet returns the information about the successful deployment to *kube-apiserver*, which saves the information in *etcd*.

> **Good to Know**
>
> This process is repeated again and again, even if there is only a small change such as increasing the memory of the container.

As you will learn in Chapter 2, there are different objects in Kubernetes, some of which build on each other. Here too, this API flow is run through again and again for each object. This may sound a bit much at first and can seem dauntingly complicated due to the amount of communication. However, the concept from Section 1.1.5 was also applied here—that is, separation of concerns.

You have probably also noticed that *kube-apiserver* is always involved in the communication. Kubernetes is based on the hub and spoke architecture (or hub and spoke API pattern). There is a hub as a central point through which all requests and messages flow. It acts as an intermediary and controls the data traffic between the various end points. The spokes are the end points that are connected to the hub. Figure 1.7 offers a simple illustration. Each spoke is responsible for a specific function or service and interacts with other spokes via the hub.

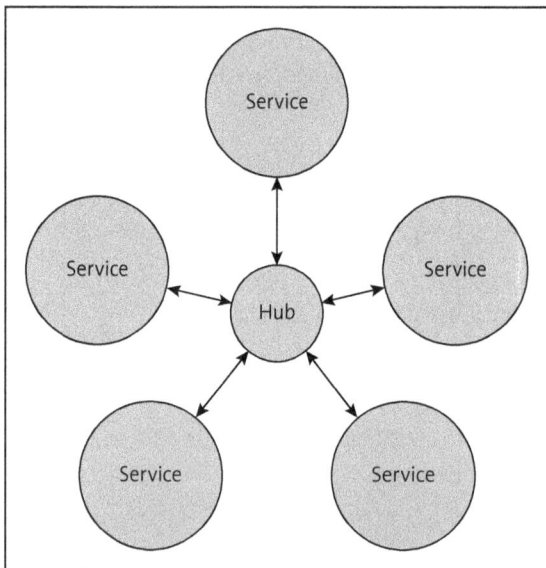

Figure 1.7 Communication through Hub and Spoke Architecture

Even if direct communication would be faster at first, this model is easier for many small services. If you take a look at Figure 1.8, you can see that there are a lot of communication channels with just five services. Each additional service increases complexity, and you need to familiarize each new service with each existing service. In the hub and spoke architecture, each spoke only communicates with the hub. The hub takes care of distribution, and if a new service is added, the hub can also receive messages from it and send them to other services.

A nice side effect of centralized communication is easy monitoring. All transactions run via the hub, which makes it easier to analyze errors. With Kubernetes, for example, you also ensure that not every service can write to the *etcd* database. This contributes to consistency and in turn reduces errors. Precisely because *etcd* is such a critical component, a well-considered communication architecture is crucial.

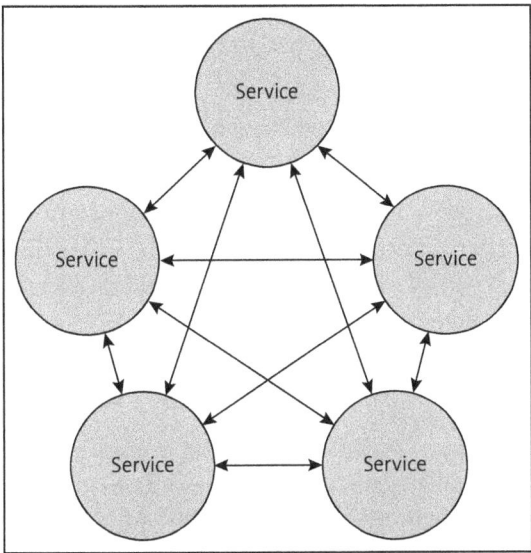

Figure 1.8 Illustration of Direct Communication

1.4 A Kubernetes Cluster on Your Computer

After all the theory, let's finally get down to the practical part. A Kubernetes cluster always consists of multiple servers (as already described). However, Minikube was developed so that you don't have to set up a server farm to learn K8s and can be able to test and play without much effort.

> **Note**
>
> In the following sections, I will introduce you to the Minikube tool, which only requires one computer. However, it is ideal if you have multiple computers available that you can use to build a small test cluster. In addition, at some points in the book, we will reach the limits of Minikube. For this reason, in addition to Minikube, I present a cost-effective way of setting up a Kubernetes cluster based on Raspberry Pis in Section 1.7.
>
> However, it is optional, and I will make it clear in the book when a demo with multiple nodes makes sense.

Minikube simulates a Kubernetes cluster on your local computer using a container or VM. This works very well for experimentation purposes, but in the end it is only a simulation of a real distributed Kubernetes setup. The performance and size of the cluster are limited to your computer, but for almost all the exercises in this book, Minikube will suffice.

In the following sections, I will guide you through the installation for the different operating systems and show you how to get Minikube up and running with Docker.

An installed Docker engine is required for the installation. I use Docker Desktop for that. You can find the installation instructions at the following address: *http://s-prs.co/v596403*.

The next chapters are all designed for the operation of Minikube in Docker. If you still want to start Minikube with a VM manager, you should take a look at the installation instructions available at *http://s-prs.co/v596404*. However, I recommend that you follow the instructions presented here to avoid possible incompatibilities.

> **Important Note for Company Computers**
>
> If you want to carry out the following instructions with a device that is managed by your employer, this can lead to problems. Most workstations have restricted rights or certain security policies that prevent the instructions provided here from working. I recommend that you use a computer that is not managed by a company and on which you have full admin rights. And of course, it makes sense perhaps not to use your own workstation with important data for such experiments.
>
> If you still want to use a company computer, then contact your company's administrator if you have any problems.
>
> If you use your company computer, you may need a license for Docker Desktop. Please check this beforehand.

1.4.1 Minikube on macOS

There are different ways to install Minikube for Mac. It is a command line tool and is also installed via the terminal. I'll show you two options, the first of which is the simplest.

Installation via the Homebrew Package Manager

The easiest way is to use a package manager called *Homebrew*. It makes installing software quick and easy, because where you would normally have to download, install, and configure packages manually, Homebrew does it for you. If you have not yet installed a package manager for your Mac, I recommend that you do so now.

Open the terminal and run the following command:

```
/bin/bash -c "$(curl -fsSL https://raw.githubusercontent.com/Homebrew/install/HEAD/install.sh)"
```

You will then be asked to enter your password and confirm the installation by pressing Enter. After the installation has been completed, you can use the package manager with the brew command.

You can now carry out the installation of Minikube. To do this, enter the following command in your terminal:

```
brew install minikube
```

This will download and install Minikube from Homebrew. Once the installation is complete, you can use the minikube version command to test whether the software has been installed and if it is ready. The command line should then output the corresponding version of Minikube. In my case, the output looks as follows:

```
minikube version
minikube version: v1.30.1
commit: 08896fd1dc362c097c925146c4a0d0dac715ace0
```

Note that you may have a newer version depending on when you read this book.

Manual Installation

If you do not wish to install the Homebrew package manager, you can also install Minikube manually. First, you need to download the installation files that match your processor architecture. The following command is suitable for Macs with Intel processors:

```
curl -LO https://storage.googleapis.com/minikube/releases/latest/minikube-darwin-amd64
```

Then, install Minikube. You need to have admin rights for this step:

```
sudo install minikube-darwin-amd64 /usr/local/bin/minikube
```

Run the following commands if your Mac has an Apple processor:

```
curl -LO https://storage.googleapis.com/minikube/releases/latest/minikube-darwin-arm64
sudo install minikube-darwin-arm64 /usr/local/bin/minikube
```

You should now also be able to test whether the installation was successful using the minikube version command.

1.4.2 Minikube on Linux

In the following sections, I will address the most common installations for Linux. You can find a complete selection at *http://s-prs.co/v596405*.

Installation on Linux with x86-64 Architecture

If you have an x86-64 architecture, you can perform the installation in three ways, depending on which distribution you are using. A Minikube package may also be available in a repository for easy installation.

If you need a Debian package, the following two commands will download the installation file and install Minikube:

```
curl -LO https://storage.googleapis.com/minikube/releases/latest/minikube_latest_amd64.deb
sudo dpkg -i minikube_latest_amd64.deb
```

If you use an RPM distribution, this will get you there:

```
curl -LO https://storage.googleapis.com/minikube/releases/latest/minikube-latest.x86_64.rpm
sudo rpm -Uvh minikube-latest.x86_64.rpm
```

You can also download and install the binary file directly:

```
curl -LO https://storage.googleapis.com/minikube/releases/latest/minikube-linux-amd64
sudo install minikube-linux-amd64 /usr/local/bin/minikube
```

Installation on Linux with ARM64 Architecture

There are also three ways to install the ARM64 architecture.

With a Debian package:

```
curl -LO https://storage.googleapis.com/minikube/releases/latest/minikube_latest_arm64.deb
sudo dpkg -i minikube_latest_arm64.deb
```

With an RPM package:

```
curl -LO https://storage.googleapis.com/minikube/releases/latest/minikube-latest.aarch64.rpm
sudo rpm -Uvh minikube-latest.aarch64.rpm
```

Downloading the binary file:

```
curl -LO https://storage.googleapis.com/minikube/releases/latest/minikube-linux-arm64
sudo install minikube-linux-arm64 /usr/local/bin/minikube
```

1.4.3 Minikube on Windows

Three options are available for the installation on Windows. If you have already installed the Chocolatey package manager or Windows Package Manager, you can skip to the corresponding instructions. The installation via a package manager is much easier, but you must install one first.

> **Package Manager for Windows**
>
> If you want to use a package manager and have installed Windows 10 or Windows 11, you should take a look at Windows Package Manager. This significantly simplifies the installation of programs such as Minikube. You can find detailed instructions from Microsoft at the following link: *http://s-prs.co/v596406*.

Installation Using Chocolatey

The installation using the Chocolatey package manager is very simple. You need to run the following command in your PowerShell:

```
choco install minikube
```

Installation Using the Windows Package Manager

The installation is also easy via Windows Package Manager. Just run the following command in PowerShell:

```
winget install minikube
```

Manual Installation

The commands for a manual installation are somewhat more complex. To avoid having to type the commands from the book, I recommend that you copy the commands from the Minikube installation page. To do this, go to *http://s-prs.co/v596405* and select the **Windows** operating system and **.exe download** in **Installer Type**.

Then you can copy the following command and paste it into the PowerShell to download Minikube:

```
New-Item -Path 'c:\' -Name 'minikube' -ItemType Directory
  -Force Invoke-WebRequest -OutFile 'c:\minikube\minikube.exe'
  -Uri 'https://github.com/kubernetes/minikube/releases/latest/download/
        minikube-windows-amd64.exe'
  -UseBasicParsing
```

For Minikube to be executable in PowerShell, the program must be entered in the PATH variable. The following command takes care of this (note that this command requires admin rights). Then you must start PowerShell as an administrator:

1 Introduction to Kubernetes

```
$oldPath = [Environment]::GetEnvironmentVariable('Path',
                     [EnvironmentVariableTarget]::Machine)
if ($oldPath.Split(';') -inotcontains 'C:\minikube'){
  [Environment]::SetEnvironmentVariable('Path', $('{0};C:\minikube' -f
                     $oldPath), [EnvironmentVariableTarget]::Machine)
}
```

You can now close and reopen PowerShell and use the `minikube version` command to test whether the software has been installed and if it is ready to use. The command line should then output the corresponding version of Minikube. In my case, the output looks as follows:

```
minikube version
minikube version: v1.30.1
commit: 08896fd1dc362c097c925146c4a0d0dac715ace0
```

Note that you may have a newer version depending on when you read this book.

1.4.4 Launching Minikube

Once you have installed Minikube, you can easily launch it from your command line. To do this, you want to run the following command:

```
minikube start
```

Minikube then creates some activity and documents all processes in logs, which will be output directly. Because it is based on Docker, Minikube will download the latest container first. Minikube also indicates which version of Kubernetes is being started. You will then see information about additional add-ons that are not relevant for the time being. You do not need to make a note of any of this as we will come back to it in due course.

You now have a Kubernetes cluster running in a Docker container on your computer.

1.4.5 Controlling Minikube

Before we get to the interaction with the Kubernetes cluster, I would like to briefly explain how you can use Minikube. We have already executed the `minikube start` command. You have used this command to start the cluster in a Docker container.

If you want to pause the containers running in the cluster, you can use the `minikube pause` and `minikube unpause` commands. These will both pause your containers, which we will deploy in later chapters, and stop the system containers that make up Kubernetes. You should use these commands when you are not using your test cluster as doing so saves resources.

You can also stop Minikube by running `minikube stop`. This stops the Minikube container completely. However, the state remains the same. This command is good to run before you shut down your computer. The same container is restarted via the start command and continues in the same state as before it was stopped.

Finally, you can use the `minikube delete` command to delete your cluster entirely. This command is particularly useful if you need a fresh cluster and want to get rid of your old tests.

These commands enable you to control Minikube. There are a few other commands that are relevant, and we will take a closer look at them in the corresponding chapters.

1.4.6 Possible Errors when Starting Minikube

You may encounter two small errors when you start Minikube. Let's take a brief look at how this happens and how you can solve it if you receive the corresponding error message.

If you run the `minikube start` command, the following error may occur:

```
Exiting due to PROVIDER_DOCKER_NOT_RUNNING:
"docker version --format -:" exit status 1: Cannot connect to the
Docker daemon at unix:///Users/kevinwelter/.docker/run/docker.sock.
Is the docker daemon running?
```

This error may look slightly different on Windows, but the issue is the same, and the hint is already in the error message. The `Is the docker daemon running?` message indicates that Docker Desktop is not running. Start Docker and try to execute the command again. Minikube should now start. This error can occur especially after restarting your computer if you have not activated Docker via autostart.

Another error message after you run the `minikube start` command might look something like this:

```
command not found: minikube
```

This indicates that the command line interface tool cannot be found. If you have carried out the installation steps given earlier correctly, then try to restart the command line. In some cases, especially when installing manually, the tool may not yet be activated in the path.

1.4.7 Container Registry of Minikube

Throughout this book, you will do exercises that require the Minikube *container registry* add-on. Here I will show you how to install and use it. You can also skip this for now and return when you need the extension. I will point this out at the appropriate place.

1 Introduction to Kubernetes

Minikube comes with some add-ons that allow you to build a nice and simple Kubernetes test environment without the need for external dependencies.

You can use the `minikube addons list` command to get an overview of the extensions. We won't need them all, but perhaps you will come back to them at some point.

The most important add-on is the container registry. If you want to develop your own containers and deploy them in Kubernetes, there is no way around a registry, as Kubernetes only retrieves the images required for the containers from there. In production environments, you naturally need a professional registry to manage and securely store your images. For our test cluster, however, such an effort is excessive and we can revert to the useful add-on.

The add-on can be activated using the `minikube addons enable registry` command. Now the next part is important! This is because a port that you should use is displayed as the output. You do *not* need this!

Instead, you want to run the `eval $(minikube docker-env)` command on Linux or on a Mac. In PowerShell, the command is `minikube -p minikube docker-env | Invoke-Expression`.

This call makes sure that you use the Docker daemon from Minikube. You can then also access the registry via the default port 5000.

> **Note**
>
> You must run the `eval $(minikube docker-env)` command with every new command line; otherwise, you will not be able to access the registry. If you don't want to think about it every time, you can also write the command in your `.bashrc` or `.zshrc`, depending on the command line, so that it gets always executed. You can enter the command in your profile in PowerShell.
>
> Note that you then use the Docker host from Minikube.

Now let's test whether you can store containers in the registry. Use the following one-line Dockerfile for this purpose:

`FROM nginx`

Create this as a Dockerfile and run the `docker build -t localhost:5000/test-nginx .` command. Your own Nginx image will now be built and tagged with the name of the registry. Then you can store the image in the registry using the `docker push localhost:5000/test-nginx` command. From now on, Kubernetes can access the image with the image name and download it.

> **Important Convention for the Image Name**
>
> If you are familiar with Docker, then you will certainly also know the naming conventions for images. You must start the name using the URL of the registry, as this is the only way Docker can also assign the image to a registry and store it there in the event of a push.
>
> You can of course continue to name the images locally as you wish.

1.5 Interaction with Kubernetes via the Command Line and Dashboard

Kubernetes comes with two options for interacting with the cluster: kubectl as a tool for the command line, and the *Kubernetes dashboard*, which can be accessed via the browser.

Both tools use the Kubernetes API and make their requests to *kube-apiserver* in the control plane.

1.5.1 Minikube Comes With kubectl

The easiest way to communicate with your Kubernetes cluster via kubectl is to use the kubectl instance that comes with Minikube. This instance is always compatible with the corresponding cluster version and can simply be used via the minikube kubectl command, which of course is particularly useful if you need an older version of kubectl because the clusters in your company have a different version.

And that's all. However, you should set an alias in your command line so that you do not always need to type the entire minikube kubectl command.

1.5.2 Installing kubectl

The kubectl instance provided with Minikube is very helpful and easy to use for the development environment. In a production environment, you should not rely on kubectl supplied by Minikube. First, the dependency between kubectl and Minikube is unnecessary, and second, you cannot install an independent version of kubectl.

For this reason, we will now take a closer look at how to install the "right" CLI tool on your computer. As with Minikube, this depends on the operating system.

> **Version and Version Conflicts**
>
> In the following sections, I will show you how to install the latest version of kubectl. At the time of writing this chapter, that's version 1.27, the same version as the Minikube cluster. If you also use the latest version of Minikube, there should be no conflicts. However, you must check that the versions are the same.
>
> The kubectl instance is always one version upward and one version downward compatible. Thus, if you install version 1.27, you can control clusters with versions 1.26, 1.27, and 1.28. In case you use an older version in your company, you will find a link to the Kubernetes documentation for each operating system at the end of the relevant section. There you can read how to install an older version of kubectl.

kubectl on macOS

The easiest way to install kubectl on macOS is also via the Homebrew package manager. If you installed this in Section 1.4.1, you can simply run the following command:

```
brew install kubectl
```

That's it. Homebrew will then install the appropriate package.

If you want to install without the package manager or install an older version, you can also install it manually. The manual installation is somewhat more complex with kubectl and differs depending on the processor type. You can find the current commands at the following link: *http://s-prs.co/v596407*.

kubectl on Linux

You can also use the native package manager for Linux to install kubectl.

Installation Using the apt Package Manager

The default apt package manager is available for operating systems based on Debian. To use it, run the following commands:

```
sudo apt update
sudo apt install -y ca-certificates curl
curl -fsSL https://packages.cloud.google.com/apt/doc/apt-key.gpg | \
    sudo gpg --dearmor -o /etc/apt/keyrings/kubernetes-archive-keyring.gpg
echo "deb [signed-by=/etc/apt/keyrings/kubernetes-archive-keyring.gpg]
    https://apt.kubernetes.io/ kubernetes-xenial main" | \
    sudo tee /etc/apt/sources.list.d/kubernetes.list
sudo apt update
sudo apt install -y kubectl
```

For Debian version 9 or older, you need the following package:

```
sudo apt install -y apt-transport-https
```

For versions older than Debian 12 and Ubuntu 22.04, you may have to create the */etc/apt/keyrings* folder manually.

Installation Using the yum Package Manager

The default `yum` package manager is available for operating systems based on RedHat. To use it, run the following commands:

```
cat <<EOF | sudo tee /etc/yum.repos.d/kubernetes.repo
[kubernetes]
name=Kubernetes
baseurl=https://packages.cloud.google.com/yum/repos/kubernetes-el7-\$basearch
enabled=1
gpgcheck=1
gpgkey=https://packages.cloud.google.com/yum/doc/yum-key.gpg https://packages.cloud.google.com/yum/doc/rpm-package-key.gpg
EOF
sudo yum install -y kubectl
```

You can also install the packages for Linux without a package manager. The instructions for this as well as the instructions for installing older versions can be found at the following address: *http://s-prs.co/v596408*.

kubectl on Windows

As with Minikube, you can use the Chocolatey and Winget package managers on Windows systems.

Run the following command in PowerShell for Chocolatey:

```
choco install kubernetes-cli
```

For Winget, run the following command in PowerShell:

```
winget install -e --id Kubernetes.kubectl
```

You can also install the packages for Windows without a package manager. The instructions for this as well as the instructions for installing older versions can be found at the following address: *http://s-prs.co/v596409*.

Function Test for kubectl

You can now check whether you can run `kubectl`:

```
kubectl version --client
```

The output should read something like this:

```
Client Version: version.Info{Major:"1", Minor:"27", GitVersion:"v1.27.3",
GitCommit:"25b4e43193bcda6c7328a6d147b1fb73a33f1598", […] }
```

Depending on the operating system, further system information is also displayed. However, the important thing here is that you can run kubectl and that you have the latest version. You should also check again against the Minikube cluster for whether there could be version conflicts, as mentioned earlier.

Activating Autocompletion for kubectl

For your Linux or macOS command lines (Bash, Zsh, or Fish), and for your Windows PowerShell, kubectl provides very useful autocompletion options by pressing Tab. Because calls using kubectl can become very long and complex, working without these completions is really no fun. To enable autocompletion, follow these steps:

1. To install the completion, run the appropriate command for your package manager:

```
# For macOS
brew install bash-completion@2
# For Linux
apt install bash-completion
yum install bash-completion
```

Note the output after the installation and add the corresponding line to the *~/.bash_profile* file so that Bash completion is permanently activated. This should look something like this:

```
Add the following line to your ~/.bash_profile:
  [[ -r "/usr/local/etc/profile.d/bash_completion.sh" ]] && . "/usr/
  local/etc/profile.d/bash_completion.sh"
```

2. Finally, run one of the following commands:

Installing for Bash:

```
echo 'source <(kubectl completion bash)' >>~/.bash_profile
```

Installing for Zsh:

```
echo 'source <(kubectl completion zsh)' >>~/.zshrc
```

Installing for Fish:

```
echo 'kubectl completion fish | source' >>~/.config/fish/config.fish
```

Installing for PowerShell:

```
kubectl completion powershell | Out-String | Invoke-Expression
```

Once you have activated autocompletion, you need to restart your command line. Now the autocompletion of kubectl should work.

> There are also other little helpers in the shell that make working with Kubernetes clusters easier. For example, take a look at the ZSH plugin for kubectl, which you can find at *http://s-prs.co/v596410*. It comes with a large number of aliases that make your work much easier.

1.5.3 Accessing the Cluster Using Kubeconfig

To be able to access Kubernetes clusters using the kubectl CLI tool, you need to configure the tool. First, the tool needs to know which cluster it should address and how it can reach it. Second, kubectl must authenticate itself against the Kubernetes API. This is defined in the so-called Kubeconfig file, or Kubeconfig for short. kubectl searches for the file

1. either automatically in *~/.kube/config*,
2. or it expects the paths to several files as environment parameters such as KUBECONFIG=~/.kube/config:/path/to/other/config.

This is the same for Linux, macOS, and Windows PowerShell.

Minikube configures kubectl automatically when you execute the minikube start command. If you installed kubectl in the previous section, you should run minikube stop again and then minikube start to be on the safe side. Minikube should also have created the Kubeconfig file correctly.

To check that Kubeconfig has been successfully created and that you can reach your Kubernetes cluster, run the kubectl get namespaces command. This then returns all namespaces that were created with the cluster by default. The output should read something like this:

```
NAME              STATUS    AGE
default           Active    16s
kube-node-lease   Active    17s
kube-public       Active    17s
kube-system       Active    18s
```

Let's analyze a Kubeconfig file in more detail to understand how Kubernetes can use it to connect to your cluster. As an example, we'll look at the Kubeconfig file generated by Minikube. This should look similar to the one shown in Listing 1.1.

```
apiVersion: v1
clusters:
- cluster:
    certificate-authority: /Users/kevinwelter/.minikube/ca.crt
    extensions:
    - extension:
```

```
      last-update: Sun, 16 Jul 2023 18:48:50 CEST
      provider: minikube.sigs.k8s.io
      version: v1.30.1
    name: cluster_info
   server: https://127.0.0.1:59746
  name: minikube
contexts:
- context:
    cluster: minikube
    extensions:
    - extension:
        last-update: Sun, 16 Jul 2023 18:48:50 CEST
        provider: minikube.sigs.k8s.io
        version: v1.30.1
      name: context_info
    namespace: default
    user: minikube
  name: minikube
current-context: minikube
kind: Config
preferences: {}
users:
- name: minikube
  user:
    client-certificate: /Users/kevinwelter/.minikube/profiles/minikube/
client.crt
    client-key: /Users/kevinwelter/.minikube/profiles/minikube/client.key
```

Listing 1.1 Kubeconfig from Minikube

The configuration is simple and quickly explained. It is divided into the following blocks:

- clusters
- contexts
- users

Information on the cluster itself can be found under the cluster item. For example, my Minikube cluster can be reached at the address `https://127.0.0.1:59746`. Information about certificates is also stored in this section. For example, you will find out that these are located under the path `/Users/kevinwelter/.minikube/ca.crt`.

A specific context for a cluster is saved under `contexts`, which `kubectl` uses when logging in. For this reason, the user with which you log in and which namespace is active after the start is stored here. You can see the active context under `current-context`.

Information about your user is stored under user, and the certificates you need for authentication are linked.

> **Readable Certificates**
> Kubeconfig often contains the certificates in the text. This allows anyone who has access to Kubeconfig to connect to the cluster.

In a company, you usually have multiple clusters for different environments. This also means that you have multiple Kubeconfig files. In addition to using the KUBECONFIG environment variable, you can also use the --kubeconfig option to provide kubectl with the path to Kubeconfig. However, this route is somewhat tedious in everyday life. Another option is to merge multiple Kubeconfig files into one Kubeconfig, in which case you do not need to set the environment variable. You can find an example of this in Section 1.7.3. I will show you the best way to work with multiple clusters in Section 1.5.6.

> **Note**
> You should discuss with the cluster admins how you want to obtain the Kubeconfig file for a cluster in your company. This varies slightly depending on the structure.

1.5.4 Namespaces

I have used the term *namespace* a few times now without explaining it properly. So let me make up for that at this point.

With Kubernetes, *namespace* is used in a similar way as it is in programming languages. A namespace is a separate area that allows you to isolate resources.

A namespace in Kubernetes is like a country in the real world. Each country (namespace) has its own government (resource management), laws (access rules), and inhabitants (pods, services, etc.). The resources and administration are isolated within a country (namespace) so that activities in one country cannot affect the other countries.

This means that a namespace offers you the options

- to assign access rights to users,
- to allocate quotas for resources, and
- to avoid name conflicts.

A classic use case on a shared development cluster occurs when you and a developer colleague are working on the same software in different branches. You both want to deploy a version on Kubernetes, but the names of the resources would lead to a conflict. If you create your own namespace, you won't get in each other's way.

1 Introduction to Kubernetes

However, not every resource is bound to a namespace in Kubernetes. In Section 1.5.5, you will get to know a command from `kubectl` to query the resources that are bound to namespaces.

At some point you may ask yourself how you can best cut namespaces and which applications should come together in a namespace. There is no universal answer to this question, as it depends in part on the cluster structure. I always look at the following points:

- Are the applications part of a larger coherent component?
- Do the applications need each other?
- Are the containers loosely coupled, but do they belong together?

If you answer *yes* to these questions, then you should deploy the applications to the same namespace. The good thing is that your decision does not have to be final. In the future, you will also be able to quickly deploy an application to a separate namespace.

Good to Know

If you set up a fresh cluster, Kubernetes will start with four initial namespaces:

- `default`
  This namespace is created so that you can start directly without having to create your own namespace.
- `kube-node-lease`
  This contains the lease objects that are connected to the Kubernetes nodes. Kubelet sends the heartbeats over it.
- `kube-public`
  This namespace is readable for everyone and is usually only used by the cluster.
- `kube-system`
  This is used for objects created by Kubernetes.

Note

You should only use the `default` namespace for quick tests. It is better to create a separate namespace for your application. Otherwise, conflicts can arise, especially in clusters that have multiple users.

1.5.5 kubectl Commands

Now that you know how to configure `kubectl` for a cluster, I would like to briefly introduce you to the structure of the tool and the most important commands. We will use the individual commands in detail in the following chapters.

> **Note**
>
> This section is perfectly suited for reference. If you have never used `kubectl` before, please follow the instructions. Otherwise, come back if you need one command or another again in the course of the book.

If you have used CLI tools before, you will quickly get used to `kubectl`. Like every CLI tool, `kubectl` also has a help function that you can always consult for the syntax. To do so, enter the following command:

```
kubectl --help
```

The shortened output will look as follows:

```
kubectl controls the Kubernetes-Cluster manager.
Find more information at: https://kubernetes.io/docs/reference/kubectl/
Basic Commands (Beginner):
    create         Create a resource from a file or from stdin
…
Basic Commands (Intermediate):
…
Deploy Commands:
…
Cluster Management Commands:
…
```

You can see the commands supported by `kubectl`. These are grouped into topics, and behind each command you will find an additional explanation. If you need help for a specific command, you can also call it for each command. If you want to get more information on the `create` command, you must run `kubectl create --help`.

You will then receive examples of the application in the output, other commands that can be combined with `create`, and options that you can use. The effect of the commands and options is also displayed.

This goes even further with most commands. If you now want to create a namespace using `create`, but do not know the exact syntax, then you can simply enter `kubectl create namespace --help` to receive the information you need in order to enter the name of your new namespace.

The really good thing about `kubectl` is that the commands have meaningful names, so it's easy if you know what you want to do. You want to create a pod? Then the command you need to enter is `create pod`.

If you want to have all pods in one output, you must use `get pods`.

If you want to delete a pod, you should enter `delete pod [PODNAME]`.

We will now go through the most important commands of kubectl, and then you will see how easy it is to use Kubernetes.

As a developer, you will hardly need some of the commands, as you can also use kubectl to administrate the cluster. For this reason, we will not go through all the commands, but you will get to know the most important ones.

Note

You can also create an alias for Kubernetes to avoid all the typing work.

For Linux and Mac:

`alias k='kubectl'`

For the PowerShell:

`Set-Alias -Name k -Value kubectl`

kubectl get

Let's start with a command that allows you to see which resources are running or are active in Kubernetes. To do this, use the `kubectl get` command. If you remember Section 1.5.3, then you already know this command. As a test, you used the `kubectl get namespaces` command to display the namespaces.

In our example, we want to display the generated pod. To do this, you simply need to replace namespaces with pods in the command—thus, `kubectl get pods`. When you run the test command in your console, you will see all pods that are available in your active namespace. However, these are not all the pods, as Kubernetes also uses some system pods that are executed in other namespaces. To assign a desired namespace to the command, you can use the `-n <namespace>` option to select a specific namespace. You can also specify the `-A` option to output all pods in all namespaces. Then the command would read `kubectl get pods -A`.

If you enter the command in your console, your output should look something like this:

```
NAMESPACE     NAME                         READY   STATUS    RESTARTS   AGE
kube-system   coredns-787d4945fb-qcsvv     1/1     Running   0          8d
kube-system   etcd-minikube                1/1     Running   0          8d
kube-system   kube-apiserver-minikube      1/1     Running   0          8d
kube-system   kube-proxy-42gdl             1/1     Running   0          8d
kube-system   kube-scheduler-minikube      1/1     Running   0          8d
kube-system   storage-provisioner          1/1     Running   0          8d
...
```

If you want to view a specific pod, you can also enter the name of the pod after the command. Note, however, that you must also specify the exact namespace. Thus, if we use the information from the first output to output the pod named `etcd-minikube`, the command looks as follows: `kubectl get pods -n kube-system etcd-minikube`.

An additional option of `get` that you will need often is `-o`. This allows you to customize the output format. For example, you can use the `-o wide` command to output more information when outputting the pods or use the `-o yaml` option to output the object as YAML code.

kubectl create

Let's now take a closer look at the command from the preceding example. You can use the `create` command to create Kubernetes resources.

When you run the `kubectl create namespace my-k8s` command, the `my-k8s` namespace should be created as a result. The `kubectl get namespace` command allows you to check whether or not the namespace has been created.

Now let's deploy the first pod in the new namespace. For this purpose, we use the deployment object. I will go into more detail about the Kubernetes objects and how they are connected in Chapter 2.

In this example, you'll install a Nginx web server in your new namespace. To do this, use the following command: `kubectl create deployment nginx --image=nginx -n my-k8s`. This command creates a deployment object with named `nginx` and passes `nginx` as the image to it. Kubernetes searches for this image in Docker Hub, downloads it, and creates a pod from it.

You need the `-n my-k8s` option so that Kubernetes knows in which namespace the object is supposed to be created. If you do not add the namespace, the deployment will be created in the `default` namespace.

Now you can check what Kubernetes has created. You can use `kubectl get pods -n my-k8s` to view the pod created and whether it is in the *running* state. With `kubectl get deployment -n my-k8s`, you can view the generated deployment object.

You have now entered imperative instructions by using the `create` command to create an object, and Kubernetes has followed them. However, it is also possible to define resources in YAML and roll out this file. The command to deploy resources from a file named `deployment.yaml` is `kubectl create -f ./deployment.yaml`. You can use the `-f` option to give a file to `kubectl`.

In most cases, you want to define and deploy resources *as code* in Kubernetes, and this also follows the declarative approach. We will deal with this topic in Chapter 3.

kubectl replace

You now know the create command, which you can use to create resources in Kubernetes. However, this command has its limits: if a resource already exists, it cannot be created by kubectl. You need another command to update a resource, and that is kubectl replace. It is the counterpart to create and also follows the imperative approach.

replace has a weakness because, as the name suggests, the resource in question is *replaced*. This can lead to a pod being dismantled and the new one being set up, even with the smallest changes. replace should therefore be carried out with caution, especially in production environments.

A resource can only be replaced if the complete manifest in YAML is provided with the change. This means that you first need the YAML code of the deployment object for the current example. For this purpose, you can use the get command that you already know. The kubectl get deployment nginx -o yaml -n my-k8s command displays the YAML code on your console. The shortened result should look similar to Listing 1.2.

```
apiVersion: apps/v1
kind: Deployment
metadata:
  annotations:
    deployment.kubernetes.io/revision: "1"
  creationTimestamp: "2023-07-25T21:32:34Z"
  generation: 1
  labels:
    app: nginx
  name: nginx
  namespace: my-k8s
  resourceVersion: "225996"
  uid: 15cdcd08-37ff-4ae3-99b5-176524daf166
spec:
  progressDeadlineSeconds: 600
  replicas: 1
...
```

Listing 1.2 Output of Deployment

I will go into the YAML format in more detail in Chapter 3, Section 3.2. For now, it is sufficient to write the output to a file named *deployment.yaml*.

As shown in Listing 1.2, in that file, you can change replicas: 1 to replicas: 2. When you install this update, Kubernetes will start a second pod. To do this, you need to run the kubectl replace -f deployment.yaml command.

As a return, you obtain `deployment.apps/nginx replaced`, and Kubernetes should immediately start deploying the second pod. Then you should use the `get` command to check whether the second pod has been started in your namespace.

> **Good to Know**
>
> You may have noticed that you did not have to specify the namespace by using `-n` in the command. If you look at the YAML manifest, you will see the `namespace: my-k8s` parameter. This enables Kubernetes to find the correct assignment.

kubectl apply

The far more elegant way to create resources and import updates using `kubectl` is the `kubectl apply` command. The advantage of `apply` is that this command checks whether a resource has already been created. If not, it will create a new one. If a resource has already been created, it will import the changes as an update. The `apply` command is therefore much more flexible than `create`.

`apply` also proceeds differently than the `replace` command during an update. This is based on a declarative approach. You can find more information on this in Chapter 3, Section 3.1. Simply put, the `apply` command will try to adjust the available resources until the result matches what you have defined in the manifest.

As with `replace`, the `apply` command requires a manifest in order to roll something out—either individually via `kubectl apply -f deployment.yaml`, or for an entire folder by using `kubectl apply -f <FOLDER>`.

Because `apply` is completely different from `create`, `kubectl` will also show you a warning if you try to update the created `nginx` using `apply`. Just give it a try with the following command: `kubectl apply -f deployment.yaml -n my-k8s`. The warning looks as follows:

```
Warning: resource deployments/nginx is missing the kubectl.kubernetes.io/last-
applied-configuration annotation which is required by kubectl apply. kubectl
apply should only be used on resources created declaratively by either kubectl
create --save-config or kubectl apply.
```

The cleanest way is to generate the resource directly by using `apply` because then there are generally no conflicts. This is also the standard procedure in most projects. To enable you to test this in your cluster, you will learn about the `delete` command in the next step.

kubectl delete

The `kubectl delete` command allows you to delete resources in their entirety. You can define and delete individual resources by name as well as entire groups of resources that you have defined in files.

In the current example, you can again reference the file using `kubectl delete -f deployment.yaml`, or you can use the name of the deployment as when creating it. To display the name again, you can use the `get` command as described previously. The deployment is called `nginx` and can be deleted accordingly by using the following command: `kubectl delete deployment nginx -n my-k8s`.

There are two other options worth mentioning for the `delete` command that you may sometimes need. First, you can force the deletion by using `--force`. This is useful if, for example, a pod no longer responds at all and can no longer be shut down correctly. The complete command would then be `kubectl delete pod TestPod --force`.

Second, you can also delete multiple resources using a label. This is useful if you have also created a service or other Kubernetes resources for the pod. Let's assume that you have created a service and a pod and given them a `Name=TestApp` label. If you now want to delete both, you can simply run the `kubectl delete pods,services -l Name=TestApp` command. I will go into the topic of labels and selectors in more detail in Chapter 2, Section 2.2.

At this point, I want to refer you once again to the `kubectl help`. There you will find additional options and information on how to use them.

kubectl describe

Because you have deleted your deployment, you can now use `apply` to roll out `deployment.yaml`. You should now find the deployment and two pods in your namespace again. You know the `get` commands and can see the names of your pods. But now you'll want to get more information about your resources. The `kubectl describe` command is available for this purpose. It provides detailed information, status messages, and events.

You can try it on one of your pods by using the `kubectl describe pods nginx-748c667d99-xtljp -n my-k8s` command. (The name of the pod is generated and will be different for you.) These outputs are important during debugging. Here you can also see whether an image could be pulled and much more.

The `describe` command also enables you to simply display all available resources by specifying only the resource in the command—for example, `kubectl describe pods -n my-k8s`. Take a look at your deployment now via `describe`. There you can see information about the replicas, the annotations, and also events such as scaling events.

This allows you to display every Kubernetes resource in your cluster. In the Section subsection, I will show you how to determine which resources are available, which you can then also view via `describe`.

kubectl logs

If you want to debug your containers, it is very useful to have access to the application logs. As with Docker, you can output these very easily using the `kubectl logs` command.

In the current example, you want to view the logs of one of your nginx containers. Run the `kubectl logs nginx-748c667d99-9448b -n my-k8s` command. (Don't forget to adapt the name of the pod to your environment.) The logs of the container are now displayed.

You can also extend the command with options. One notable feature is the ability to output logs of multiple pods that are identified by a label. You can check which label your pods have by using the `describe` command. The `app=nginx` label should be on the pods. When you now use the `kubectl logs -l app=nginx --all-containers=true -n my-k8s` command, you'll get the logs from both pods. As you can see, I have added the additional `--all-containers` option. We will look at the structure of a pod in Chapter 2, Section 2.1, but this much can be said in advance: a pod can consist of multiple containers, and this option will give you the logs from each of the containers.

If you run the command, you will receive the current logs that have been written up to this point. Sometimes it is necessary to receive the logs directly without having to send the command every time. You can activate the streaming function using the `-f` option.

kubectl exec

Sometimes you may need to enter the container for debugging purposes—for example, to view the folder structure. This is also possible with the `kubectl exec` command, which enables you to execute commands in the container of a pod. For example, if you want to check which date is set in the nginx container, you can use the `kubectl exec nginx-748c667d99-9448b -n my-k8s -- date` command to call the `date` tool in the container and display the date.

You can run any command in the container in this style. If you want to display the folder structure under the /usr folder, you need to use `kubectl exec nginx-748c667d99-9448b -n my-k8s - ls /usr`. You can also interactively connect to the container and take over the shell there. The `kubectl exec --stdin --tty nginx-748c667d99-9448b -n my-k8s -- /bin/bash` command will lead you into the container, while every further command will then be executed in the container. You can terminate this again by using the `exit` command.

> **Note**
>
> Not every container comes with a Bash command line. This means that you may have to use /bin/sh instead of /bin/bash, depending on which image the container is based on.

kubectl port-forward

Now let's look at a very interesting command. You can use `kubectl port-forward` to open a tunnel from your machine to access your pods.

> **Note**
> With regard to development, this is a wonderful opportunity to test something quickly. In production environments, however, you should only use this method for debugging purposes. If it is not even restricted for your user, clarify this with your cluster admins to be on the safe side.

Let's try this out on your nginx pods right away. The `kubectl port-forward pod/nginx-748c667d99-9448b -n my-k8s 8080:80` command allows you to open port 8080 on your computer and forward it to port 80 of nginx, which naturally expects requests on port 80 as it is a web server. You can now call address 127.0.0.1:8080 in your browser and should then see the welcome page of your nginx container.

As long as forwarding is active, your command line is blocked. If you want to continue working, you can open a second window or cancel the forwarding process. You are now welcome to check the logs of your pods again. There you should find an entry from your call that looks similar to the following:

```
127.0.0.1 - - [27/Jul/2023:21:04:32 +0000] "GET / HTTP/1.1" 200 615
"-" "Mozilla/5.0 (Macintosh; Intel Mac OS X 10_15_7) AppleWebKit/537.36
(KHTML, like Gecko) Chrome/115.0.0.0 Safari/537.36" "-"
2023/07/27 21:04:32 [error] 28#28: *1 open() "/usr/share/nginx/html/
favicon.ico" failed (2: No such file or directory),
client: 127.0.0.1, server: localhost, request: "GET /favicon.ico HTTP/1.1",
host: "127.0.0.1:8080", referrer: "http://127.0.0.1:8080/"
```

kubectl api-resources

The `kubectl` tool also comes with a command that enables you to find out more about the resources. We will go into the most important components in more detail in Chapter 2, but when you use `kubectl`, it is important for you to know what you can query by using `get` or `describe`, for example.

The command is `kubectl api-resources`. When you run this command, you will receive a list of all resources offered by Kubernetes in the current version. There you can also see in which version this component is used. This is particularly relevant for Kubernetes updates, as the versions of the resources can change and thus the YAML manifests can as well.

What is also interesting to know is whether a resource is assigned to a namespace or not. As you already know, a pod is assigned to a namespace. This is why you also need the `-n` option in the commands to refer to the corresponding namespace. A *persistent volume* (PV), on the other hand, is not assigned to a namespace and is only bound to a namespace by the *persistent volume claim* (PVC). But we will come to that in Chapter 5.

1.5 Interaction with Kubernetes via the Command Line and Dashboard

> **Example**
>
> You can use the following commands to query whether resources are bound to namespaces or not:
>
> - `kubectl api-resources --namespaced=true`
> - `kubectl api-resources --namespaced=false`

1.5.6 Switching Clusters and Namespaces Easily

At this point, I would like to introduce two tools that will make your work with `kubectl` much more convenient. There are two things that become quite annoying over time:

- Specifying the namespace for each command by using `-n`
- Working with multiple clusters simultaneously

There is a nice solution for both. The `kubens` tool helps you to change namespaces, while `kubectx` helps you to change clusters. You can find simple installation instructions for your system at *http://s-prs.co/v596411*.

After the installation, you must also check the completion for your command line, because this is the big advantage and makes your work much more pleasant, and you can select namespaces very easily.

Let's go through a quick example. You are currently in the `default` namespace and want to switch to the `test` namespace. The `kubens test` command allows you to change the namespace. The output should look as shown in Listing 1.3.

```
Context "minikube" modified.
Active namespace is "ingress-nginx".
```

Listing 1.3 Output from kubens

> **Good to Know**
>
> You can change the namespace permanently with `kubectl` as follows:
>
> `kubectl config set-context --current --namespace=my-namespace`

`kubectx` works just as easily as `kubens`. The `kubectx minikube` command enables you to switch to the cluster named `minikube`. `kubectx` requires the corresponding Kubeconfig file to be integrated as described in Section 1.5.3. The tool extracts all information from the context of the Kubeconfig file.

1 Introduction to Kubernetes

> **[+] Good to Know**
>
> A permanent change of the Kubernetes cluster via `kubectl` is also possible, as follows:
>
> `kubectl config use-context minikube`

> **[»] Note**
>
> If you have many namespaces or clusters to choose from, you should take a look at the `fzf` tool. It is a fuzzy finder for the command line, which ensures that you only have to call `kubectx`, for example, and can then select the cluster interactively. You can find the tool in GitHub at *http://s-prs.co/v596412*.

1.5.7 The Kubernetes Dashboard

The Kubernetes dashboard is the graphical user interface for Kubernetes, which also uses the Kubernetes API to allow you to manage it. Minikube comes with the dashboard out of the box, and the Kubernetes dashboard can also be used in most companies.

Starting the Dashboard with Minikube

With Minikube, the dashboard can be started via a simple command. To do this, you need to run the following command in your command line.

`minikube dashboard`

After executing the command, Minikube will download the dashboard container and deploy it in your Kubernetes cluster. A connection to the dashboard is then automatically established, and the page opens in your default browser. The dashboard should look like the one shown in Figure 1.9.

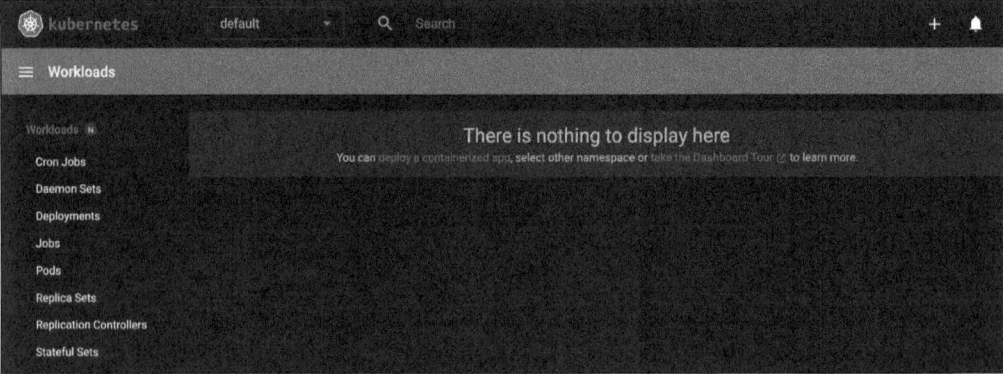

Figure 1.9 Kubernetes Dashboard: Overview

1.5 Interaction with Kubernetes via the Command Line and Dashboard

A Tour of the Dashboard

Let's explore the dashboard interface together so that you can familiarize yourself with its operation.

In the top bar, you can see a dropdown field that reads **Default**. There you can select the namespace you want your dashboard to display. If you have gone through the introduction to kubectl, you will also find your **my-k8s** namespace there. (If you have not done this, go through the examples in the previous sections again and create the resources there.) If you select the namespace, you will be shown an overview of your resources, as in Figure 1.10. You can see the deployment of nginx, the two pods, and a ReplicaSet, as well as additional information. If you recall the outputs of kubectl get, you will also see similarities there.

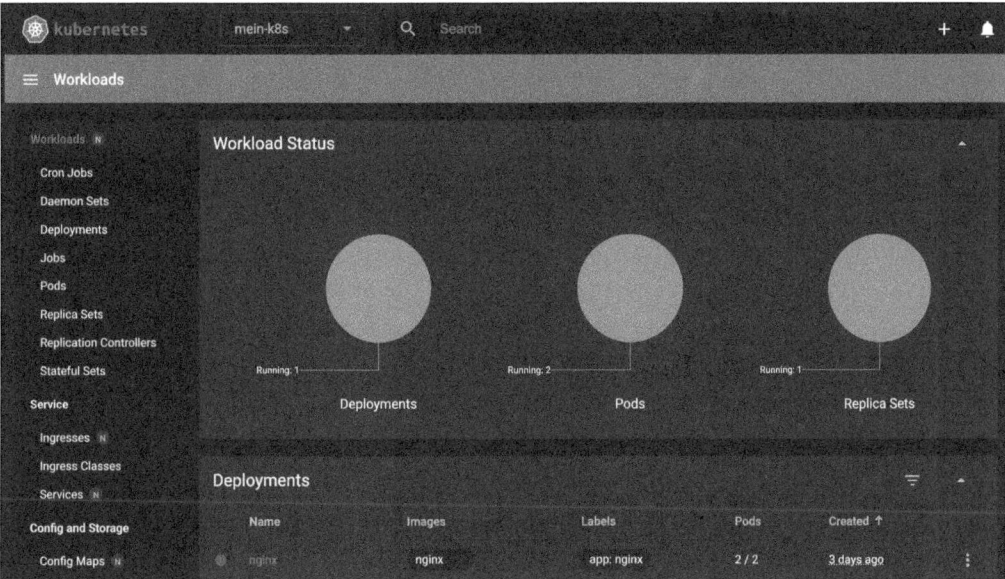

Figure 1.10 Overview of Resources in Your Namespace

On the left-hand side you will find a menu with the option to select the most important resources in Kubernetes. For example, if you click **Pods**, you will see a list of all pods running in this namespace. This allows you to navigate through the individual resources.

If you now click the name of one of your pods, you should be taken to a page where you can get all the information about this specific pod, just as you can do using kubectl describe.

Let's say you want to debug this pod. You've already collected the information you need, and now you want to look at the logs. Kubernetes provides additional menu items in the pod view in the form of icons, as shown in Figure 1.11. From left to right, there are icons for **View Logs**, **Exec into Pod**, **Edit**, and **Delete**.

Figure 1.11 Menu Bar in Pod View

Click the **View Logs** icon. This opens a window containing the logs. It is important to know that you can also select the different containers within a pod using a dropdown list, as shown in Figure 1.12.

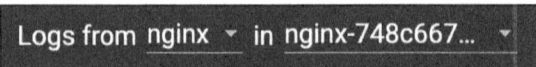

Figure 1.12 Selecting Logs for Different Containers in Pod

You have now looked at the logs and want to go into the container to check something there. To do this, return to the pod view and select the second icon from Figure 1.11. This will take you to the pod's shell window. As with `kubectl exec`, you can enter commands in the container and navigate through the file system.

The **Edit** icon opens a window containing the pod's YAML manifest, which you can customize and update using the **Update** button. An indication is given that an update via this route is equivalent to a `kubectl apply` operation. For this reason, a new manifest is passed to Kubernetes, and it attempts to adapt the resources accordingly.

You now want to make this adjustment on the basis of the deployment resource. To do this, you need to navigate to **Deployments** in the dashboard and click the **nginx** deployment. Then click the **Edit** icon and search for the definition of `replicas`, as shown in line 98 in Figure 1.13.

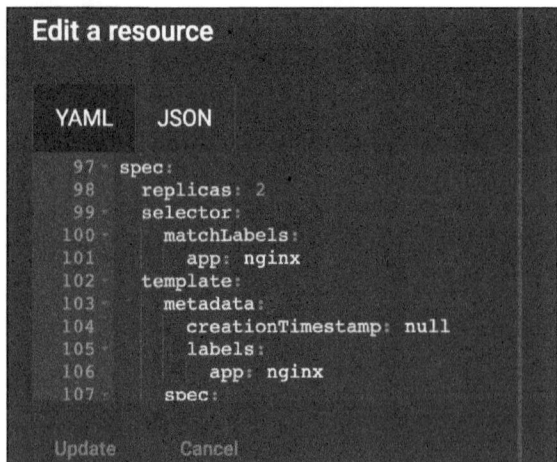

Figure 1.13 Editing Deployment in Kubernetes Dashboard

Kubernetes always extends the YAML manifest with status messages and metadata, which is why the manifest is significantly larger. However, we are only interested in the definition that is shown under spec. Change the number of replicas from 2 to 3, then click **Update**. Now the deployment will start another pod. Then click **Pods** in the navigation bar and see how the third pod is started.

So much for the Kubernetes dashboard. You now know how to navigate through the individual resources and how to view and customize resources and debug pods. You should take some time to click through the dashboard and familiarize yourself further with it, because it is a good alternative to kubectl in order to get a quick and graphical overview of your Kubernetes setup.

1.6 Lens: The IDE for Kubernetes

With kubectl and the Kubernetes dashboard, you have become familiar with the standard tools for using Kubernetes. I remember clearly that I only worked with those two until a few months ago. Most of the time, I used the command line with kubectl, because the Kubernetes dashboard is sometimes a bit slow depending on the cluster. In addition, it is difficult to switch between the individual clusters in a multicluster setup with the Kubernetes dashboard because a new website must be called each time.

I always wanted a tool that combines both worlds: a graphical user interface and high speed while working. Then I tried OpenLens. You will see shortly that this tool combines kubectl and the dashboard. The integrations with Helm, which I will introduce in Chapter 8, also make OpenLens an excellent tool for developers. It will expand your tool set and definitely make you more flexible in the use of Kubernetes. But enough raving. Let's get started now so you can see for yourself.

To install OpenLens, go to the following website: *http://s-prs.co/v596413*.

There you will find the correct installation command for your operating system and can install OpenLens.

> **Licensing Terms of Lens**
>
> While I was writing this book, Lens changed its licensing terms, with the result that it is no longer free of charge.
>
> I have looked for an alternative for you and found OpenLens. The Lens team continues to develop the core in OpenLens as an open-source product. Unfortunately, a few features are missing, which I think is a shame, but you don't have to buy a license for the exercises described in this book. kubectl is perfectly adequate.
>
> I will use the name *Lens* herein synonymously with *OpenLens*.

1.6.1 Overview of Lens

When you open Lens, you will first be taken to the start page. On the left-hand side there is a hotbar. That's where you can store your clusters for quick access, and that's exactly what we're going to do for Minikube now. Figure 1.14 shows how you can get to your cluster. Click the menu button ❶ in the hotbar to access the catalog. Select ❷ **Clusters** to see all your clusters.

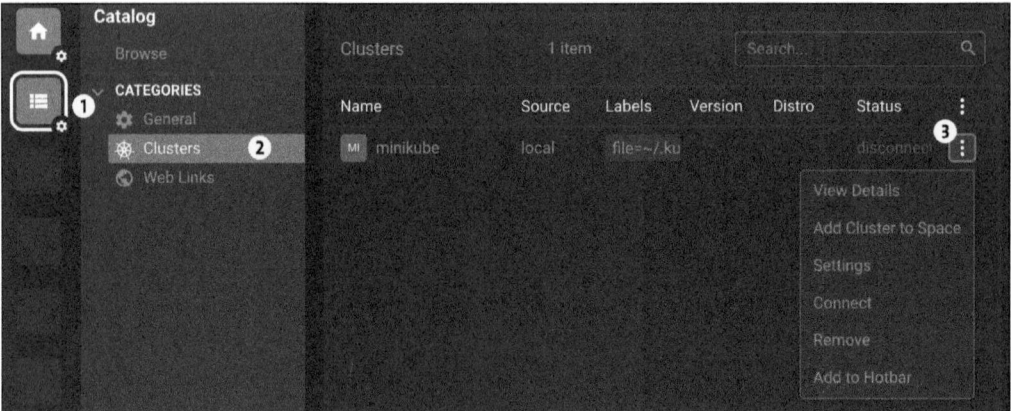

Figure 1.14 Inserting Minikube in Hotbar

Lens enables you to manage multiple clusters. The tool retrieves the access data and information on your clusters from the Kubeconfig file. This means that you do not need to configure the clusters manually; your Minikube cluster is already known, as the `minikube start` command also writes the Kubeconfig file. If other clusters are already included in your Kubeconfig file, then these are also listed here.

To add your cluster to the hotbar, click the options button ❸ and select **Add to Hotbar**. After that a new icon will appear below the catalog. When you click it, you will be logged into your cluster.

First you see the overview of the cluster, as shown in Figure 1.15. On the left, you can recognize the similarity to the Kubernetes dashboard. You have the option of selecting and displaying each resource individually. Click **Workloads · Overview** to be taken to the same page that you know from the Kubernetes dashboard.

As you can see in Figure 1.16, you can select your namespace via the dropdown menu. Now you can also navigate through the resources created in the previous chapters and familiarize yourself with Lens. At its core, it is similar to the dashboard.

Try to display a pod as described in Section 1.5.7, and open the logs. You may already notice how quickly you can navigate and how well thought-out the design and user experience are. The fact that logs and terminal windows are opened at the bottom and navigation through the resources is still possible is unbeatable in your day-to-day work.

1.6 Lens: The IDE for Kubernetes

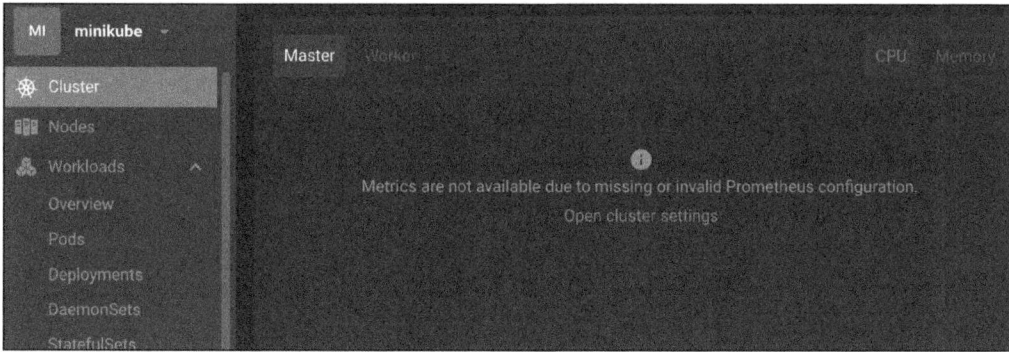

Figure 1.15 First Screen: Cluster Overview

Figure 1.16 Selecting Namespace in Lens

1.6.2 Advantages over the Kubernetes Dashboard

The design and speed are not the only advantages of Lens over the Kubernetes dashboard. I now want to show you a few more features that will be useful for you in the future.

Activating Cluster Metrics

I would like to start by showing you the metrics. In Figure 1.15, you probably noticed the following message: **Metrics are not available due to missing or invalid Prometheus configuration.** *Prometheus* is a very common monitoring tool in the world of Kubernetes. Among other things, it collects metrics as time series that can be queried and displayed using simple commands. I will go into more detail about Prometheus in Chapter 7, Section 7.4, but for now note that Lens recognizes Prometheus and can display the metrics.

Activating Simple Port Forwarding

As you learned in Section 1.5.5, you can use port forwarding with `kubectl` to reach your containers via a tunnel. This can be very convenient during development and debugging. As the Kubernetes dashboard is a web application, port forwarding is not possible there. Lens offers port forwarding and even makes forwarding very easy compared to `kubectl`.

1 Introduction to Kubernetes

Like `kubectl`, Lens can create tunnels for pods or services. A port definition is required that specifies the port under which the application can be reached. Unfortunately, we do not have these in our nginx example and must first add them to the YAML manifest.

To do this, click **Workloads · Deployments** in Lens, select **Nginx** in your namespace, and click the **Edit** icon. There, as shown in bold in Listing 1.4, you want to insert the corresponding code snippet and then click **Save & Close**.

```
...
spec:
  containers:
    - name: nginx
      image: nginx
      ports:
        - name: http
          containerPort: 80
          protocol: TCP
...
```

Listing 1.4 Adding Port Definition for Pods

Good to Know

I want to make a small addition to the port definition in Listing 1.4. You may have asked yourself whether `name: http` is necessary, as you are using port 80, and you must also set `https` as the name if you use port 443. The answer is no. The name is freely selectable.

For me, the name `http` was the most logical in this case. You can also name the port according to its function, such as `metrics` for a port on which metrics are queried.

The deployment will update and deploy the pods with the new configuration. Go to the pod overview and click a pod. The info window opens on the right. In it, you will also find information on the containers, as shown in Figure 1.17.

Click **Forward** and enter "8080" as the local port. A browser window should now open automatically, showing you the default page of your nginx. If that is not the case, you can reach the container at *http://127.0.0.1:8080/*.

Note

When you use port forwarding, you should keep in mind that a port can only be used once on your computer. For example, if you try to enter port 8080 a second time, you will receive an error message. If that happens, you can either select a different port or delete the old forwarding.

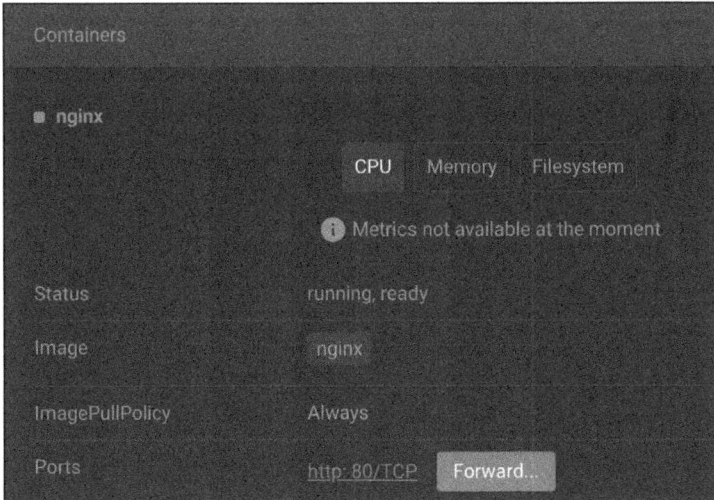

Figure 1.17 Container Info Window with Port Forward

In contrast to kubectl, port forwarding does not block the terminal, but is managed by Lens in the background. You can view and manage the overview of your tunnels in the navigation on the left under **Network • Port Forwarding**.

Helm Integration

The integration of Helm into Lens makes developing easier. Whereas I used to operate the tools on the command line, I can now do most things via Lens. We will go into Helm in more detail in Chapter 8, so a brief summary will suffice here.

Helm is a package manager for Kubernetes manifests. It allows you to provide Kubernetes objects very easily in so-called charts, which you can configure using parameters. This is pretty useful if you have different environments and also makes it convenient to use other charts. For example, it allows you to activate or deactivate autoscaling by setting a variable. This means you can always deliver your application appropriately.

If you click **Helm • Charts** in the menu on the left, you will be taken to an overview page. Lens has already activated a Helm repository from the Bitnami provider for you and shows you an overview of the software offered by Bitnami, as shown in Figure 1.18. This is very convenient for development purposes because you can find many tools, from Jenkins to databases, and roll them out in Minikube at the touch of a button.

Try it out for yourself and roll out a Jenkins. You should also familiarize yourself with this. The more you work with it, the easier it will become.

1   Introduction to Kubernetes

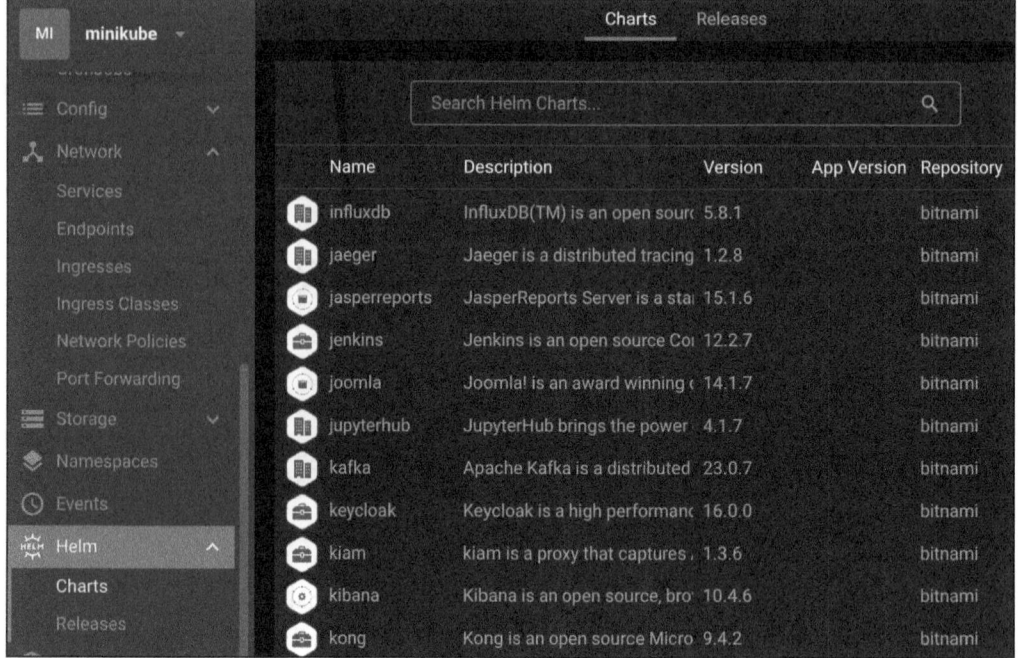

Figure 1.18 Overview of Helm Charts in Lens

1.6.3 The Lens Reference

I would like to provide you with a reference so that you can look up how to do something in Lens whenever you need to. I expanded this section whenever an exercise was added in an upcoming chapter that requires you to do something in Lens. This means that the following chapters with exercises are not unnecessarily large and you can simply look them up again if necessary.

Creating Resources

You can easily create new resources in Lens. Regardless of which menu item you are on, you will see a **+** button in the bottom line, as shown in Figure 1.19. If you click that button and select **Create Resource**, an editor opens to write YAML manifests. In the **Select Template ...** dropdown menu, you will find a selection of resources and templates that you can use and expand. This is very convenient if you want to familiarize yourself with Kubernetes or get to know new resources. Of course, you can also use your own manifests here or use the examples from this book.

Finally, click **Create** and the manifest will be transferred to Kubernetes and set up. If you have errors in the manifest, Lens will point them out and you can correct them.

1.6 Lens: The IDE for Kubernetes

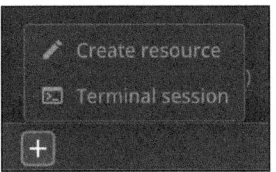

Figure 1.19 Creating Resources and Opening Terminals in Lens

Terminal within Lens

In Figure 1.19, you can also see the **Terminal Session** option, which allows you to open your own command line in Lens. This is very helpful as you can then simply use kubectl within Lens.

This integration makes your work even easier because you don't need to switch between the different windows.

> **Good to Know**
> Especially since OpenLens has removed some functions, such as exec or logs, the terminal is worth its weight in gold.

The Pod Action Bar

With kubectl, you have already learned some commands to perform actions with pods. At this point, I want to show you how you can use Lens to perform actions on pods.

To do this, go to the pod overview under **Workloads · Pods** and click the pod you want to use. This opens the pod overview on the right, and at the top you will see the action bar, which looks like the one shown in Figure 1.20.

Figure 1.20 Action Bar for Pods in Lens

> **Note**
> The action bar is slightly smaller in OpenLens. Unfortunately, you can perform only edit and delete actions here. I have nevertheless covered the complete action bar to give you a full overview.

Let's go from left to right and try the options out directly:

1. The first icon is the equivalent of kubectl attach. It allows you to connect to the running container, and you are in the running process. This enables you to receive all

1 Introduction to Kubernetes

log messages directly on the console. In my daily work, I don't use this as I prefer using the logs.

2. You can use the second icon to run `kubectl exec`. In Lens, this function is called *pod shell* because you can use it to open a new command line in the corresponding container and execute commands there. This is useful if you want to debug something or search for files in the file system.

3. You can view the pod's logs by clicking the third icon. It does the same thing as `kubectl logs` and shows you the current logs of the container in a new window.

4. Icon number 4 performs a deletion by eviction. *Eviction* is a special way of "expelling" pods from nodes. I do not use this option either and instead recommend the normal delete function. Eviction can lead to problems as it does not delete the pod cleanly from the *etcd* database.

5. The pen icon is intended for editing the pod. When you click it, a window opens and displays the YAML manifest, which you can edit and save.

6. The sixth icon performs the normal deletion of the pod via `kubectl delete`. The pod is then terminated and the pod manifest is deleted from *etcd*. However, if it is part of a deployment with a ReplicaSet, for example, Kubernetes will create a new pod again.

Custom Resources and Custom Resource Definitions

You will learn about the concept of custom resources (CRs) and custom resource definitions (CRDs) in Chapter 4. Here, I'll show you where you can find them later in Lens.

At the bottom of the left-hand menu, you will find the custom resources menu, as shown in Figure 1.21. As soon as you start creating CRs and CRDs, these will also be displayed in the left-hand menu in dropdowns corresponding to the CRD group. You therefore have the option of navigating either via the left-hand menu or via the **Definitions** menu item. As usual, all CRDs are then displayed in the main window, which you can select to display the corresponding CRs.

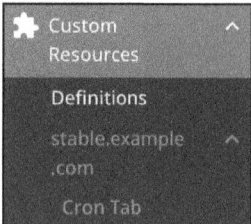

Figure 1.21 Custom Resources Menu in Lens

Adding a New Cluster

Usually, Lens simply retrieves the information about your Kubernetes clusters from all Kubeconfig files in your home directory under *~/.kube*. You can therefore simply extend your Kubeconfig file with a new cluster or store another Kubeconfig file there and thus also have access to it in Lens.

In addition, Lens provides the option of adding a new Kubeconfig file and thus a new cluster without having to adapt the Kubeconfig file in your home directory. To do this, you need to go to the catalog in Lens and click **Clusters** in the categories. As shown in Figure 1.22, you will find the **+** button there; click it and select **Add from Kubeconfig**. This opens a text window into which you can copy the Kubeconfig file. Then, click **Add Cluster**.

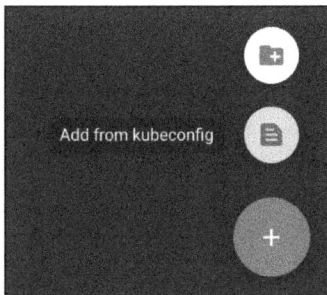

Figure 1.22 Adding New Kubeconfig File to Lens

The second option shown in Figure 1.22 is to use **Sync Kubeconfig(s)**, which allows you to select a folder or a Kubeconfig file. The folder is then searched for Kubeconfig files in the same way as the ~/.kube path. You can also manage the Kubeconfig syncs later under **General** • **Preferences** • **Kubernetes** and remove or add folders.

In the catalog, you can click **Settings** via the button with three dots and adjust the name of the cluster. In addition, you can add a Pi cluster to the hotbar.

> **Note**
>
> *Remember:* If you do not store the Kubeconfig file in the ~/.kube directory, you cannot control the new cluster via `kubectl`. Another option is to merge two Kubeconfig files. I will describe this in Section 1.7.3.

1.7 The Kubernetes Cluster from Raspberry Pis

You have already taken your first steps with Kubernetes in Minikube in the previous sections. In this section, I will introduce a simple and inexpensive way of setting up a real cluster. Minikube itself is virtualized in a Docker container and therefore quickly reaches its limits, especially with network demos.

You are also welcome to use this section at a later stage and set up a cluster if you need it in an exercise. The hardware requirements of a "real" Kubernetes cluster naturally depend on the number of containers you want to manage and can become very large. Our test environment is much more modest in this respect; a few small Raspberry Pis are sufficient for the examples in this book. However, if you want to run a large number of different containers at the same time, the cluster will reach its limits. But that is precisely the strength of Kubernetes: you have two nodes and can try out networks such as a service and an ingress, which we'll discuss in Chapter 2, Section 2.5. If that's not enough, you can simply add more devices and let your cluster grow.

In this section, we will go through the following steps:

1. I will introduce the hardware I use.
2. You will set up the Raspberry Pis.
3. We will install Kubernetes together.

Note

I followed various instructions on the internet to set up the Raspberry Pis. A good website to consult if you have problems is *https://tutorials-raspberrypi.com/*.

1.7.1 Choosing the Right Raspberry Pis

At the beginning, I was a little undecided about which Raspberry Pis would work best for this use case. There are different models and versions available, which of course also cost different amounts of money. In the end, I decided on the following setup, and am more than happy with it:

- 2 × Raspberry Pi 4s, 4 GB memory ($61.75 each)
- 2 × SD cards, 64 GB each ($15.00 for two)
- 2 × official Raspberry Pi power supply units ($7.99 each)
- Five-port switch + network cable ($17.99)
- Stackable acrylic case ($6.50)

In my experience, the longer I research, the more expensive it gets, because the tinkerer in me becomes active. There are many possibilities to buy even better housings for the Pis or to install fans in the housings. I have been running the cluster for several days in a row now and do not see any sign of overheating. You will find a small example of temperature monitoring later in Chapter 7, Section 7.4.2. I also bought a heat sink set. This lowers the temperature by a few degrees—but in the end, this is all optional and not necessary for a Kubernetes test cluster.

Of course, you can also use memory cards and power supply units that you still have at home, and a 32 GB memory card will probably suffice. In this case, I have opted for a setup that can also withstand larger requirements, but you can also start a little smaller.

The great thing about Kubernetes is that if you push your cluster to the limit, you can simply add another Pi.

> **Note**
>
> The Pis all have a built-in Wi-Fi module. However, I immediately added the switch so that I have a more stable connection between the master and the worker. A Wi-Fi connection would of course be sufficient for your test cluster, and you can save yourself the switch.

The Pis arrive without anything loaded. To be able to use the Raspberry Pis, you must install an operating system on an SD card. There are specially developed operating systems available, which are based on Debian. To record to SD card, you can use Raspberry Pi Imager, which you can find at *https://www.raspberrypi.com/software/*; select the download for your computer.

Once you have installed the imager, you can select your suitable Raspberry Pi version, as shown in Figure 1.23. You'll want to use Raspberry Pi OS Lite (64 bit) as the operating system, as this comes without a desktop, which you won't need anyway. You can find it in the selections under **Raspberry Pi OS (Other)**. Then select the SD card and click **Next**. You will be asked if you want to make OS customizations; click **Edit Settings** to set some important elements.

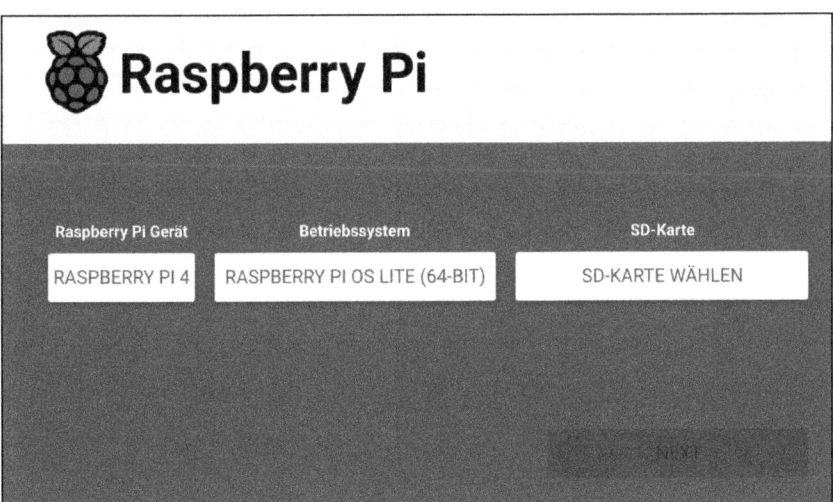

Figure 1.23 Raspberry Pi Imager

I have chosen `raspberrypi1` as the host name for the master and `raspberrypi2` for the worker. The user's name is `kevinwelter`. I have stored my public key in the **Services** tab so that I can later access the Pis via SSH. Next, click **Save**, then select **Yes** to apply the customizations.

1 Introduction to Kubernetes

Once you have installed the image on both SD cards, you can insert them into the Pis and connect them with a power supply unit. The Pis will then start automatically. To log in, I use two command line windows and the commands from Listing 1.5.

You have now set up the Pis initially. In the next step, you will install Kubernetes.

```
ssh -i .ssh/pi_key kevinwelter@raspberrypi1.local
ssh -i .ssh/pi_key kevinwelter@raspberrypi2.local
```

Listing 1.5 Login to Pis via SSH

1.7.2 Installation of Kubernetes

We suggest K3S (*https://docs.k3s.io*) as your Kubernetes installation. This is a slightly adapted version that is very lightweight and therefore perfect for your Pi cluster. However, you must first make small adjustments to the hosts.

Commands You Need to Execute on Both Raspberry Pis

Open the *cmdline.txt* file by using the `sudo nano /boot/cmdline.txt` command, then insert a space and the following values at the end:

```
cgroup_enable=cpuset cgroup_memory=1 cgroup_enable=memory
```

If you have not used Nano previously, note that you can save and close the changes via `Ctrl`+`X`.

Then you must change `iptables` to `legacy`, because there is a small problem with the Debian of the Pis. Finally, restart the devices:

```
sudo update-alternatives --set iptables /usr/sbin/iptables-legacy
sudo update-alternatives --set ip6tables /usr/sbin/ip6tables-legacy
sudo reboot
```

Commands for raspberrypi1

The first Pi will be the K3s master. For this purpose, you need to run the following commands:

```
export K3S_KUBECONFIG_MODE="644"
curl -sfL https://get.k3s.io | sh -
```

This command downloads the script for installation and executes it directly. At the end, you will receive the information that K3s has been started. You can also check the current status via the `sudo systemctl status k3s` command.

Now you still need the node token with which worker nodes can register with the master. You can display it via `sudo cat /var/lib/rancher/k3s/server/node-token`.

1.7 The Kubernetes Cluster from Raspberry Pis

Commands for raspberrypi2

You can now install the worker and register it with the master. You also need two other environmental parameters. The first one is the URL for the master, which in this case is simply the host name. The second one is the node token that you issued in the previous step. Run the commands as follows and adjust the values for your setup if necessary:

```
export K3S_KUBECONFIG_MODE="644"
export K3S_URL="https://raspberrypi1:6443"
export K3S_TOKEN="K101a…560"
curl -sfL https://get.k3s.io | sh -
```

You can then check on the master whether the worker has been installed correctly. Run `sudo k3s kubectl get node` and you should get an output similar to the one shown in Listing 1.6. You have now installed Kubernetes on your Raspberry Pis.

```
NAME           STATUS   ROLES                  AGE
raspberrypi2   Ready    <none>                 18d
raspberrypi1   Ready    control-plane,master   18d
```

Listing 1.6 Displaying Nodes of Raspberry Pi Cluster

> **Note**
> These instructions make sure that you will always install the latest version of K3s. However, this should not be a problem for the examples in this book.

1.7.3 Using the Kubeconfig File of the Pi Cluster

So long as you are logged in on the master, you can use the `kubectl` installed there, but in the next step you need to get access to the Pi cluster with Lens and with `kubectl` on your machine. To enable this, you need the Kubeconfig file, which you can obtain using the `cat /etc/rancher/k3s/k3s.yaml` command on the master. It is best to first copy the file into an editor of your choice, as you must change the `server: https://127.0.0.1:6443` parameter to the name of the master—in this case, `server: https://raspberrypi1:6443`. You can now use the customized Kubeconfig file and add it to your system in three ways:

1. You can add the Kubeconfig file only in Lens, as described in Section 1.6.3.
2. If you store the Kubeconfig file under the *~/.kube* path and adjust the environment parameters as described in Section 1.5.3, you can access it via both Lens and `kubectl`.
3. Alternatively, you can transfer both Kubeconfig files to a single file under *~/.kube/config*; this means that you do not have to set any environment parameters.

1 Introduction to Kubernetes

If you have selected option 2 or 3, you can use `kubectx` to switch clusters. I introduced this tool in Section 1.5.6.

> **Example**
>
> There are several ways to merge Kubeconfigs. I want to describe the most elegant one now. For PowerShell, you have to adapt the commands slightly:
>
> 1. First, you need to make a backup of your Kubeconfig file:
>   `cp ~/.kube/config ~/.kube/config-backup`
> 2. Add all Kubeconfig files to the `KUBECONFIG` environment variable so that they can be found by `kubectl`:
>   `export KUBECONFIG=~/.kube/config:/path/one:/path/two`
> 3. You can merge the Kubeconfig files using `kubectl`:
>   `kubectl config view --flatten > config`
> 4. Then replace the old Kubeconfig file with the new one:
>   `mv config ~/.kube/config`
> 5. You can now check in a new terminal window whether all clusters are found:
>   `kubectl config get-clusters`

Once you have integrated the Kubeconfig file, you should carry out a short test to check whether you also have access to the cluster. If everything works, you are ready to carry out the exercises from the book on your own Raspberry Pi cluster.

> **Good to Know**
>
> You can switch the Raspberry Pis on and off without any problems. The cluster should rebuild itself correctly as you are only using one master.
>
> If you remember Section 1.3, you will know that a connection failure with more than one master would be fatal. By breaking the connection, the *etcd* quorum would no longer know which dataset is the correct one. In that case, you would have to perform a recovery or rebuild the cluster.

Chapter 2
Basic Objects and Concepts in Kubernetes

I never dreamed about success. I worked for it.
—Estée Lauder

In the preceding chapter, you got to know Kubernetes at a high level. You now know the architecture and have your own test cluster. Along the way, you became familiar with some resources and executed your first commands in the cluster. Let's now take a look at the theory and the concepts behind it. The theory is usually the hardest part of a book, but I will provide some practical examples that you can try out using Lens and Minikube.

In the first step, you will get an overview of the most important objects, which are also referred to as *API objects*, and then I will introduce them in more detail thereafter. The sections build on each other.

> **Note**
> I recommend that you really try out everything. Every exercise, no matter how small it may seem, will help you to anchor your new knowledge. You will see how your brain links the knowledge directly to your own use cases and how this gives you new ideas.

Let's start by looking at which components are active in Kubernetes when you make a request to a website that gets its data from a backend. You will get to know all these components in this chapter.

Figure 2.1 shows how a user sends a request that is accepted by an *ingress*. The ingress redirects the request to a service that has the frontend as a selector. The request then gets redirected from the service to a frontend pod, which in turn retrieves the data from the backend via a service.

This looks like a lot of communication effort at first, but you will see that each of the objects fulfills a purpose, especially for a distributed system.

2 Basic Objects and Concepts in Kubernetes

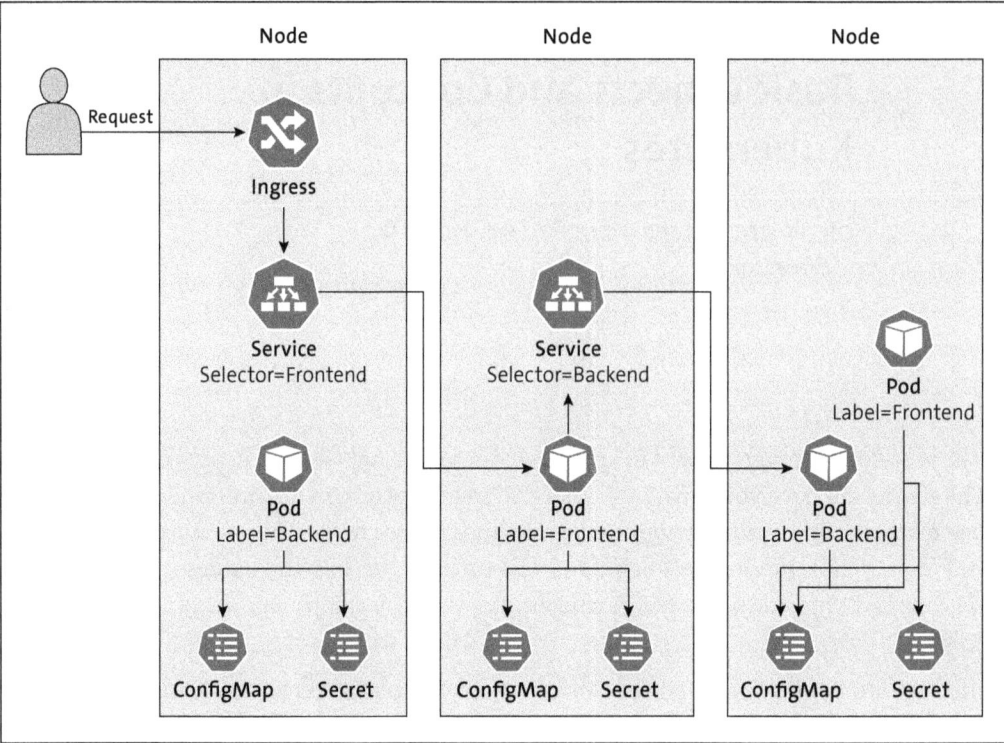

Figure 2.1 Sequence of Request in Kubernetes

> **Note**
>
> Feel free to take a look at the overall picture from time to time during the course of the chapter. Think about how what you have learned fits into this picture. By the end of the chapter, you will understand how everything fits together.

Let's walk through the components:

- Let's start with the *pod*, which can contain and manage one or more containers. When you use Kubernetes, you always create and manage pods and cannot manage the containers they contain by yourself. This is why the pod is the smallest unit in Kubernetes. So when you roll out your application in Kubernetes, it always runs under the control of a pod. The pod is responsible for the execution of containers, provides resources such as network configurations and storage, and can share these between several containers within the pod.

  Pods form the basis for other resources, such as deployments or StatefulSets, and are generally not created in production systems without them because they manage and monitor the individual pods.

- One of the most important concepts in Kubernetes is that of *labels* and *annotations*. These are not only there to identify Kubernetes resources; you can also use them to control the entire cluster as you wish. You can link resources to each other, group objects, or assign pods to specific nodes. Although simple, this concept represents a very strong implementation.

- You already used a Kubernetes *deployment* in Chapter 1, Section 1.5.5. I have not included the deployment in Figure 2.1 as it is more of an object in the background. You can consider deployments to be a higher-level abstraction for the provision of applications. They allow you to update your applications, perform rollbacks, and scale your pods. A deployment always creates a ReplicaSet, which in turn creates and manages the pods. A deployment is the perfect resource if your application is stateless.

- The *ReplicaSet* ensures that you always have the number of pods available that you actually need. Like the deployment, it is not contained in Figure 2.1 as it manages your pods in the background. For example, if your application is supposed to have two pods for reasons of reliability, the ReplicaSet takes over the monitoring. If it detects that a pod has been terminated, it will scale a new one. At the same time, it will terminate a pod if you manually scale a third one. This means that it continuously monitors the number of pods running and automatically creates or deletes pods to reach the desired number.

    ReplicaSets are actually used only in combination with deployments. Like pods, you can also create ReplicaSets individually, but this is not useful in practice as you would lose the additional rollout mechanisms of the deployment.

- Now let's move to *ConfigMaps* and *secrets*. These two objects allow you to configure your applications externally, which enables you to separate the configuration data and sensitive data from the pods. As its name suggests, sensitive data such as passwords or API keys are provided in a secret, and you can provide any configurations for your applications in ConfigMaps.

- The Kubernetes *service* enables network communication with pods and takes care of load balancing. It provides stable IP addresses and DNS names to reach your pods. Thanks to load balancing, the service makes it easier for you to scale your applications. It always directs the requests to the right pods. Services are a crucial component for the communication of your applications.

- Using *ingress* resources, you can implement external access to your application. These allow you to route HTTP and HTTPS requests to the service, which then redirects the traffic to your pods. This can be done on the basis of host or path rules, which we will look at in more detail later. To apply the rules, Kubernetes uses the so-called ingress controller, which is usually Nginx. The controller then redirects the traffic based on the ingress rules.

2.1   Pod and Container Management

Let's now take a closer look at the concept of the Kubernetes pod. The pod is the smallest unit that you can create and manage using Kubernetes, and its task is to manage one or multiple containers. Kubernetes calls this unit a *pod* and thus remains true to Docker's whale metaphor as in the world of whales, a pod is a unit made up of several whales that live together.

Think of the pod as a small team of employees working together to accomplish a specific task. Each member of the team has a distinct role, but they share resources and communicate with each other to achieve their common goals. Each employee on the team corresponds to a container.

If you work according to the separation of concerns principle discussed in Chapter 1, then a container has a specific task that is mapped and executed in a single process. In addition, the container with your application only executes exactly this application. If your application requires additional logging or proxy functionalities, then it makes sense to outsource these to separate containers in order to keep your application lean. However, this additional functionality is necessary for your application and belongs to it, which is why it makes sense to link these two containers in a logical unit.

The question you need to answer as a developer is as follows: When do containers belong together in a pod and when do they not?

The advantages of a single container per pod are as follows:

- More flexible deployments
- More granular scaling
- Higher degree of decoupling

The advantages of multiple containers per pod are as follows:

- Faster communication between containers
- Improved separation of tasks
- Further development of the individual containers independently of each other

The team metaphor for a pod makes the concept a little more tangible, but the additional pods are actually mostly little helpers that support the main application. Table 2.1 describes typical use cases for these helpers. This may make some decisions easier for you.

The important thing here is that the little helper must be decoupled from the main application. For example, if adding a sidecar limits scalability, then you need to move this functionality to a separate pod.

Name	Function
Ambassador	These little helpers perform proxy tasks, which allows you to outsource the authentication to an API, for example. Your application simply sends the data to the ambassador, which in turn takes care of setting up the HTTPS connection and managing the access tokens.
Adapter	An adapter modifies data that goes into or out of the main application. This is useful if you need to convert data into a specific format but do not want to implement this in the main application. This is a perfect use case for legacy applications.
Sidecar	A sidecar adds additional functionality to your application. Monitoring or logging is usually implemented via a sidecar.

Table 2.1 Terminology of Helper Containers within Pods

> **Note**
> Here's a little help to decide whether your containers belong in the same pod. Simply ask yourself the following questions:
>
> - **Do the containers have to share a common resource?**
>   For example, if you have two containers that share a common file system, then they belong in a pod. A use case could look as follows: A container makes files available to users. Another container is responsible for updating this data. According to the *separation of concerns* concept, this would be a separation that makes sense, and yet the two containers belong inseparably together.
> - **Do the pods have different scaling requirements?**
>   For example, if you have a database and a web server, both applications have different scaling requirements. While the web server is stateless, the database is stateful and cannot be replaced that easily. Accordingly, you must move these two containers to individual pods.
>   In addition, a high load on the web server does not directly mean that you also need more database instances. The goal is to have a good utilization for each pod so that you don't waste any resources.
> - **Can the containers run on separate machines?**
>   If your answer to this question is yes, then you can move the containers to different pods with a clear conscience.

Figure 2.2 illustrates the inner workings of a pod. In this example, the pod has
- two containers,
- two volumes, and
- a pause container.

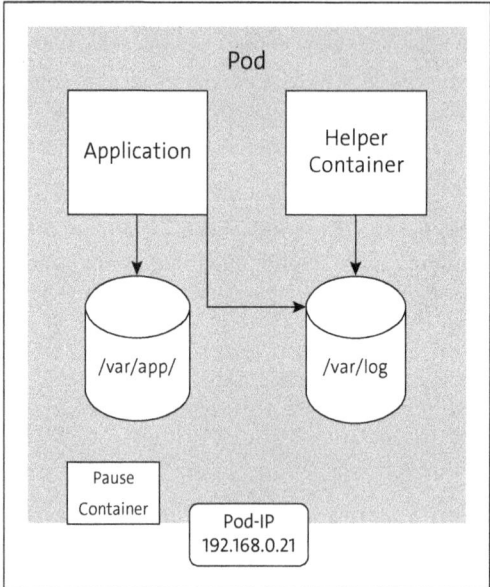

Figure 2.2 Structure of Pod

If a pod is created, the *pause container* starts first. It initializes the IP address, the namespace, and the *cgroups* and holds them until the pod gets terminated. Each additional container within the pod can rely on the initialized resources and receive incoming traffic via the IP address, for example. The pause container is useful because even if your application containers are restarted, the IP address remains.

> **Good to Know**
>
> The pause container is invisible and is never displayed in the Kubernetes CLI. However, it runs on the node and you can see it via the container runtime.

The volumes can also be used by both containers. You can even access it at the same time and use it to exchange data, for example. However, you must be aware that this can cause conflicts that Kubernetes does not prevent or monitor for you. For this reason, you should pay close attention to which container can write and read during implementation. It is best if one container writes and the other is read-only. However, we will return to the topic of shared storage in Section 2.1.3, and you will learn about a use case that illustrates how you can use it.

> **Good to Know**
>
> As Kubernetes manages only the pod object, all containers within a pod are of course started on the same node. This is the only way that the containers within a pod can also share resources such as network and storage.

2.1.1 Container Engines

When you first came into contact with containers, you probably used Docker. At least that was what I did, and for me Docker is still synonymous with containers today. Docker may not be the first one, but it is currently one of the best-known container engines on the market. But Docker is much more than a simple container runtime. It provides an easy way to develop containers, and with Docker Swarm it has developed a competitor product to Kubernetes.

> **Good to Know**
>
> Docker is a *container engine* that runs the Docker containers via a *container runtime*. To distinguish between these two terms, remember the following: A container engine processes the user requests, interprets them and commands the container runtime. The container runtime only takes care of the execution of the containers and what is necessary for this.

As Kubernetes wanted to define a more open and independent standard, Docker was marked as *deprecated* in Kubernetes version 1.20 and can no longer be used as a runtime in the newer versions as before. This was not a decision that was made overnight.

Container Runtime Interface (CRI) was developed so that Kubernetes could open up to other runtimes in addition to Docker.

> **Good to Know**
>
> CRI is a plug-in interface that can be implemented by container runtimes. It defines the communication between Kubelet and the container runtime, which is implemented using the *gRPC* protocol. This interface allows Kubernetes to communicate with any runtime that has implemented the interface.

A lot has happened on the market in recent years, and driven by Docker, an industry standard for containers was defined in 2015. The standard of the Open Container Initiative (OCI) comprises three specifications that can be used as a guide for projects and developers. The specifications describe what containers are and how they should be used, and from this what a runtime must look like or how images can be saved in a registry is derived. You can find an overview of the specifications in Table 2.2. The OCI standard allows you to build your application with Docker, Buildah, or Podman and still be able to deploy it on Kubernetes. It is therefore definitely worth taking a look at the OCI website at *https://opencontainers.org/*.

Specification	Description
runtime-spec	Describes what a runtime must be able to do and what it should look like
image-spec	Describes the standard of a container image and what its manifest should look like
distribution-spec	Is closely related to runtime-spec and image-spec and describes, for example, how the images are to be uploaded to a registry

Table 2.2 Overview of OCI Specifications

Docker has contributed a lot to the independence and openness of containers. For example, Docker has extracted its own runtime and transferred it to an independent project called *containerd*. Another component called *libcontainer* was also donated, from which *runC* was developed.

That should be enough history. Now, how is a runtime structured?

If you search for container runtimes, you will always find a division into

- low-level runtimes, and
- high-level runtimes.

runC, for example, is a low-level runtime that creates and executes containers. However, it is controlled by a high-level runtime such as containerd.

containerd takes care of everything as a high-level runtime. It downloads the image from the repository, manages the storage, and passes runC the container specification it needs to start the container. It also monitors runC during the execution of the containers. In Kubernetes, starting a container would be similar to Figure 2.3.

The CRI interface allows you to exchange containerd and runC with any other runtime. In addition to containerd, there is another runtime that currently plays an important role in the Kubernetes context. *CRI-O* is a runtime that is specifically developed for Kubernetes. CRI-O also uses runC to execute the container, but like containerd, it has the option of using other low-level runtimes.

Note

CRI-O is sometimes also referred to as the *container engine*. According to my research, however, it is only a container runtime. The terms are often used interchangeably on the internet, which sometimes makes it difficult to determine what is right or wrong.

For questions like this, I prefer to look at the GitHub repo, because that's where you'll find the truth: *https://github.com/cri-o/cri-o*.

2.1 Pod and Container Management

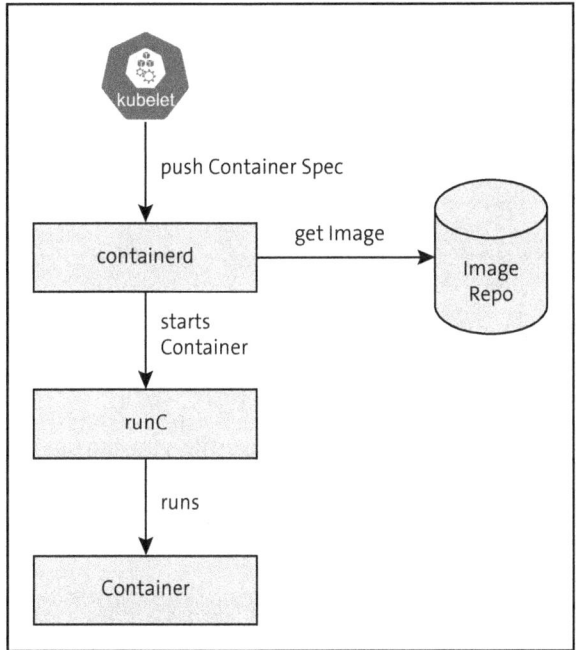

Figure 2.3 Communication of Runtimes

Ultimately, which runtime you want to use under the hood of Kubernetes is a matter of taste. But let me be honest with you: I have never dealt with the runtime as intensively as I did for this book, and it will make no difference to you as a developer. For some basic understanding, it's good that you know how containers are executed, but you don't need to dive much deeper.

> **Good to Know**
>
> Let's have a brief digression on how Kubernetes starts a container through CRI-O.
>
> As you learned in Chapter 1, Section 1.3.3 about the API flow, Kubernetes passes the pod manifest to the kubelet. Here's what happens next:
>
> 1. The kubelet forwards the request to CRI-O via the CRI interface of Kubernetes.
> 2. CRI-O extracts the container image from the specified registry, unpacks it, and creates a root file system.
> 3. CRI-O then generates an OCI runtime specification that describes how the container must run.
> 4. It transfers these to the low-level runtime runC.
> 5. The container is then monitored via a separate *conmon* process. This is a monitoring process that ensures communication between CRI-O and runC. It redirects logs and exit codes and thus passes on the status to the top.

2.1.2 Your First Own Pod

That's enough theory! Let's now deploy your own first container in a pod. We will use a simple example and extend an Nginx web server. To do this, you can use the Dockerfile from Listing 2.1. In Chapter 1, Section 1.4.7, I described how you can create a container registry using Minikube. We need that container registry now, and if you haven't activated it yet, you can do so now.

> **Note**
>
> I use Docker here for the build. You can of course also use any other tool for this purpose.

If your command line is located in the same folder as your Dockerfile, you can use the `docker build -t localhost:5000/my-nginx` . command to build your image. The `docker push localhost:5000/my-nginx` command allows you to load the image into the registry.

```
FROM nginx
RUN echo '<!DOCTYPE html><html><body><h1>Hello, World!</h1></body></html>'
    >/usr/share/nginx/html/index.html
```

Listing 2.1 Dockerfile for Nginx Container

> **Note**
>
> If pushing the image causes problems, you should compare the respective error message with the following:
>
> `Get "http://localhost:5000/v2/": dial tcp [::1]:5000: connect: connection refused`
>
> This is an indication that you need to run the `eval $(minikube docker-env)` command again. You must execute this in every new command line so that Minikube uses the correct Docker environment. Only then will you reach the registry.

Now that your Nginx image is already available in the registry, you can use it to start the pod in Kubernetes. To do this, you can use the manifest from Listing 2.2 and generate it via Lens. Take another look at the Lens reference guide (Chapter 1, Section 1.6.3), where you will find the instructions for creating resources. Under **Image**, you enter the container image that you have built and pushed into the registry.

> **Good to Know**
>
> The `apiVersion` in the manifest references the version of the corresponding Kubernetes object. In Chapter 4, Section 4.6, you will learn how objects are versioned in Kubernetes.

```yaml
apiVersion: v1
kind: Pod
metadata:
  name: my-nginx
spec:
  containers:
  - name: my-container
    image: localhost:5000/my-nginx
    ports:
    - containerPort: 80
```

Listing 2.2 Pod Manifest for Your Own Nginx

To check, open Lens and look at your pod under **Workloads • Pods**. This should look like Figure 2.4. Activate port forwarding as described in Chapter 1, Section 1.6.2, and open the Nginx page in the browser. You should now be greeted by a **Hello World** message.

> **Note**
> You are not dependent on Lens in these examples. You can also use `kubectl`. In this case, you can see the following output via `kubectl get pod`:
>
> ```
> NAME       READY   STATUS    RESTARTS   AGE
> my-nginx   1/1     Running   0          12s
> ```
>
> Then you can use the `kubectl describe pod my-nginx` command to obtain detailed information. Here's the abridged version:
>
> ```
> Name:              my-nginx
> Namespace:         default
> Priority:          0
> Service Account:   default
> Node:              minikube/192.168.49.2
> Start Time:        Tue, 13 Feb 2024 23:34:32 +0100
> Labels:            <none>
> Annotations:       <none>
> Status:            Running
> IP:                10.244.1.6
> ...
> ```
>
> Even if I don't always use both tools in the book, you can simply try it out with both. It will help you to familiarize yourself with Lens and `kubectl`.

Congratulations! You have your first container running on your Kubernetes cluster. Let's be honest, it's actually not that complicated. Try it out with one of your own applications: the quicker you take what you have learned into your own projects, the better it will anchor itself in your subconscious.

2 Basic Objects and Concepts in Kubernetes

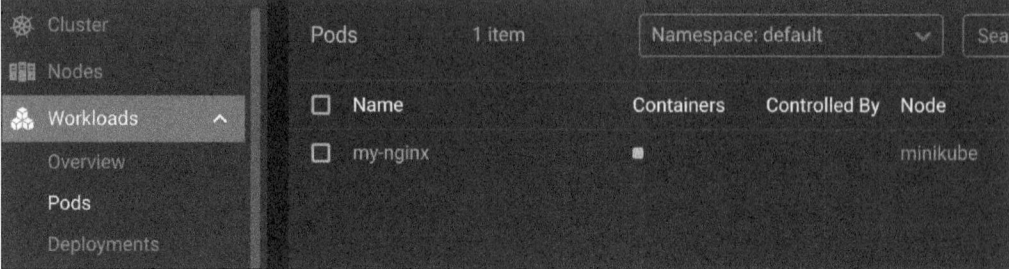

Figure 2.4 First Pod in Lens

2.1.3 Multiple Containers within a Pod

You have already learned that a pod can contain more than one container. Let's take a look at this concept in practice.

You now have your own nginx web server that delivers your *Hello, World* page. As a little helper, we want to provide the web server with a sidecar container that collects and processes the logs. This means that the main container does not also have to take care of redirecting logs to a central server; instead, we swap this task out to the sidecar container. You can see the target image of the pod in Figure 2.5.

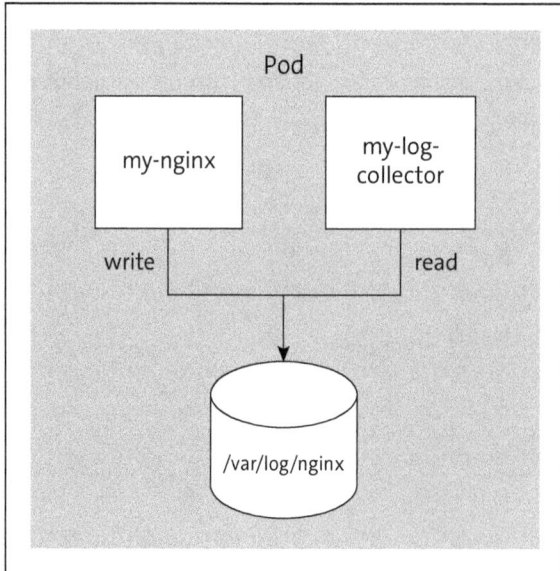

Figure 2.5 Nginx with Log Collector

Use the Dockerfile from Listing 2.3 to create your own log collector. The Dockerfile contains a small shell script that accesses the *access.log* file of Nginx in an endless loop and transfers it to a separate file named *sidecaraccess.log*. Once you have adopted the code, build an image again and store it in the registry. I have named the image `localhost:5000/my-log-collector`.

> **Note**
> Don't be surprised by the example. Of course, it is not thought through to the end, and lines that have already been transferred are simply bluntly transferred to the file again. However, the example gives you a good first impression of how a sidecar container should function.

```
FROM busybox
RUN echo -e '#!/bin/sh\nwhile true; do cat /var/log/nginx/access.log >>
    /var/log/nginx/sidecaraccess.log; sleep 5; done' > /bin/log-collector.sh
    && chmod +x /bin/log-collector.sh
ENTRYPOINT ["/bin/log-collector.sh"]
```

Listing 2.3 Dockerfile for Log CollectorSidecar Container

> **Good to Know**
> You will come across the busybox image often in the Kubernetes world. It is known as the Swiss Army knife of embedded Linux because it contains most of the standard Linux tools. At the same time, the image is very compact and extremely lightweight at less than 5 MB. So if you want to do something with Linux tools, think of the busybox image.

Now you can expand the pod manifest from Listing 2.2 as shown in Listing 2.4. You can see that two containers are now specified, and the same log volume has been assigned to both. This is also the transfer point of the data at the end. Both containers can access the volume. Nginx writes its logs to this path, and the log collector picks them up and processes them further.

Deploy the manifest via Lens and set up port forwarding again. Now open the Nginx page so that the log collector also has logs to collect. You can now use `kubectl exec` to open a shell in one of the containers. You can find the command in Chapter 1, Section 1.5.5. In the */var/log/nginx* path, you need to check whether the log collector extracts the logs correctly and writes them to its own file.

> **Note**
> Use an `emptyDir` as the volume in Listing 2.4. This is an empty folder that you mount under */var/log/nginx*. The lifecycle of this volume ends after the pod is terminated. We will look at volumes in more detail in Chapter 5.

```
apiVersion: v1
kind: Pod
metadata:
  name: my-nginx
spec:
  containers:
  - name: my-container
    image: localhost:5000/my-nginx
    ports:
    - containerPort: 80
    volumeMounts:
    - name: log-volume
      mountPath: /var/log/nginx
  - name: my-log-collector
    image: localhost:5000/my-log-collector
    volumeMounts:
    - name: log-volume
      mountPath: /var/log/nginx
  volumes:
  - name: log-volume
    emptyDir: {}
```

Listing 2.4 Pod Manifest for Log Collector as Sidecar

> **Note**
>
> In case something does not work as it should, I would like to point out the possibilities for error analysis. I always proceed according to a similar pattern.
>
> In this example, it would be as follows:
>
> 1. Check in Lens or `kubectl` whether the containers are running and, if not, analyze the error messages. Possible issues include the following:
>   – The image name is incorrect.
>   – The image is not in the registry.
> 2. Check the manifest:
>   – Are the indentations correct?
>   – Is the volume stored on the same path on both containers?
> 3. Check the logs of the individual containers:
>   – Perhaps one of the paths is not correct.
>   – The script has an error.

2.1.4 Communication between Containers

You have now developed and deployed your first pod with a sidecar container. Perhaps you are now asking yourself the following question: Do the containers always have to talk to each other via the file system?

There are different strategies for how the containers can interact with each other. You can select the appropriate strategy depending on the function of the second container. You have familiarized yourself with the standard helper containers and their functions in Table 2.1. You can see the corresponding communication channels in Figure 2.6.

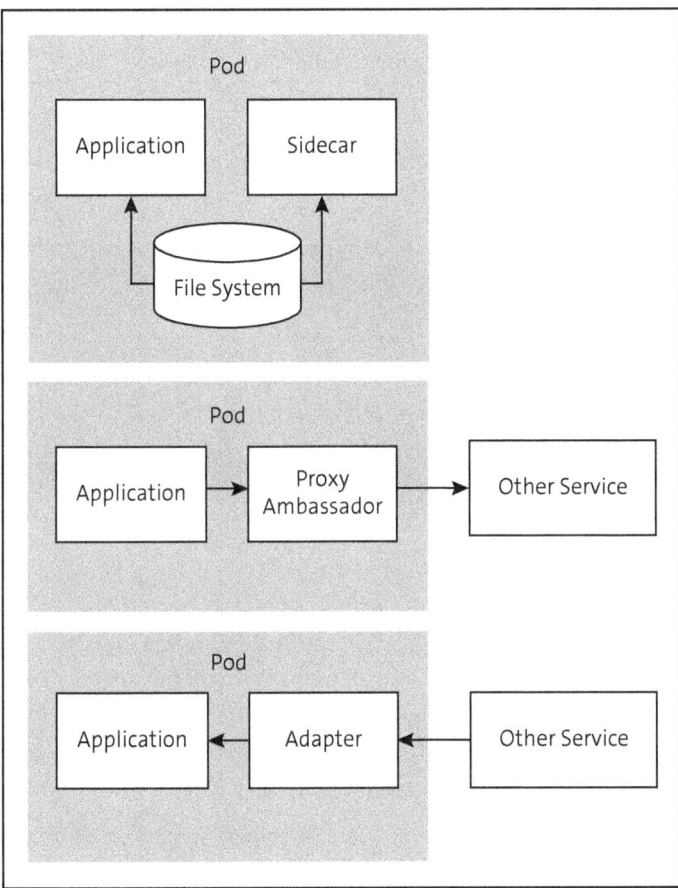

Figure 2.6 How Helper Containers Communicate with Each Other

The classic sidecar container communicates via the file system, just like your log collector. A log collector is even the typical use case for this: your main application produces data, and the helper container processes it further.

The second option is the ambassador, which is used as a proxy to the outside world. Because the containers within a pod share an IP address, the main application can easily access the ambassador container via `localhost` and send data. Possible tasks for the ambassador are as follows:

- Act as a service broker to other services
- Apply authentication to APIs

In each of these cases, your main application does not have to take care of a certain part of the logic itself. It knows only the ambassador container, sends its requests to it, and the rest is done by the ambassador.

The third strategy is the adapter container, which provides communication from the outside to the inside. Here too the communication is established via `localhost`, but in the other direction. Possible use cases can include the following:

- Validation of incoming data
- Transformation of data before it reaches the main application
- Transformation of communication protocols

Here the incoming traffic is simply transformed or prepared for the main application, and only then does it go into the main application.

> **Good to Know**
>
> The containers in a pod all start at the same time. It is therefore impossible to predict which container will be the first to boot up. This means that in your applications, you cannot rely on the sidecar container being started after the application.

2.1.5 Init Container

The more complex an application in a pod is, the more demanding deployment becomes. When you deploy the pod, you have little control over the speed of the startup. Sometimes this can be a little frustrating because the database is not yet ready or other prerequisites are missing. You could of course implement such checks in your application, but there is a much nicer method: the init container.

Each pod can define one or more init containers, which are always started before the main application. The sequence is strict, and one init container starts after the other. Your application container can only start once all of them have been successful. This means you can always be sure that your main application will not start until all init containers have successfully completed their tasks. The following use cases for this could be conceivable:

- The init container checks whether interfaces such as databases are accessible.
- Init containers can prepare the file system and assign or restrict rights.
- An init container can start on the basis of a different image with other tools that do not belong in the main application but are needed for preparation.

A major advantage of the init container is that you can assign your own authorizations to it. This allows you to assign secrets to the init container that are not required by the main application, which allows you, for example, to assign rights that enable you to access resources or APIs during initialization that you do not need during the execution. This increases the degree of security, as you assign rights according to the *least privilege* principle.

Let's now extend your pod with an init container in Listing 2.5, which creates two files in the file system for preparation. First it initializes the *sidecaraccess.log* file, and so that you can see that it is actually creating something, it also creates the *initcontainer.log* file. In the end, the pod will look as shown in Figure 2.7.

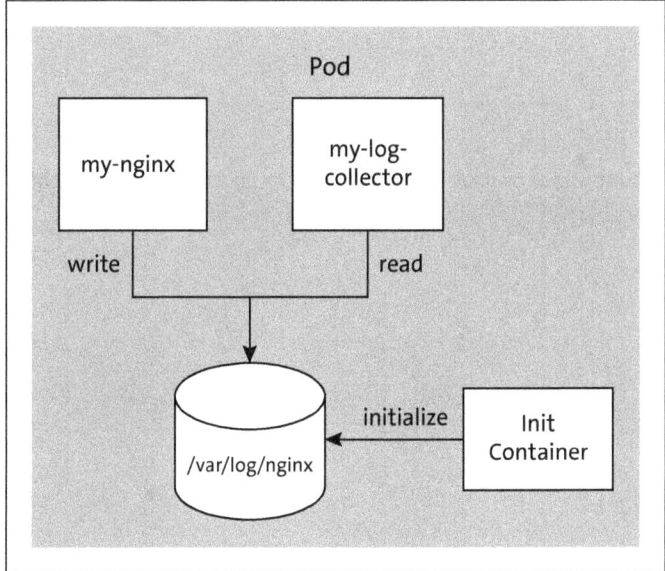

Figure 2.7 Pod Extension with Init Container

Note

You can already see that the pods will fill up with many containers over time, but they are inseparable. This is why Kubernetes manages pods and not the individual containers.

You can now roll out the new pod manifest via Lens and see how the containers behave. The pod now contains three containers: it successfully completes the init container first and then starts the other two containers. In the end, the pod in the Lens overview should look as shown in Figure 2.8. As the init container has been successfully completed, it is only displayed as an empty box at the end. Now use the pod shell to check whether the two files have been created correctly.

Figure 2.8 Pod with Init Container in Lens

As you can see, the concept of the init container is also easy to use. Can you think of an example from your company that is perfect for init containers? Try it out for yourself.

> **Note**
>
> You have a great deal of freedom when you develop an init container. You can use your own code of your choice, whether Python, Java, or other. The only important thing is that the container must complete successfully at the end; otherwise, the main application won't be able to start.

> **Note**
>
> If the init container fails, Kubernetes will start it again and again by default until the attempt is successful. In Section 2.1.7, you will learn how to set the restart policy.

```
apiVersion: v1
kind: Pod
metadata:
  name: my-nginx
spec:
  initContainers:
  - name: init-my-log-collector
    image: busybox
    command: ['sh', '-c', 'touch /var/log/nginx/sidecaraccess.log && touch /var/log/nginx/initcontainer.log']
    volumeMounts:
    - name: log-volume
      mountPath: /var/log/nginx
  containers:
  - name: my-container
    image: localhost:5000/my-nginx
    ports:
    - containerPort: 80
    volumeMounts:
    - name: log-volume
      mountPath: /var/log/nginx
  - name: my-log-collector
    image: localhost:5000/my-log-collector
    volumeMounts:
```

```
  - name: log-volume
    mountPath: /var/log/nginx
volumes:
- name: log-volume
  emptyDir: {}
```

Listing 2.5 Log Collector Pod Manifest with Init Container

2.1.6 Pod Phases and Container Statuses

You have already started and run several containers on Kubernetes. Finally, I would like to introduce you to the different phases that containers or pods can be in.

Let's start with the phases of a pod that are listed in Table 2.3. You can already see the phases in Lens in the pod overview in the **Status** column. These are a good indication of the current state of your pod. For example, if you have a pod that has been in the pending phase for a long time, then you should take action and see why it cannot switch to running.

Phase	Description
Running	The pod is running on a node and all containers have been successfully created. At least one container is running or is in the start process.
Succeeded	All containers in a pod have been successfully terminated and will not be restarted. This is particularly visible in Kubernetes jobs, which you will learn more about in Chapter 4, Section 4.2.
Pending	This status indicates that the pod is currently being created. During this time, Kubernetes loads the images, it gives the pod access to resources such as volumes and secrets, and the containers are waiting to be started. If any of these cannot be provided to the pod or the scheduler cannot find a suitable node, the pod remains in this status. It waits until all preconditions have been met.
Failed	All containers of a pod are terminated and at least one of them is faulty. This can happen through the system or through a status code that is not 0.
Unknown	Kubernetes cannot determine the state of the pod. This happens, for example, if the node on which the pod is running is no longer accessible.

Table 2.3 Possible Pod Phases

> **Good to Know**
>
> The statuses of pods do not always indicate existing issues. A pod in pending status may have problems pulling the image or mounting a volume. A pod in running status

> may be in a crash loop where the container in the pod is constantly restarting due to a problem.
>
> In most cases, you will need to take a closer look at the pod to understand how it is doing.

In addition to the pod phases, there are also the container statuses, which you can find in Table 2.4. These are not always so obvious, but are very meaningful, especially in the case of errors.

Status	Description
Running	A container that is in this phase was able to start without any problems and is currently being executed.
Terminated	A container is terminated when it has either been successfully completed or something has gone wrong for some reason. You can use the exit code to find out why the container was terminated.
Waiting	If a container is not running or terminated, it is in the waiting phase. Here, for example, it waits for a secret or for the container image to be downloaded.

Table 2.4 Possible Container Phases

In Lens, you can either see these in the pod overview as in Figure 2.9, or you can click a pod to see more information under **Containers**, including exit codes. This will help you with debugging, because sometimes you can only find the decisive clue as to why a container is not starting or has been terminated in the "last state."

Figure 2.9 Container Status Display in Lens

You can also output this information by using the `kubectl describe pod` command.

2.1.7 The Restart Policy of Pods

For init containers, I stated that they start again and again until they complete successfully. However, this is only half the truth, as you have the option of configuring this behavior. The setting for this is called the *restart policy* and applies to all containers in a pod.

The default setting ensures that containers that fail are always restarted. This is suitable in most cases, but sometimes there are applications where this is not the best option. Via the restart policy, you can determine under which circumstances a pod should be restarted. You can choose from the three options listed in Table 2.5.

Option	Description	Area of Use
Always	This is the default value and always restarts a container, regardless of why it was terminated.	This is suitable for applications that need to run continuously.
OnFailure	Only restarts the container if the exit code is not equal to 0. This means that the container has not been successfully completed.	This is useful, for example, for batch jobs that are not restarted until they have completed successfully.
Never	The container is never restarted, regardless of why it was terminated.	If you need full control over the running of the container, then this option is the best.

Table 2.5 Restart Policy Options

In Listing 2.6, you can see the extension of the pod manifest from Listing 2.2 with a restart policy. You can see that this policy is defined at the pod level, which is why it applies to all containers within the pod. For example, if you define restartPolicy: Never, the init container will not restart if it fails.

```
apiVersion: v1
kind: Pod
metadata:
  name: my-nginx
spec:
  containers:
  - name: my-container
    image: localhost:5000/my-nginx
  restartPolicy: Always
```

Listing 2.6 Setting Restart Policy in Pod Manifest

2.1.8 When the Pod Comes to an End

An important process that you should know as a developer is the scheduling of pods. In Kubernetes, pods should be scaled according to requirements, which means that pods are constantly being started and terminated. This is part of your daily work, which is why it is important that your application can be shut down properly (a *graceful shutdown*).

> **Good to Know**
>
> A graceful shutdown includes, for example,
> - the conclusion of current transactions,
> - closing a database session, and
> - exiting your application via exit code 0.

For your application to even realize that the pod is supposed to be terminated, it must respond to signals sent by the kubelet. The kubelet gives your application a certain amount of time to shut down. This is referred to as the *grace period* and is set to 30 seconds by default. If your application cannot respond to signals or takes longer than the grace period, the container is simply terminated by means of a hard shutdown. This must be avoided so that it does not lead to unwanted inconsistencies in the data or disconnections for your users.

Let's take a look at how scheduling works with Kubernetes. It does not matter whether you have triggered the scheduling manually using `kubectl` or whether Kubernetes does it itself for some reason:

1. In the first step, the pod is updated in the API server with the time at which the pod should be deleted. For this to happen, the grace period is simply added to the current time.
2. If a `preStop` hook has been defined, it will now be executed.
3. The kubelet will then send a `TERM` signal to each container in the pod. Only the process with ID 1 receives the signal.
4. If all containers have been terminated, the pod can also be completely terminated.

While the signals are being sent to your application, the kubelet will start to shut down the pod properly. For example, it must be removed from connected objects such as ReplicaSets or services.

If the grace period expires at any point in the process, the pod and all containers in it are forcibly deleted (*force delete*). For this purpose, the following is true:

- Each process in a container receives the `SIGKILL` signal and is thus simply terminated hard.
- The pod object is immediately deleted from the API server by setting the grace period to 0, which means that the pod can no longer be found by any client.

Finally, the final remnants such as the pause container are tidied up.

2.1 Pod and Container Management

Good to Know

There are two container hooks that you can define:

1. postStart
2. preStop

Here you can either use exec to execute a command or a script in the container or call another endpoint via HTTP. You can find out more about this at *http://s-prs.co/ v596414*.

Note

The *terminating* status is not one of the classic phases of a pod that you learned about in Section 2.1.6. This status is displayed in some kubectl commands if the pod is currently in the scheduling process but is still waiting for the grace period.

The grace period is the most important option in the scheduling process. You should consider whether the default value of 30 is sufficient for your application. The grace period is defined via the terminationGracePeriodSeconds option, as you can see in Listing 2.7.

```
apiVersion: v1
kind: Pod
metadata:
  name: my-nginx
spec:
  containers:
  - name: my-container
    image: localhost:5000/my-nginx
  terminationGracePeriodSeconds: 60
```

Listing 2.7 Grace Period Set in Pod Manifesto

You always have the option to override the grace period using kubectl and force the deletion of a pod via kubectl delete --force or kubectl delete --grace-period=0.

Good to Know

If the grace period is set to 0, the termination will be forced immediately.

Now that you know how scheduling works in Kubernetes, you need to prepare your application for a graceful shutdown. For this to work, two things must be fulfilled:

- Your application runs on PID 1 of the container.
- Your application should be able to handle signals.

To get your application to PID 1, there are several ways to structure your Dockerfile. You can take a look at the following article, for example, which explains this in detail: *http://s-prs.co/v596415*.

How your application can handle signals depends on the programming language. You'll have to see how this can be implemented in your application or framework. Consider which steps are necessary for your application in the signal handler in order to be switched off without negative side effects.

2.2 Annotations and Labels

Each object in Kubernetes is initially independent of all others. However, each one has specific functions that help you to operate your application, although there are some objects that can support and manage others. Pods can, for example, be managed by other objects, such as deployments. There must be a simple way to link these more or less independent objects together.

Another challenge is the number of objects in a cluster. The more objects are running in a cluster, the more important it is to keep an overview of them.

But it's not just the technical components that need clarity about individual objects. Precisely because there is a lot going on in a cluster, you also want to be able to quickly see which pods belong together and how you need to handle them.

For this reason, the concept of labels and annotations was introduced:

- *Labels* are like labels on a folder in your office cabinet that provide information about the contents of the folder. They make it easier for you to find the right folder or group of folders. Labels are flexible key-value pairs, and you can customize them according to your needs to add context-related information.
- *Annotations* are, as the name suggests, notes that provide additional information about an object. They are also key-value pairs, but are not used to group or identify objects. An annotation could be a table of contents in a folder. You cannot use it to find the right folder, but it gives you more information about the contents of the folder.

But not only pods are organized with labels. Kubernetes nodes also have labels, and scheduling decisions are made based on these labels. The principle is very simple, and yet you can set up complex rules that allow you to organize your entire cluster.

> **Good to Know**
> There are always matching selectors for the labels, which allow you to select and filter resources based on the labels.

2.2.1 Using Labels and Selectors

Let's jump straight into an example. In Kubernetes, labels are a way of adding metadata to your resources. This metadata can contain various information about an object. In the simplest case, this is the name of a group or the environment. But it can also be anything else that helps you organize your resources, such as

- the assignment to a larger application,
- the version of your application, or
- the owner of your application.

Listing 2.8 offers an example. You can see an Nginx pod to which the app and environment labels are assigned. You can use those labels now to select this pod.

The concept of selectors is not new. You refer to a specific label and can thus make it clear which resources you want to use. In the example, you could display the group of all production resources. Or you can display all resources that belong to the nginx application.

```
apiVersion: v1
kind: Pod
metadata:
  name: nginx example
  namespace: default
  labels:
    environment: production
    app: nginx
spec:
  containers:
  - name: nginx
    image: nginx:1.14.2
    ports:
    - containerPort: 80
```

Listing 2.8 Labels on Pod

There are different ways to write a selector. For example, there are the equality-based selectors:

- Equality = or == selects resources whose labels correspond exactly to the specified value:

  kubectl get pods -l app=nginx

- Inequality != selects resources whose labels do not correspond to the specified value:

  kubectl get pods -l app!=nginx

You can also use set-based selectors. These can be filtered using a set of values, whereby only one of these values needs to apply:

- In selects resources whose labels are contained in a specified list of values:

  kubectl get pods -l 'environment In (production,test)'

- NotIn selects resources whose labels are not contained in a specified list of values, whereby resources that have not set the label at all are also output:

  kubectl get pods -l 'environment NotIn (test)'

- Exists selects resources that have a specific label, regardless of value:

  kubectl get pods -l 'environment'

There are also selectors that you define for Kubernetes objects. The Kubernetes service, which you will learn about in Section 2.5, uses the selectors to distribute the network traffic to the correct pods. You can see an adequate example in Listing 2.9.

```
apiVersion: v1
kind: Service
metadata:
  name: nginx example
  namespace: default
  labels:
    app: nginx
    environment: production
spec:
  ports:
    - protocol: TCP
      port: 80
      targetPort: 80
  selector:
    app: nginx
    environment: production
```

Listing 2.9 Selector in Service

2.2 Annotations and Labels

You are completely free to choose your labels. Kubernetes does not provide any specifications; there are only recommended labels that you can use. Some of these are also used by third-party tools such as Helm. An overview is shown in Table 2.6.

Key	Description	Example
app.kubernetes.io/name	Name of your application	nginx
app.kubernetes.io/instance	Unique name that can identify a single instance of your application	Nginx-1337
app.kubernetes.io/version	Version of your application; you are free to use whatever version you like	1.0.5
app.kubernetes.io/component	Which component takes up the resource in the context of a large application	Web server
app.kubernetes.io/part-of	Where you can add the name of the larger application	Sales platform
app.kubernetes.io/managed-by	Set automatically if you use a tool such as Helm	helm

Table 2.6 Recommended Default Labels

Good to Know

In general, all labels without a prefix are private, which means there are no specifications. However, Kubernetes uses certain prefixes to mark system resources. You can separate a prefix from the actual label by using a / character.

All labels starting with `kubernetes.io` and `k8s.io` are created and managed by Kubernetes. You can find a complete overview at *http://s-prs.co/v596416*.

2.2.2 Field Selectors

Field selectors are a nice addition to select resources based on specific field values. They make it possible to make precise queries on objects based on the values of their fields, such as `metadata.namespace` or `metadata.name`. You could query the pod in Listing 2.8, for example, using `kubectl get pods --field-selector metadata.name=nginx-example`.

Thus, not only are you not limited to labels, but you can also make more interesting queries, such as to view all pods with a certain status. Using a command like `kubectl get pods --field-selector status.phase=Running`, you can query all running pods.

> **Note**
> The field selectors also have their limits, and you cannot query each and every field. The available fields vary depending on the resource. For more information on this, refer to the documentation at *http://s-prs.co/v596417*.

The field selectors only support the = and != operators. You can also connect several selectors together to make more complex queries—for example:

```
kubectl get pods --field-selector= \
   status.phase!=Running,spec.restartPolicy=Always
```

Field selectors are a nice extension that can help you to quickly query certain resources, especially on the command line.

2.2.3 NodeSelector

As you know, one of the purposes of Kubernetes is to make server management easier for you. Usually, you don't want to have to worry about where your application is currently running. However, there are times when it is important to be able to control the scheduling, such as in the following cases:

- Your pods require certain resources.
- Not all replications of a pod should run on the same node.
- Certain pods such as the backend and frontend should run on the same node.

For this purpose, you have three settings available that you can adjust in your pod manifests:

- **NodeSelector**
  You select the node on which you want your pod to run.
- **Node affinity and antiaffinity**
  You provide your pod with rules that tell it which node it should run on.
- **Pod affinity and antiaffinity:**
  You provide your pod with rules for which pods may or may not run on the same node.

All of these adjustments are also based on labels. You can assign labels to each node, which can then be processed with selectors to create such rules. But let's start with *NodeSelector*.

Let's take an application that needs an Nvidia GPU. It should only be trained on a node that also has an Nvidia GPU; otherwise, the application will not run. Listing 2.10 shows an example in which the nodeSelector option is used.

```
apiVersion: v1
kind: Pod
metadata:
  name: cuda-test
spec:
  containers:
    - name: cuda-test
      image: "registry.k8s.io/cuda-vector-add:v0.1"
      resources:
        limits:
          nvidia.com/gpu: 1
  nodeSelector:
    graphic: nvidia
```

Listing 2.10 NodeSelector Example

The pod will then only run on nodes that have the `graphic: nvidia` label assigned. If there is no node with this requirement, Kubernetes will not be able to start the pod.

> **Good to Know**
> NodeSelector is the easiest way to assign pods to specific nodes. If a node with the desired label is available in the cluster, the pod can start. If there is none, the container remains in pending status until a corresponding node is available.

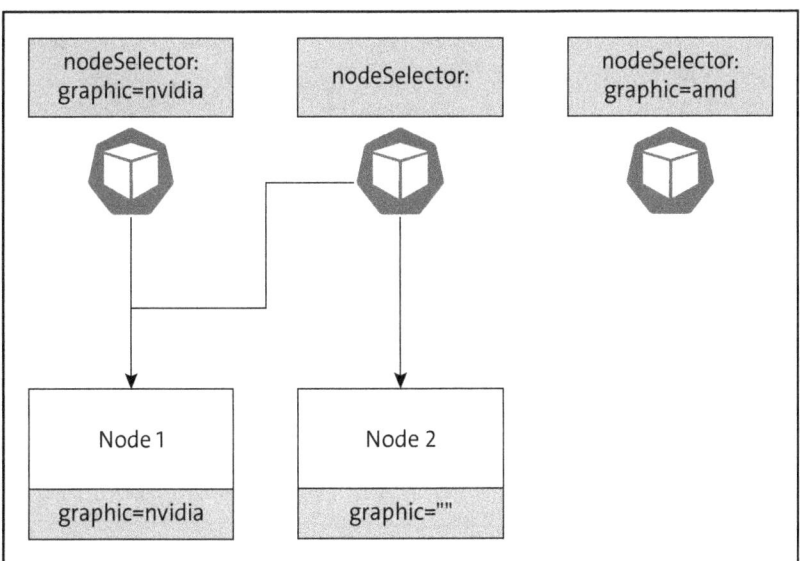

Figure 2.10 Function of NodeSelector

Figure 2.10 shows how the scheduler would distribute pods:

- The first pod has the `graphic=nvidia` NodeSelector defined and can therefore only be trained on node 1.
- The second pod does not have a NodeSelector and can be trained on both node 1 and node 2.
- The third pod has a NodeSelector on `graphic=amd` and does not find a node.

In Lens, you can easily display the labels of the nodes. To do this, go to your cluster, click **Nodes**, and then click the name of the node you want to use. You should be able to see the labels, as shown in Figure 2.11. On the Raspberry Pi master, for example, you can see the `node-role.kubernetes.io/master=true` tag, which you can use to identify the master nodes. As always, however, you are free to add more labels.

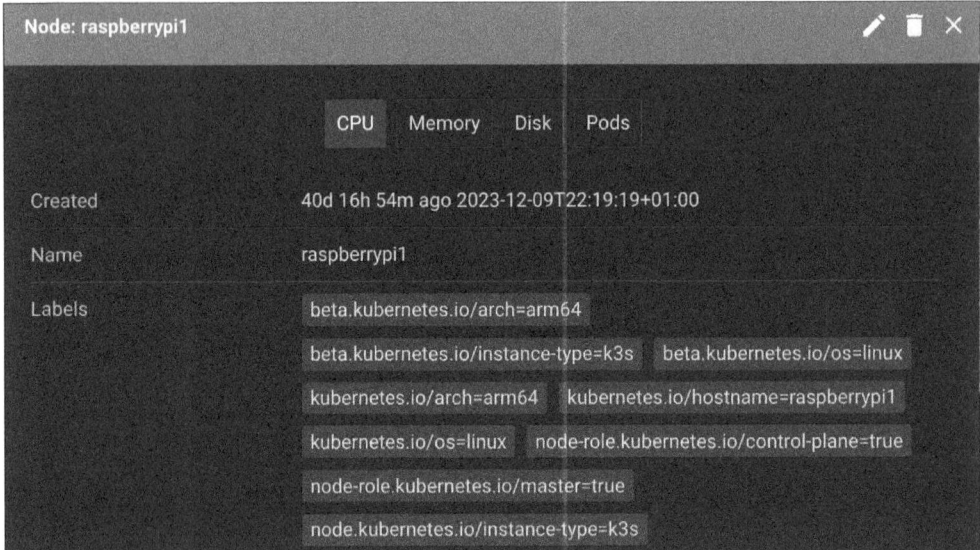

Figure 2.11 Labels of Raspberry Pi Master Node

> **Good to Know**
>
> Depending on the infrastructure on which the cluster is based, you will also see other useful labels. For example, a cluster in AWS should also have the availability zones or the region as a label. This allows you to make very fine-grained decisions about where your pods should run.

2.2.4 Node Affinity and Antiaffinity

An extension to NodeSelector is the *node affinity* and *node antiaffinity* concept. It allows you to define significantly better specifications and more complex rules. For example, you can formulate an affinity so that the pod would rather run on a specific node. However, if there is no space on the node, the pod can also start on another node. Where the NodeSelector simply decides that "the pod will not be started," with affinity, you can say that it's "better to be on the wrong node than on no node at all."

Good to Know

Affinity is particularly interesting if you have clusters with many nodes where you need to be able to control very precisely how the scheduling should work. For production systems, you should at least be familiar with the concept.

You are welcome to read this section and come back to it if it is of interest to your application.

You can configure affinity in the manifest as follows:

- `requiredDuringSchedulingIgnoredDuringExecution`
  The rule must be fulfilled. As with NodeSelector, this forces the default setting.
- `preferredDuringSchedulingIgnoredDuringExecution`
  The scheduler tries to comply with the rule, but can also schedule if it cannot adhere to it.

Note

The names of the two affinity options give the impression that there should also be a `requiredDuringExecution` option due to the `IgnoredDuringExecution` part. However, this option is not currently available. The topic has been discussed in the community for some time, but there is no longer any reference to it in the official Kubernetes documentation.

From my point of view, it would be a special case anyway. The option would only become active if the label of a node were to change at runtime, and I can't think of any use case in which this happens.

An example could look like the one shown in Listing 2.11.

```
apiVersion: v1
kind: Pod
metadata:
  name: region-pod
spec:
  affinity:
    nodeAffinity:
      requiredDuringSchedulingIgnoredDuringExecution:
        nodeSelectorTerms:
        - matchExpressions:
```

```
        - key: "region"
          operator: "In"
          values:
          - "USA"
          - "Europe"
  containers:
  - name: example-container
    image: nginx
```

Listing 2.11 Pod with Necessary Node Affinity

Here you have a cluster with nodes in different regions, and you want a pod to be allowed to launch only in Europe or the USA. By using the `requiredDuringScheduling` option you can enforce the rule and the pod cannot be started in China by mistake.

The rules can be extended in two ways:

- You can add another condition in `matchExpressions`. In that case, a pod can only be scheduled on the node if *all* rules apply. For example, in the affinity in Listing 2.12, you enter the hard disk type as an additional condition. This means that only nodes from Europe or the USA that use SSDs can be used by the pod.

- However, you can also add another condition in `nodeSelectorTerms`. This means that the node must fulfill either one or the other in order to be eligible for the pod. In Listing 2.13, this rule means that the node must either be located in the USA or Europe or have the `production` tag as its environment.

```
nodeAffinity:
  requiredDuringSchedulingIgnoredDuringExecution:
    nodeSelectorTerms:
    - matchExpressions:
      - key: "region"
        operator: "In"
        values:
        - "USA"
        - "Europe"
      - key: "disktype"
        operator: "In"
        values:
        - "ssd"
```

Listing 2.12 Additional AND Condition within Affinity

> **Note**
> You can use `In`, `NotIn`, `Exists`, `DoesNotExist`, `Gt`, and `Lt` as operators for affinities.

```
nodeAffinity:
  requiredDuringSchedulingIgnoredDuringExecution:
    nodeSelectorTerms:
    - matchExpressions:
      - key: "region"
        operator: "In"
        values:
        - "USA"
        - "Europe"
    - matchExpressions:
      - key: "environment"
        operator: "In"
        values:
        - "production"
```

Listing 2.13 Additional OR Condition within Affinity

The syntax for `preferredDuringScheduling` is slightly different. You can also assign a weighting for each rule, which helps to decide where the pod should prefer to run. For this purpose, Kubernetes will check each node for the rules, and if one of them applies, the node is assigned the weighting in terms of points. At the end, the pod on the node with the most points is scheduled. In Listing 2.14, the pod prefers a node from the USA.

```
apiVersion: v1
kind: Pod
metadata:
  name: region-pod
spec:
  affinity:
    nodeAffinity:
      preferredDuringSchedulingIgnoredDuringExecution:
      - weight: 2
        preference:
          matchExpressions:
            - key: "region"
              operator: "In"
              values:
              - "USA"
      - weight: 1
        preference:
          matchExpressions:
            - key: "region"
              operator: "In"
```

```
              values:
              - "Europe"
  containers:
  - name: example-container
    image: nginx
```

Listing 2.14 Pod with Preferred Node Affinity

In comparison, Listing 2.15 prefers a node in Europe with an SSD over a node in the USA without an SSD. However, a node in the USA with an SSD beats all others because it has the highest number of points.

```
nodeAffinity:
  preferredDuringSchedulingIgnoredDuringExecution:
  - weight: 2
    preference:
      matchExpressions:
        - key: "region"
          operator: "In"
          values:
          - "USA"
  - weight: 1
    preference:
      matchExpressions:
        - key: "region"
          operator: "In"
          values:
          - "Europe"
  - weight: 2
    preference:
      matchExpressions:
        - key: "disktype"
          operator: "In"
          values:
          - "ssd"
```

Listing 2.15 Pod with Another Node Affinity Rule

2.2.5 Pod Affinity and Antiaffinity

Another way to tell the pod where it should prefer to run is *pod affinity* and *pod antiaffinity*. You can use this to tell a pod which other pods should run on a node and which should not run there. Use cases for this include the following:

- One of your applications should run on the same node as the database in order to speed up communication.
- Two replications of the same pod should not run on the same node to ensure reliability.

If you look at Figure 2.12, you will see three pods from Nginx and three pods from the backend. The distribution is poor: if a node fails, either all backend pods or all Nginx pods are gone. In addition, requests from Nginx to the backend must always be routed to another node.

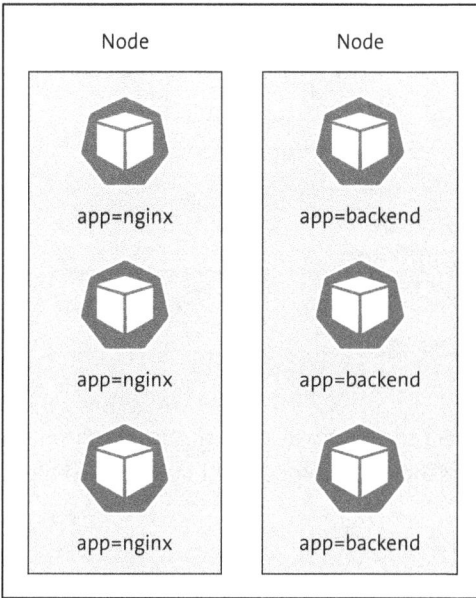

Figure 2.12 Pod Distribution without Pod Affinity

You can see a much nicer distribution in Figure 2.13. There is one Nginx pod and one backend pod on each node, but there is another node, and the utilization of the other nodes may be lower than before.

> **Note**
> Higher reliability and availability can lead to higher costs.

> **Note**
> As is the case with many other settings, you should also proceed iteratively with affinities. Set a hypothesis such as "I want to reduce latency," then try to verify it. Try to approach the sweet spot bit by bit.

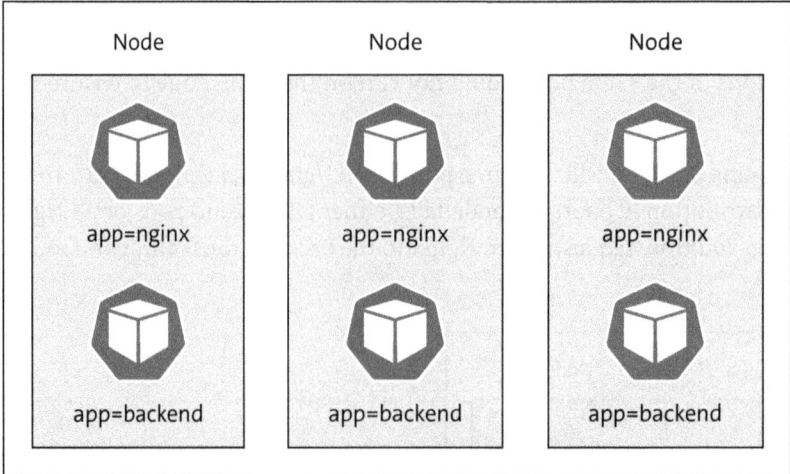

Figure 2.13 Pod Distribution with Pod Affinity

As with node affinity, you have the following options:

- requiredDuringSchedulingIgnoredDuringExecution
- preferredDuringSchedulingIgnoredDuringExecution

In Listing 2.16, an Nginx pod is forced by pod affinity to be deployed on a node with a backend pod. If you roll out this manifest via Lens, you will see that the status simply remains set to **Pending**. Kubernetes cannot schedule the pod until you roll out a pod with the app=backend label.

```
apiVersion: v1
kind: Pod
metadata:
  name: nginx-pod
spec:
  affinity:
    podAffinity:
      requiredDuringSchedulingIgnoredDuringExecution:
        - labelSelector:
            matchLabels:
              app: backend
          topologyKey: "kubernetes.io/hostname"
  containers:
  - name: example-container
    image: nginx
```

Listing 2.16 Pod Affinity: app=backend

2.2 Annotations and Labels

You can see an example of antiaffinity in Listing 2.17. There you force Nginx not to run on nodes on which another Nginx pod is already running. If you only have two nodes but want to deploy three pods, Kubernetes cannot roll out the third pod. On the Raspberry Pi cluster, it then looks as shown in Figure 2.14.

```
apiVersion: v1
kind: Pod
metadata:
  name: nginx-pod
  labels:
    app: nginx
spec:
  affinity:
    podAntiAffinity:
      requiredDuringSchedulingIgnoredDuringExecution:
        - labelSelector:
            matchLabels:
              app: nginx
          topologyKey: "kubernetes.io/hostname"
  containers:
  - name: example-container
    image: nginx
```

Listing 2.17 Pod Antiaffinity to Itself

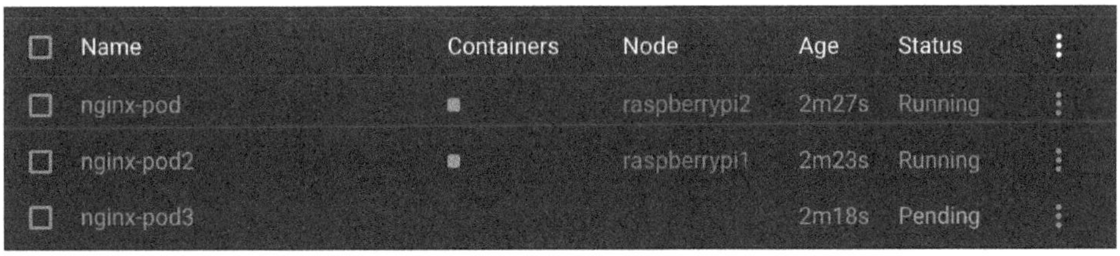

Name	Containers	Node	Age	Status
nginx-pod	■	raspberrypi2	2m27s	Running
nginx-pod2	■	raspberrypi1	2m23s	Running
nginx-pod3			2m18s	Pending

Figure 2.14 Pod Antiaffinity on Raspberry Pi Cluster

If you want to make the rules a little less strict, you can do this in the same way as with the node affinities. Listing 2.18 shows the syntax for a `preferredDuringScheduling` rule. Just try it out!

```
affinity:
  podAntiAffinity:
    preferredDuringSchedulingIgnoredDuringExecution:
      - weight: 100
        podAffinityTerm:
          labelSelector:
```

```
      matchLabels:
        app: nginx
    topologyKey: "kubernetes.io/hostname"
```

Listing 2.18 Pod Antiaffinity with preferredDuringScheduling Rule

> **[+] Good to Know**
>
> You can also use `matchExpressions` as with the node affinities.

You have probably already noticed the `topologyKey` option. This option defines the label according to which the nodes are grouped. For example, if you have nodes in different data centers or in the cloud in different availability zones, your nodes should be marked with the `topology.kubernetes.io/zone` label.

If you then write an antiaffinity rule, you can specify the following: "The pod should not run in a zone where a pod of the same application is already running." This will remove all nodes that are running in the same zone as the node on which one of your pods is already running. I have recorded this for you in Figure 2.15. The second pod selects the node from zone B because one is already running in zone A.

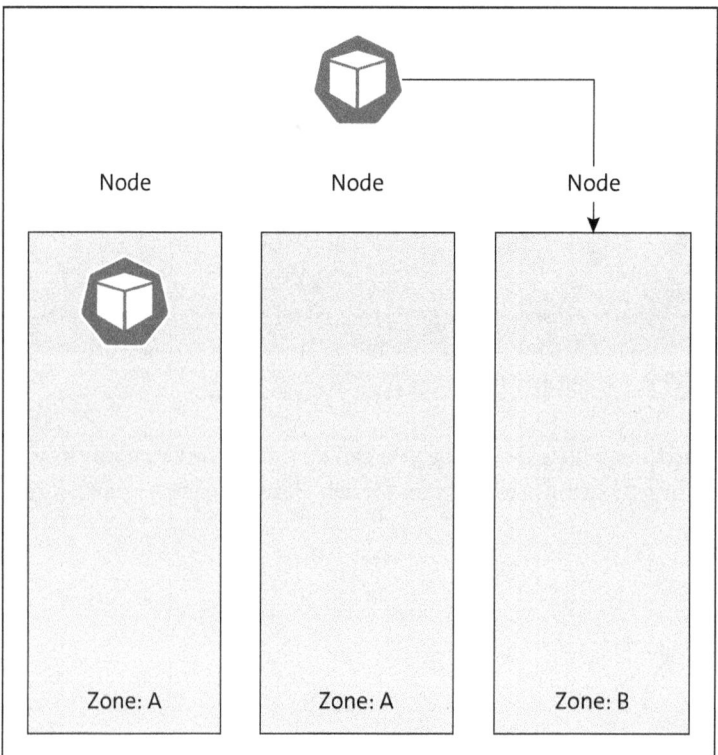

Figure 2.15 Grouping Nodes by topologyKey

This gives you even more control over where your pods should run and can increase reliability.

> **Note**
> Depending on the number of nodes in a cluster, pod affinities require a high computing effort. Kubernetes recommends not using pod affinity for clusters with several hundred nodes.

2.2.6 Taints and Tolerations

You now know how to use NodeSelector and affinities to make pods favor specific nodes. However, there are situations in which you will also want to allow the node to reject certain pods. The reasons for this include the following:

- **Reservation of nodes**
  Certain nodes should be reserved for special purposes or special groups of pods, such as the masters.
- **Dealing with special hardware**
  In clusters where some nodes have specialized hardware such as GPUs, you want to prevent ordinary pods from blocking these resources.
- **Controlled pod management**
  During maintenance work or upgrades of a node, you want to prevent new pods from being started there.
- **Error handling**
  In the event of node errors such as network issues or resource bottlenecks, you want to prevent new pods from being started on the faulty nodes.

For this purpose, you can use the concept of taints and tolerations. You assign a *taint* to a node, and any pod that does not tolerate this taint will be rejected. You set the *tolerations* on the pods that you want to allow to run on these nodes.

For example, if your master has set a taint and you want to run one of your applications on it, then it only needs the toleration and can thus be deployed by the scheduler on the master. Taints and tolerations work together in this way to ensure that pods are not placed on unsuitable nodes.

You can assign one or more taints to a node. Kubernetes also uses taints automatically to control the scheduling of pods. For example, if a node is not yet ready, a taint prevents pods from being started on the node.

You can assign three effects to your taints, with which you can control how Kubernetes should handle pods. A detailed list is provided in Table 2.7.

> **Good to Know**
>
> The *eviction* of pods is the process by which a pod on a node must be deleted. If possible, it will then get restarted on another node. There are different scenarios when this is advantageous. The node controller automatically sets taints if, for example, the node is no longer accessible or has other problems.
>
> You can find more about this topic at *http://s-prs.co/v596418*.

Taint Effect	Effects on Running Pods	Effects on New Pods
NoExecute	<ul><li>Pods without toleration are evicted immediately.</li><li>Pods with toleration, but without tolerationSeconds, remain on the node indefinitely.</li><li>Pods with toleration and tolerationSeconds are evicted after the specified time has elapsed.</li></ul>	The pod cannot start without toleration.
NoSchedule	No eviction of existing pods.	The pod cannot start without toleration.
PreferNoSchedule	No eviction of existing pods.	New pods without toleration are avoided if possible, but that is not guaranteed.

Table 2.7 Taints and Their Effects

To set a taint, you need a key, a value, and an effect. For example, if you want to define a NoSchedule taint for the master, you could run the following command:

```
kubectl taint nodes node1 nodeType=master:NoSchedule
```

nodeType is the key, while master is the value. To remove a taint, you simply need to add a hyphen (-) after the command:

```
kubectl taint nodes node1 nodeType=master:NoSchedule-
```

On the other hand, you need to assign tolerations to your pods if you want them to accept taints. The tolerations look similar to affinities, and you can use the Exists and Equal operators. Listing 2.19 shows a toleration that accepts a taint with the nodeType key. The value of the taint is irrelevant to this toleration.

```
apiVersion: v1
kind: Pod
metadata:
  name: nginx
spec:
  containers:
  - name: nginx
    image: nginx
    tolerations:
    - key: "nodeType"
      operator: "Exists"
      effect: "NoSchedule"
```

Listing 2.19 Pod with Toleration Operator "Exists"

In Listing 2.20, however, only one taint is tolerated where the `nodeType` has the `master` value. Consequently, if you have a node with the `worker` value, the pod will not be able to start there.

Of course, you can also add multiple tolerations to a pod, such as to tolerate other effects or completely different taints.

> **Good to Know**
> An empty value for `key` with `operator: Exists` matches all keys, values and effects. This means that your pod will simply tolerate all taints.
> An empty value for `effect` can be used with all effects.
> The term *empty value* simply refers to passing an empty string with "".

```
apiVersion: v1
kind: Pod
metadata:
  name: nginx
spec:
  containers:
  - name: nginx
    image: nginx
    tolerations:
    - key: "nodeType"
      operator: "Equal"
      value: "master"
      effect: "NoSchedule"
```

Listing 2.20 Pod with Toleration Operator "Equal"

2.2.7 Annotations

Like labels, annotations are an essential component of Kubernetes. They provide a flexible method for enriching objects with additional information. While labels are used to identify and organize objects within Kubernetes, annotations allow you to store additional information that goes beyond the core functionality of Kubernetes. Classic use cases include the following:

- **Storage of complex data**
  For example, this includes data that can contain entire JSON objects.

- **Additional information**
  You can include contact information, release notes, or auditing information. For example, one of my customers has stored the application's protection requirements there.

- **Extensions and integrations**
  Kubernetes tools and extensions use annotations to enable specific functions or provide information.

You can view examples of annotations and what is saved there in each object that you have created. Take a look at an Nginx pod that you created in the previous sections. There you will find the `kubectl.kubernetes.io/last-applied-configuration` annotation and, as a value, a JSON object with the complete manifest that you rolled out last.

You will also find annotations in other objects. In Section 2.3, you will learn about deployments. The revision of the deployment is counted under the `deployment.kubernetes.io/revision` value. With each new rollout, Kubernetes counts up by one, and you can see which deployment is currently active.

> **[+] Good to Know**
>
> An annotation can store data of up to 256 KB.

Let's take a look at how you can add annotations to a pod yourself. In Listing 2.21, we give the pod a JSON object under the `build` annotation.

```
apiVersion: v1
kind: Pod
metadata:
  name: nginx-example
  namespace: default
  annotations:
    build: |
      {
        "repo": "nginx-example"
        "hash": "afj34iweo",
        "timestamp": "2023-12-12T14:38:23Z"
```

```
    }
spec:
  containers:
  - name: nginx
    image: nginx:1.14.2
    ports:
    - containerPort: 80
```

Listing 2.21 Pod with Annotations

The nice thing about this is that you can pass a multiline string in YAML, which makes it very easy to read in the code. We will go deeper into the YAML syntax in Chapter 3, Section 3.2.

The advantage of JSON is that you can easily see the current status of your application's code yourself. You can also have the data read automatically and process it further. The options for using this are enormous, and this can make it easier for you to manage your applications.

Figure 2.16 shows how the annotations are displayed in Lens. Simply click your pod, and you will see the annotations directly in the menu that pops up.

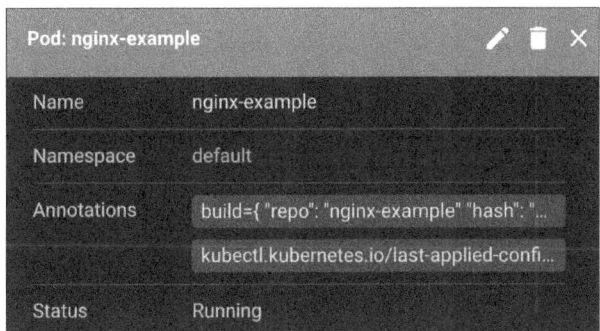

Figure 2.16 Displaying Annotations in Lens

An example of how the Prometheus annotations-monitoring application uses annotations is shown in Listing 2.22. The tool you will learn about in Chapter 7, Section 7.4 uses the annotations to detect whether a pod provides metrics and, if so, on which port. In Chapter 8, you will learn that Helm also uses annotations to store and read information about objects.

```
metadata:
  annotations:
    prometheus.io/scrape: "true"
    prometheus.io/port: "8080"
```

Listing 2.22 Annotation for Prometheus

2.3 Deployments and ReplicaSets

At this point, you know how to run your application in Kubernetes as a pod. You already know the label concepts and how the Kubernetes resources can be controlled and coupled. But now we come to the real magic of Kubernetes.

As you know, Kubernetes is primarily designed for managing a large number of pods. In a production environment, you will want to create multiple pods of the same type so that you can distribute the load or increase reliability. Now you don't need to manually create and deploy multiple pod manifests, because this is exactly what the deployment and ReplicaSet Kubernetes objects are responsible for.

You can imagine these as the manager and the foreman of a craft business. The manager (deployment) knows the plan of the project and its goals, and they can also make strategic decisions to replace workers in a project. The manager always sees the big picture. The manager instructs the foreman (ReplicaSet) to supervise the workers. The foreman ensures that the work is carried out in accordance with the manager's instructions and that there is always a sufficient number of workers on the project. If a worker is absent due to illness, the foreman takes care of a replacement.

Note

Although you can also use ReplicaSet without a deployment, the two objects simply belong together. Kubernetes recommends never using ReplicaSet without a deployment, as the additional features make running your application much easier. However, we will take a close look at both properties.

Figure 2.17 shows how the objects are connected. The deployment is the top-level object, manages a ReplicaSet, and instructs the ReplicaSet to manage and monitor the pods.

The deployment

- manages the lifecycle of pods and ReplicaSets,
- enables controlled updates through rolling updates, and
- enables rollbacks to previous versions.

The ReplicaSet

- ensures that the specified number of pods are running and
- monitors the pods and replaces faulty ones.

Both objects always check the current status with regard to the desired status. If something is not as desired, the objects try to restore the state.

2.3 Deployments and ReplicaSets

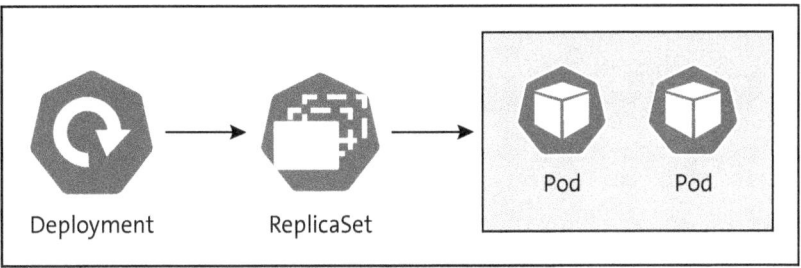

Figure 2.17 Architecture of Deployments and ReplicaSets

2.3.1 The Role of ReplicaSets

ReplicaSet is one of the key objects for running your application stably in Kubernetes. It makes sure that a certain number of pods of your container are always running. To ensure this, the ReplicaSet also uses labels and selectors, which you learned about in Section 2.2. These are also referred to as *MatchLabels* in the ReplicaSet and serve as a selector. This is how the ReplicaSet knows which pods belong to it and will manage them accordingly. MatchLabels are also key-value pairs that you define in the specifications of a ReplicaSet and the pods.

Listing 2.23 shows the extension of your Nginx pod from Section 2.1. Under `template`, the ReplicaSet finds everything it needs to know to create the pods. In addition, we have added the `app: nginx` label, and under `selector.matchLabels` you can see the selector used to select the set label. Another difference from the pod manifest is the `replicas` option. There you tell ReplicaSet how many pods you want to run.

Now you can roll out the manifest from Listing 2.23 and view the pods created in Lens.

```
apiVersion: apps/v1
kind: ReplicaSet
metadata:
  name: my-nginx-replicaset
  labels:
    app: nginx
spec:
  replicas: 2
  selector:
    matchLabels:
      app: nginx
  template:
    metadata:
      labels:
        app: nginx
```

```
    spec:
      containers:
      - name: my-container
        image: localhost:5000/my-nginx
        ports:
        - containerPort: 80
```

Listing 2.23 ReplicaSet Manifest

As shown in Figure 2.18, you can see that ReplicaSet creates two pods as desired. For each pod, ReplicaSet will add a unique ID to the name so that there is no name conflict. You can also see in the overview that the pod is controlled by a ReplicaSet.

Figure 2.18 Pods Generated by ReplicaSet

You can view the ReplicaSet itself in Lens under **Workloads · ReplicaSets**. In the detail view, you can then see which pods are managed by the ReplicaSet and what the status of the pods is, as shown in Figure 2.19.

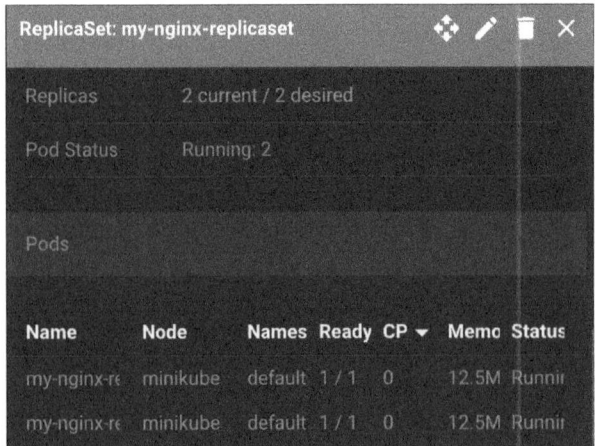

Figure 2.19 Detail View of ReplicaSet in Lens

Good to Know

I mentioned that ReplicaSet will generate the pod manifest. This makes ReplicaSet the owner of the pods. In Kubernetes, this concept is referred to as *owners* and *dependents*. In this case, the ReplicaSet is the owner of the pods, and the pods are the dependents of the ReplicaSet.

In the deployed pod manifesto, this looks as follows:

```
ownerReferences:
  - apiVersion: apps/v1
    kind: ReplicaSet
    name: my-nginx-replicaset
    uid: b6238a34-7656-45e4-a377-e864e8ad99f9
    controller: true
    blockOwnerDeletion: true
```

The nice thing about this approach is that you do not need to specify and link each resource individually because Kubernetes is smart enough to generate the pods from the ReplicaSet manifest. There is also a kind of garbage collection: if you delete the owner, Kubernetes will delete its dependents first. You can control this option using `blockOwnerDeletion`.

There are other owner and dependency connections that you will get to know. For example, the deployment will be the owner of the ReplicaSet. You can find more information about this topic at the following address: *http://s-prs.co/v596419*.

Take some time to play around with ReplicaSet:

- Delete one of the pods.
- Increase the replicas using `kubectl scale`.
- Try to reduce the number of replicas using Lens.

Try it out and see how ReplicaSet reacts. The principle behind ReplicaSet is very simple and yet extremely powerful. By regularly comparing the desired and current status, it always keeps the number of pods at the right level. This is the first step toward a system that is able to perform a self-healing process. A very interesting extension is autoscaling, which allows Kubernetes to increase or decrease the number of replicas itself depending on the load. We will look at this in more detail in Chapter 7, Section 7.3.

> **Note**
>
> If you also want to start a single pod in addition to ReplicaSet with the `app: nginx` MatchLabel, you should make sure that the pod does not have the same label. ReplicaSet will immediately see the pod as its task and include it in its management process.
>
> Try rolling out the following manifest and see what ReplicaSet does to the pod:
>
> ```
> apiVersion: v1
> kind: Pod
> metadata:
>   name: my-nginx
>   labels:
>     app: nginx
> ```

```
spec:
  containers:
  - name: my-container
    image: localhost:5000/my-nginx
    ports:
    - containerPort: 80
```

The pod is terminated immediately, as ReplicaSet already has the required number of pods running.

2.3.2 Creating Deployments

You now know how ReplicaSet works and how it monitors pods and keeps them stable at a desired number of replications. The functionality of ReplicaSet is kept simple, but it needs a little more logic, especially when it comes to new rollouts. You can use the Kubernetes deployment object for this purpose.

While ReplicaSet takes care of the pods, the deployment takes care of ReplicaSet, monitors its status, and provides additional logic. A deployment can

- create, monitor, and clean up ReplicaSets;
- roll out a new version of pods by creating a new ReplicaSet; and
- monitor and pause rollouts and perform rollbacks.

If you look at the manifest in Listing 2.24, you can see that it is almost exactly the same as Listing 2.23. You provide the pod template, the replications, and the MatchLabels. Of course, the deployment manifest needs this information because it has to use it to create the ReplicaSet and the pods.

```
apiVersion: apps/v1
kind: Deployment
metadata:
  name: my-nginx-deployment
  labels:
    app: nginx
spec:
  replicas: 2
  selector:
    matchLabels:
      app: nginx
  template:
    metadata:
      labels:
        app: nginx
```

2.3 Deployments and ReplicaSets

```
spec:
  containers:
  - name: my-container
    image: localhost:5000/my-nginx
    ports:
    - containerPort: 80
```

Listing 2.24 Deployment Manifest

Now try to roll out the manifest and take a look at the deployment in Lens. To do this, select **Workloads · Deployments** and click the **my-nginx-deployment** you have created. Compared to ReplicaSet, you can see two special features, as shown in Figure 2.20: the strategy type and the deploy revisions. These properties allow you to control the rollout and perform a rollback. Let us take a closer look at this in an example.

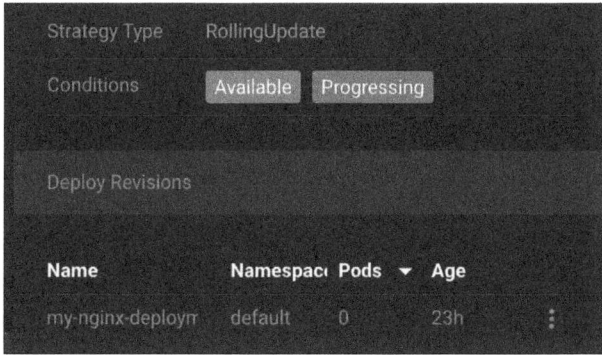

Figure 2.20 Detailed View of Deployment in Lens

Take a look at what exactly the deployment has generated. If you click the ReplicaSet and the pods, you can see that the pods still belong to the ReplicaSet. However, the ReplicaSet now has an owner reference to your deployment. The structure is therefore exactly the same as you saw in Figure 2.17. The deployment takes care of the ReplicaSet, and the ReplicaSet takes care of the pods.

> **Good to Know**
>
> As the deployment manages the ReplicaSet, any changes you make to the ReplicaSet will be overwritten by the deployment. Try changing the number of replicas in the ReplicaSet. The change will not take effect because the deployment will overwrite your value again.
>
> The deployment is the single point of truth. In this way, Kubernetes avoids inconsistencies.

143

2.3.3 Rolling Updates via the Deployment Object

There are two update strategies you can choose between:

- `Recreate`
  All existing pods are first deleted and then replaced by new ones.
- `RollingUpdate`
  You can control pod replacement and thus minimize downtime.

A use case for the `recreate` option could be a development cluster where you have few resources available. If new pods cannot be trained, the old pods must first be deleted. However, this strategy leads to failures in any case. You should therefore do without it in production clusters.

The second strategy represents the standard for Kubernetes. The `RollingUpdate` option replaces one pod after the other for you. This allows you to avoid application downtimes. Ideally, your application should have the following features to support rolling updates:

- **Horizontal scaling**
  Your application should be able to run multiple instances simultaneously.
- **Statelessness**
  Your application should be stateless or save the state externally to enable seamless updates.
- **Readiness checks**
  You should implement readiness checks as described in Chapter 7, Section 7.2 to ensure that new pods are healthy before they start managing the traffic.

> **Note**
>
> Kubernetes can also perform a rolling update without a readiness check. However, the problem with this is that Kubernetes does not know when the new pods will be available and able to process requests. This is why you need readiness checks, because this is the only way you can be sure that your application is fully up and running before Kubernetes replaces the next pod.
>
> Another advantage is that you can cancel the update of your application if the new version has a problem and cannot be started up. Without a readiness check, Kubernetes would not notice the problem.

Figure 2.21 shows what it will look like when you install an update. The deployment has the following tasks during an update:

1. Creating a new ReplicaSet with a new pod specification
2. Upscaling pods in the new ReplicaSet
3. Downscaling pods in the old ReplicaSet

A rollout is successful if all pods are running in the new ReplicaSet and the old ReplicaSet is set to 0. But let's just give it a try. If you have not yet rolled out the deployment from Listing 2.24, please do so now. The next step will be to update the image. To do this, swap your own Nginx image `image: localhost:5000/my-nginx` with the official one. Set the value to `image: nginx` and roll out the update.

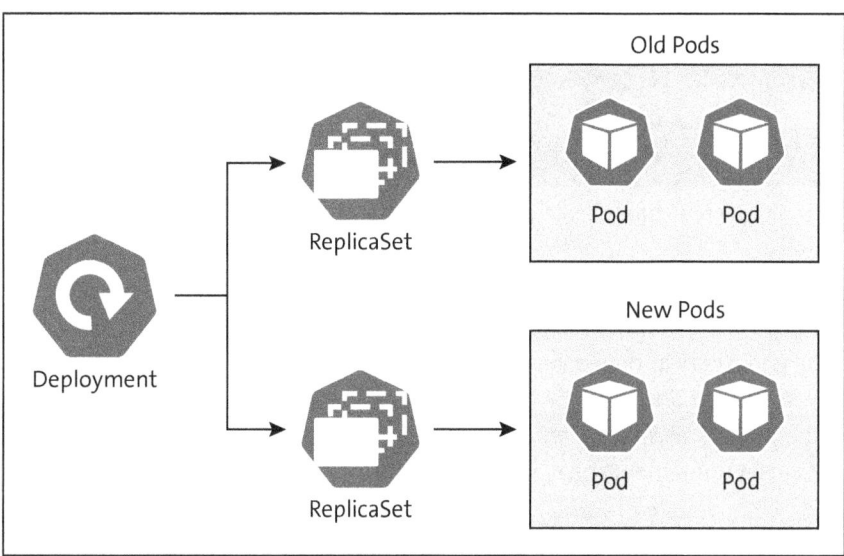

Figure 2.21 Rolling Update Managed by Deployment

Observe the rollout process. This will happen very quickly in this case, as the Nginx pods do not need long to report that they are ready. In Lens, click **Workloads · ReplicaSets**, then take a look at the way the deployment works. You now have two ReplicaSets there, and the pods are terminated on the old one and scaled on the new one.

If you compare the old ReplicaSet with the new one, you can see that your old image is still stored in the old one. This makes it easy to roll back. In the deployment itself, you can find the ReplicaSets under **Deploy Revisions**, as shown in Figure 2.22.

> **Good to Know**
> When you update a deployment, the old ReplicaSet remains in place. You can use the `.spec.revisionHistoryLimit` option to control the maximum number of ReplicaSets that can be kept. The default value is 10, but if you set the value to 0, you can no longer perform a rollback.

> **Good to Know**
> Imagine the deployment is currently in the update process and you want to make a quick change. Kubernetes can handle this use case without any problems. When you

install the new update, the deployment will create another ReplicaSet, scale up pods there, and scale down all old ReplicaSets.

Figure 2.22 ReplicaSets as Deploy Revisions in Deployment

There is another nice option that allows you to see how the deployment proceeds. Run the `kubectl get events` command. Your output will look similar to Listing 2.25, where you will find it in a very abridged version. It is nice to see how the new pods are started by the new ReplicaSet and how the old pods are terminated at the end. In our Nginx example, the process completes within just a few seconds. You can use one of your applications here, implement the readiness check from Chapter 7, Section 7.2, and then observe the rolling update process.

> **[+] Good to Know**
>
> You can pause and resume a rollout using the `kubectl rollout pause` and `kubectl rollout resume` commands. This can be particularly useful for larger deployments with a large number of replicas—for example, if you see an error in the new version that you want to analyze or resolve.
>
> While the rollout is paused, you can make updates to the deployment. However, these updates do not become active until the rollout is reactivated.

```
REASON        OBJECT       MESSAGE
Scheduled     pod/799      Successfully assigned default/799
Pulling       pod/799      Pulling image "nginx"
Pulled        pod/799      Successfully pulled image "nginx" in 1.13s
Created       pod/799      Created container my-container
Started       pod/799      Started container my-container
Scheduled     pod/7zm      Successfully assigned default/7zm
Pulling       pod/7zm      Pulling image "nginx"
Pulled        pod/7zm      Successfully pulled image "nginx" in 1.2s
Created       pod/7zm      Created container my-container
Started       pod/7zm      Started container my-container
Succ...te     rs/8d        Created pod: 7zm
Succ...te     rs/8d        Created pod: 8d-brvnt
```

```
Killing      pod/c7        Stopping container my-container
Killing      pod/s6        Stopping container my-container
Suc…Delete   rs/94         Deleted pod: s6
Suc…Delete   rs/94         Deleted pod: c7
ScalingRS    deployment    Scaled up replica set 8d to 1
ScalingRS    deployment    Scaled down replica set 94 to 1
ScalingRS    deployment    Scaled up replica set 8d to 2
ScalingRS    deployment    Scaled down replica set 94 to 0
```

Listing 2.25 Procedure of Rolling Update

For rolling updates, you have two setting levers, maxUnavailable and maxSurge, which you can use for configuration.

> **maxUnavailable**
>
> - **What it does**
>   It determines the maximum number of pods that may not be available during the update.
> - **How it works**
>   It can be a fixed number or a percentage, such as 5 pods or 10% of the pods. When calculating the integer from the percentages, Kubernetes rounds down.
> - **Default value**
>   25%
> - **Example**
>   If you set the value to 30%, then at least 70% of the pods must be running during the update. New pods start, old ones are terminated, but Kubernetes makes sure that at least 70% of the pods are always available and can answer requests.

> **maxSurge**
>
> - **What it does**
>   It determines how many additional pods can be created during the update.
> - **How it works**
>   It can also be a fixed number or a percentage. When calculating the integer from the percentages, Kubernetes rounds up.
> - **Default value**
>   25%
> - **Example**
>   With the 30% setting, 30% more pods can be started during the update than you specified in the manifest. Once old pods have terminated, Kubernetes can continue to scale, but never more than 130% of the desired pods in total.

Let's start with an example that illustrates these two levers. Listing 2.26 uses a deployment that runs with three replicas of busybox. The busybox container is called via several simple commands:

- First it waits for 20 seconds.
- Then it creates the /tmp/ready file.
- After that, it waits for another 3,600 seconds so that it does not terminate.

A readiness check is set for the */tmp/ready* file, which checks whether the file exists. Thus, the readiness check will run successfully after about 20 seconds and mark the pod as *ready*. This will give you some time to watch the rolling update.

We have configured 1 for maxUnavailable and maxSurge. In this case, this means that at least two pods should always be available and a maximum of four pods may run simultaneously. Before you get started with the demo: How do you think the deployment will proceed with a rolling update?

Roll out the deployment manifest in Kubernetes. Observe the pods and how much time it takes them to get ready. You can also see in the deployment itself, as in Figure 2.23, that it takes a while for **0/3** to become **3/3**. This is due to the readiness check, which is only successful after 20 seconds. To enable you to now trigger a rolling update, you need to change sleep 20; to sleep 30; and roll out the manifest again. Observe exactly what happens.

Name	Namespace	Pods	Replicas	Age
rolling-deployment	default	0/3	3	12s

Figure 2.23 Deployment after Creation

As you can see in Figure 2.24, the pod display jumps to **2/4**. If you display the pods, you will see that a pod switches directly to **Terminating** status and two new pods are scaled at the same time. The two remaining pods remain untouched for the time being. Not until one of the new pods is ready will the next pod switch to **Terminating** status. The third of the new pods is started immediately. At about the same time, the second pod is ready and the last of the old pods is terminated.

Name	Namespace	Pods	Replicas	Age
rolling-deployment	default	2/4	3	5m44s

Figure 2.24 Deployment after Start of Rolling Update

We were therefore able to observe that at least two of the pods were always available and a maximum of four pods were running at the same time. The pods in **Terminating** status were not included here.

2.3 Deployments and ReplicaSets

Try out the rolling update a few more times and observe the process. You can try out the following exercises:

- Use `kubectl get events` to take a look at the individual steps performed by Kubernetes.
- Update the deployment twice in quick succession with different values.
- Try pausing and resuming the rollout.
- Finally, you can play around with the `replicas`, `maxUnavailable`, and `maxSurge` values.

You should really take some time for this, because this is a core process in a production system. You should be able to roll out your application with every release without any downtime. The better you understand this process and the levers, the better you can adapt Kubernetes to your application.

> **Good to Know**
>
> Kubernetes only distinguishes between the *rolling update* and *recreate* rollout strategies in the deployment. There are a few other strategies such as *blue-green* or *canary deployment*. These types of rollout are also possible with Kubernetes, but sometimes require a little more effort. You can find a good article on this here: *http://s-prs.co/v596420*.

```
apiVersion: apps/v1
kind: Deployment
metadata:
  name: rolling-deployment
spec:
  replicas: 3
  strategy:
    type: RollingUpdate
    rollingUpdate:
      maxUnavailable: 1
      maxSurge: 1
  selector:
    matchLabels:
      app: rolling
  template:
    metadata:
      labels:
        app: rolling
    spec:
      containers:
      - name: busybox
        image: busybox
```

2 Basic Objects and Concepts in Kubernetes

```
      args:
        - /bin/sh
        - -c
        - >
          sleep 20;
          touch /tmp/ready;
          sleep 3600;
      readinessProbe:
        exec:
          command:
            - cat
            - /tmp/ready
        initialDelaySeconds: 5
        periodSeconds: 5
```

Listing 2.26 Example of Rolling Update

2.3.4 Rollback via Deployment

As described previously, the deployment provides a rollback mechanism. In addition, ReplicaSets from previous versions simply remain and continue to hold the old pod manifest. Accordingly, you still have all the manifests you need for a rollback in the *etcd* database.

For our sample rollback, let's extend the example from Section 2.3.3. If you have experimented a lot, your ReplicaSet overview could look exactly like Figure 2.25. Now you'll use `kubectl` to roll back to the desired version.

Name	Namespace	Desired	Current	Ready	Age
rolling-deployment-55d66	default	0	0	0	21m
rolling-deployment-74466	default	0	0	0	6m51s
rolling-deployment-76588	default	0	0	0	26m
rolling-deployment-b8f5c	default	3	3	3	6m28s

Figure 2.25 Overview of ReplicaSets from Previous Versions

Good to Know

If you have played a lot with the rolling updates, you will have noticed the following: Kubernetes recognizes when you import the same YAML manifest from a previous version and will not create a new ReplicaSet for it. The old ReplicaSet then becomes the current one again, and the pods are scaled there.

2.3 Deployments and ReplicaSets

To get an idea of the latest versions, you can use the `kubectl rollout history deployment/rolling-deployment` command. Depending on how much you have tried, your output should look similar to that shown in Listing 2.27. Unfortunately, we have not provided a change cause, which is why it is somewhat more difficult to recognize which version we want to go back to. You can display the exact details via `kubectl rollout history deployment/rolling-deployment --revision 2`. The entire pod template is displayed, and you can choose which one you want. In this example, the changes were only minor.

> **Good to Know**
>
> The change cause from Listing 2.27 is taken from the `kubernetes.io/change-cause` deployment annotation. This means that if you set this annotation during the rollout with info that indicates what has been changed, you can better recognize this in the history.
>
> Think carefully about whether you want to use this type of information. I personally do not use the change cause, as the change is usually always available via a CI/CD pipeline release and is contained in the Git commit. It must fit your processes because duplicate maintenance is unnecessary.

Once you have decided on a revision, you can use either `kubectl rollout undo deployment/rolling-deployment` to return to the last revision or `kubectl rollout undo deployment/rolling-deployment --to-revision=2` to return to revision 2. When you run the command, you will see that Kubernetes performs the same procedure as with the normal rolling update.

Take another look at the history. In my case, I rolled back to revision 2. This is no longer displayed in the history, but is now the latest one—in my case, revision 5. Kubernetes is smart enough to recognize that both revisions have the same manifest, so it will only display one of them.

```
REVISION     CHANGE-CAUSE
1            <none>
2            <none>
3            <none>
4            <none>
```

Listing 2.27 Deployment History

Finally, I want to show you a good way of easily recognizing differences between the revisions. Especially if no change cause is provided and you need to debug quickly, you can simply redirect the output of the revision details to a file and then compare it via a diff—for example:

```
kubectl rollout history deployment/rolling-deployment --revision 1 \
  > first.out
kubectl rollout history deployment/rolling-deployment --revision 3 \
  > second.out
diff one.out two.out
```

The output of the diff command should look similar to Listing 2.28. In the example, you can see that we have switched from 30 seconds to 20 seconds and that the pod template hash has changed as a result. Even major changes can be analyzed very well.

```
1c1
< deployment.apps/rolling-deployment with revision #1
---
> deployment.apps/rolling-deployment with revision #3
4c4
<         pod-template-hash=76588598d6
---
>         pod-template-hash=55d6687d47
13c13
<         sleep 30; touch /tmp/ready; sleep 3600;
---
>         sleep 20; touch /tmp/ready; sleep 3600;
```

Listing 2.28 Diff of Two Revision Details

> **Note**
> Deployment rollbacks are a great tool for responding quickly to problems. However, I hardly ever use them in production systems. The problem with this is the traceability of changes because if you perform a rollback, your manifests in the Git repository are no longer up to date, and other team members will find it more difficult to track the changes.
>
> In the repo, you have the history and can also directly see the changes to the code. Suitable CI/CD pipelines will also enable you to roll back quickly, and you will have tracked all changes to your production system.

2.4 ConfigMaps and Secrets

When you develop and build containers, everything your application needs belongs in the container, with a few exceptions. One of them is secrets such as passwords or certificates, and the other is configurations. The reason is that you don't want to build a separate image for each environment, but you want to be able to deploy a container in different environments and configure it as in Figure 2.26 using environment-specific configuration.

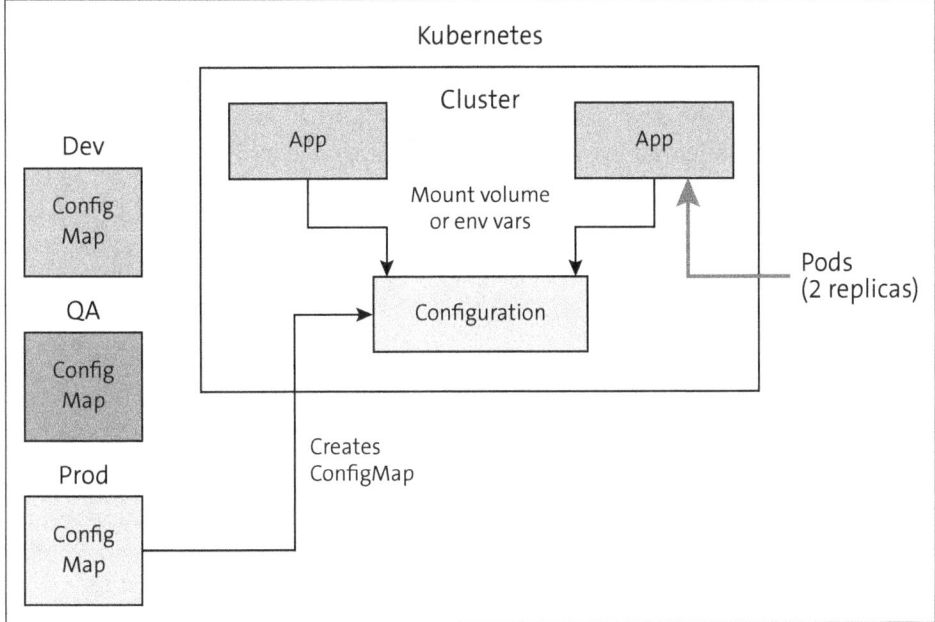

Figure 2.26 Configurations per Environment

Kubernetes provides the *ConfigMap* and *secret* objects for this purpose. These two objects are like notebooks that you have on your desk. Anyone from your company can take a look inside, but the notebook is locked with your passwords and can only be read by you.

> **Secrets in etcd**
>
> Even if secrets provide a higher level of security, they can be found unencrypted in the *etcd* database by default. This means that anyone with API access can read them. Every cluster admin also has access to your secrets. It is therefore important that you think in advance about the secrets you want to store there and how you want to restrict access to them. Of course, you have options to increase security:
>
> - Setting encryption at rest
> - Implementing role-based access control (RBAC) rules
> - Restricting secret access to specific containers
>
> Keep these points in mind when using Kubernetes secrets. It is best to discuss them with your cluster admins. Instructions for activating encryption at rest can be found at the following address: *http://s-prs.co/v596421*.

2.4.1 What Are ConfigMaps?

You can use the ConfigMap object to implement the configuration of your applications using Kubernetes. This decouples the configuration and the container image. ConfigMap primarily stores data as key-value pairs, which you can then inject into your container and use in your application. In addition to the classic key-value pairs, you can also store an entire file, such as a configuration in JSON, in a ConfigMap. This is very useful for larger configurations and can be easily integrated into an application.

> [!] **Secrets instead of ConfigMaps**
>
> ConfigMaps provide no protection at all for sensitive data; it is better to use the Kubernetes secrets object for this.

A ConfigMap has a very simple structure. In Listing 2.29, you can see that the relevant fields for your data are `data` and `binaryData`. Normal key-value pairs are stored under `data`, and `binaryData` was developed for Base64-encoded strings.

But to really understand what ConfigMap can do, we will try it out in different ways. There are a total of four ways to make data from ConfigMap available for your application. The content of ConfigMaps can be

- mounted by pods as a file system so that the application can read the file,
- set as environment parameters for the application within the pod manifest,
- passed as a command line argument for the container, or
- read within your application using the Kubernetes API.

As always, there is no perfect way to integrate ConfigMaps. Depending on the use case and the type of data you need in your application, one of these four approaches will make the most sense. In the following sections, I will describe each of the four approaches so that you can then choose the right one for your application.

```
apiVersion: v1
kind: ConfigMap
metadata:
  name: example-configmap
data:
  simpleKey: simpleValue
binaryData:
  binaryKey: dGVzdCBiaW5hcnkgZGF0YQ==
```

Listing 2.29 Simple ConfigMap Manifest

2.4 ConfigMaps and Secrets

Good to Know

You cannot import an infinite amount of data via ConfigMap. A ConfigMap must not exceed the maximum size of 1 MiB.

Integrating ConfigMaps as a Volume

Let's start with a ConfigMap that you will integrate as a volume. To do this, you need to create the ConfigMap from Listing 2.30. In it, we have defined a complete JSON object under the `config.json` key. If you roll out this ConfigMap and display it in Lens, it should look like the one shown in Figure 2.27. You can see that the JSON object has been imported correctly.

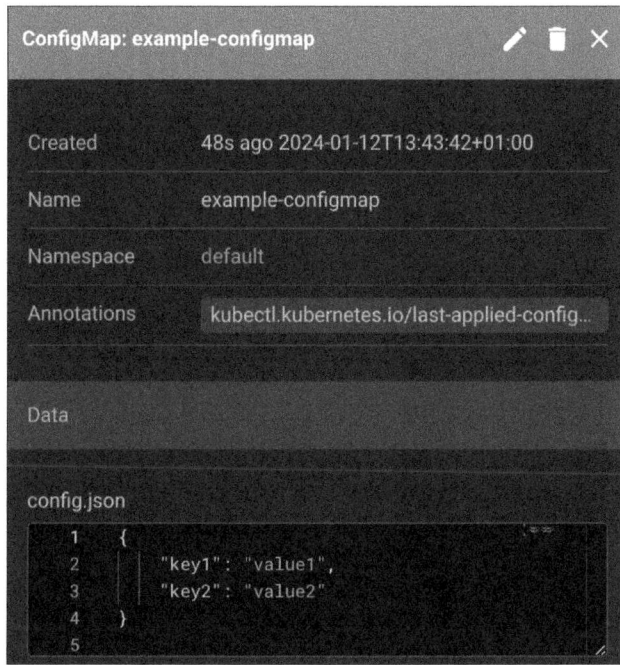

Figure 2.27 Displaying ConfigMap in Lens

```
apiVersion: v1
kind: ConfigMap
metadata:
  name: example-configmap
data:
  config.json: |
    {
```

```
        "key1": "value1",
        "key2": "value2"
}
```

Listing 2.30 ConfigMap Manifest with One File

You will learn more about the topic of volumes in Chapter 5, but at this point I want to explain what you need to know about ConfigMaps. You can use the pod manifest from Listing 2.31 for the example.

Here we define a volume named `config-volume` under `volumes` and link to the ConfigMap from Listing 2.30 and the `config.json` item. This volume is mounted under `volumeMounts` in the `mountPath: /etc/config`, and you should then find the `config.json` item in the container under this path.

```
apiVersion: v1
kind: Pod
metadata:
  name: example-pod
spec:
  containers:
    - name: example-container
      image: nginx
      volumeMounts:
        - name: config-volume
          mountPath: /etc/config
  volumes:
    - name: config-volume
      configMap:
        name: example-configmap
        items:
        - key: "config.json"
          path: "config.json"
```

Listing 2.31 Pod Manifest with ConfigMap Volume

In Figure 2.28, you can see how the volume is created by the kubelet using the ConfigMap. The volume contains the data you have defined in ConfigMap.

Roll out the pod manifest and connect to the pod via `kubectl exec`. Now you can check the contents of the volume. The nice thing about integrating as a volume is that a separate file containing the value is stored for each key in the ConfigMap. This means that you will find the file in the `/etc/config/config.json` path. You can also use `cat` to check whether the content is correct.

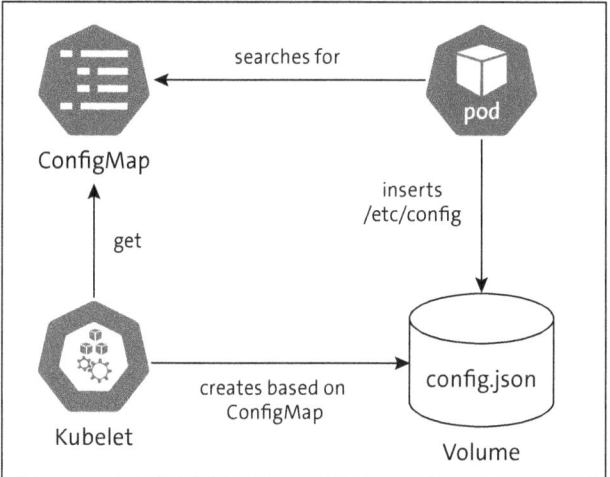

Figure 2.28 ConfigMap Provided as Volume

Note
As you can see in Listing 2.30, you can also design the values in multiple lines. You can use the full range of YAML. You will learn more about YAML in Chapter 3, Section 3.2.

Next, let's add another key-value pair to the ConfigMap from Listing 2.30. I have added test: "next one" underneath and rolled out the update in Lens. It will take a little time, but then you will find a new file named test with the content next one in the volume mount.

Good to Know
When you update a ConfigMap, the values in a volume mount get updated as well. The kubelet periodically checks whether anything has changed and will then also import the update into the volume mount. It can therefore take a few seconds for the update to arrive in the pod.

When you integrate ConfigMaps as environment parameters, you will not receive an update. In this case, you must restart the pod.

In the current pod definition, the kubelet will also mount each new parameter of the ConfigMap in the volume as soon as it recognizes an update. However, you can already restrict which items you want to provide for your application in the pod manifest. You can find the corresponding extension in Listing 2.32. This gives you significantly more control over the mount, for example, if you share a ConfigMap between multiple applications.

```
volumes:
  - name: config-volume
    configMap:
      name: example-configmap
      items:
        - key: "config.json"
          path: "config.json"
```

Listing 2.32 Special Selection of Keys in ConfigMap

> **Note**
>
> If you want to prevent an update of ConfigMap, you can achieve this by using an `immutable` tag as in the following example:
>
> ```
> apiVersion: v1
> kind: ConfigMap
> metadata:
>   name: example-configmap
> immutable: true
> data:
>   config.json: |
>     {
>         "key1": "value1",
>         "key2": "value2"
>     }
>   test: "next one"
> ```

Integrating ConfigMaps as Environment Parameters

Another common method of carrying out configurations in an application is to set environment parameters. The values of a ConfigMap can also be used to set environment parameters. I would not pass an entire JSON object as a parameter in this case, but it is perfect for classic configurations such as the log level or the host name of a database.

For the example, we have prepared the ConfigMap from Listing 2.33 and the pod manifest from Listing 2.34 for you. In the pod manifest, a reference to the ConfigMap is transferred via `envFrom`. Kubernetes will then set all key-value pairs as environment parameters.

Just try it out and log in to the pod via `kubectl exec`. You can use the `env` command to view all environment parameters and also find `LOG_LEVEL` and `DB_HOST` there. You can now also have the environment parameters imported by your application.

```
apiVersion: v1
kind: ConfigMap
metadata:
  name: example-env-configmap
data:
  LOG_LEVEL: "debug"
  DB_HOST: "localhost"
```

Listing 2.33 ConfigMap Example for Environment Parameters

```
apiVersion: v1
kind: Pod
metadata:
  name: example-pod-env
spec:
  containers:
    - name: example-container
      image: nginx
      envFrom:
        - configMapRef:
            name: example-env-configmap
```

Listing 2.34 Pod Uses ConfigMap as Environment Parameter

This is a good example if you have a ConfigMap for an application. As with mounting as a volume, you can specify the selection of environment parameters even more precisely. For example, you can load the log level from a different ConfigMap than the database host. This allows you to design your ConfigMaps more freely.

In Listing 2.35, we have turned one ConfigMap into two, and in Listing 2.36 you can see the update of the pod manifest. In this example, you use `valueFrom` to reference an explicit value that is to be taken from a specific ConfigMap, and as you do not need the value of `PORT` from the database ConfigMap in Listing 2.35, you do not have to drag it along unnecessarily.

> **Good to Know**
>
> You can only define string values in ConfigMap. For this reason, the `PORT` parameter from Listing 2.35 is not an integer.

```
apiVersion: v1
kind: ConfigMap
metadata:
  name: example-env-configmap-log
```

```
data:
  LOG_LEVEL: "debug"
---
apiVersion: v1
kind: ConfigMap
metadata:
  name: example-env-configmap-db
data:
  DB_HOST: "localhost"
  PORT: "1234"
```

Listing 2.35 Split Environment Parameter ConfigMaps

```
apiVersion: v1
kind: Pod
metadata:
  name: example-pod-env
spec:
  containers:
    - name: example-container
      image: nginx
      env:
        - name: DB_HOST
          valueFrom:
            configMapKeyRef:
              name: example-env-configmap-db
              key: DB_HOST
        - name: LOG_LEVEL
          valueFrom:
            configMapKeyRef:
              name: example-env-configmap-log
              key: LOG_LEVEL
```

Listing 2.36 Pod Manifest with Selected Parameters from Various ConfigMaps

> **Note**
>
> In addition, you can always include a ConfigMap as an option. If the ConfigMap or the parameter does not exist, the mounted volume or the environment parameters remain empty. See Listing 2.37 and Listing 2.38 for these examples.
>
> ```
> - name: LOG_LEVEL
>   valueFrom:
>     configMapKeyRef:
>       name: example-env-configmap-log
> ```

```
          key: LOG_LEVEL
        optional: true
```

Listing 2.37 Example from Listing 2.36

```
volumes:
  - name: config-volume
    configMap:
      name: example-configmap
      optional: true
```

Listing 2.38 Example from Listing 2.31

Transferring a Container Command via a ConfigMap

I am showing you this option for the sake of completeness. There are probably use cases where you want to transfer a command to a pod that can be configured by a ConfigMap. But so far, I have never come across this in real life.

You can also parameterize the command for the container by setting an environment parameter as in the previous section. For the example, we use the ConfigMap from Listing 2.33 and pass an echo on the parameter to a busybox image as a command in Listing 2.39. Kubernetes will start the pod, run the echo, and then the pod will terminate.

The `kubectl logs example-pod` command enables you to see that `localhost` gets output. This is exactly the content you have defined in the ConfigMap under the `DB_HOST` parameter.

```
apiVersion: v1
kind: Pod
metadata:
  name: example-pod
spec:
  containers:
    - name: example-container
      image: busybox
      command: ["/bin/sh", "-c", "echo $HOST"]
      env:
        - name: HOST
          valueFrom:
            configMapKeyRef:
              name: example-env-configmap
              key: DB_HOST
  restartPolicy: Never
```

Listing 2.39 Transferring Container Command as ConfigMap Parameter

Querying ConfigMaps via Kubernetes API

Finally, I want to describe a brief example of how you can query a ConfigMap via the Kubernetes API. You could do this directly in your application or implement it as a sidecar.

> **Note**
>
> If you want to use the Kubernetes API directly from the application, you also need to think about authorizations and access data in a production environment. The application then requires a technical user who is authorized to read the ConfigMap.

For this purpose, we have prepared a Python script for you in Listing 2.40.

```
from kubernetes import client, config
config.load_kube_config()
v1 = client.CoreV1Api()
configmap_name = 'example-configmap'
namespace = 'default'
config_map = v1.read_namespaced_config_map(configmap_name, namespace)
print(config_map)
```

Listing 2.40 Python Script for Reading ConfigMap

The script uses the Kubernetes Python package, which you can download either using `pip install kubernetes` or within your IDE, for example. The script does the following:

1. It loads your Kubeconfig file.
2. It creates an API client.
3. It tries to load the ConfigMap from Listing 2.30 from the `default` namespace.
4. It displays the ConfigMap.

As a result, you receive the entire ConfigMap as a JSON payload, which you can now continue to use and load the data from it.

2.4.2 What Are Secrets?

Kubernetes provides additional protection for secrets compared to ConfigMaps, as secrets often contain sensitive data such as passwords, tokens, or certificates. However, as we get into the examples, you will see that the YAML syntax is very similar to ConfigMaps, which makes it very easy to use. To ensure that the data in secrets is well protected, Kubernetes handles it very carefully. For example, the secret

- is sent only to the node on which a pod that relies on it is running;
- is stored by the kubelet in a temporary file system to prevent confidential data from being stored permanently;

- is deleted from the node as soon as the pod no longer needs the secret; and
- is only assigned to the containers in a pod to which you grant explicit access.

You can also contribute to the security of a secret's data by treating it as a secret in your application even after it has been read. When writing a manifest, you should make sure that only the container that needs the secret has access.

Good to Know

In addition to Kubernetes secrets, there are also other products available that make managing secrets easier for you. For example, I've worked with customers who used AWS Secrets Manager or HashiCorp Vault. Every product has its justification, and here too I can only say: it needs to fit in with your process.

However, if you want to use a vault cluster to merely inject passwords, then this would be a case of overengineering. It is best to talk to your cluster admins to find a suitable solution.

Full Access through Privileged Containers

Usually, a secret can only be read by a pod to which you have explicitly granted access. However, every container that is started with the `privileged: true` option can read all the secrets that are stored on its Kubernetes node.

If you look at the manifest from Listing 2.41, you will see that the structure is the same as for ConfigMaps.

```
apiVersion: v1
kind: Secret
metadata:
  name: example-secret
type: Opaque
data:
  username: YWRtaW4=
  password: cGFzc3dvcmQ=
```

Listing 2.41 Kubernetes Secret Manifest

However, there are three small differences:

1. As you can see in the overview in Table 2.8, there are different types of secrets.
2. The values under `data` must always be Base64 encoded.
3. If you want to transfer strings, you must store them under `stringData`.

In general, like ConfigMaps, you can also use secrets in multiple ways and inject them into your pods. Secrets can be

2 Basic Objects and Concepts in Kubernetes

- set as an environment parameter in a pod,
- mounted as a volume, and
- used as a pull secret for private container registries.

Type	Description
Opaque	Default type for your data. Not subject to any specifications.
kubernetes.io/dockercfg	Contains the serialized ~/.dockercfg.
kubernetes.io/dockerconfigjson	Contains the serialized ~/.docker/config.json.
kubernetes.io/tls	Here you can save TLS certificates. A common use case is to store the certificates for an ingress.
kubernetes.io/ssh-auth	Contains access data required for an SSH connection.
bootstrap.kubernetes.io/token	Secrets that are used when setting up new Kubernetes nodes.
kubernetes.io/basic-auth	Kubernetes checks whether the username and password keys are set during creation; otherwise, there are no further advantages over the opaque type. It makes it immediately clear to other developers what the secret is intended for.
kubernetes.io/service-account-token	Tokens from service accounts are stored here. This token can be used by a pod to authenticate itself to the Kubernetes API.

Table 2.8 Types of Secrets

Good to Know

Like ConfigMaps, secrets can also be

- optionally integrated,
- created as immutable, and
- created with a maximum size of 1 MiB.

Secrets in the Repo

You should never check unencrypted secret manifests into a version management system such as Git. Tools such as *sops* (*https://github.com/getsops/sops*) allow you to encrypt the data beforehand and decrypt it during deployment. Clarify these procedures with your company or your cluster admins in advance.

2.4 ConfigMaps and Secrets

Integrating a Secret as an Environment Parameter

Setting secrets as environment parameters works in the same way as with ConfigMaps except that the syntax is slightly different. Let's look at a small example. For the example, we'll use the secret from Listing 2.41 and the pod manifest from Listing 2.42. I have highlighted the changes to ConfigMap in bold.

```
apiVersion: v1
kind: Pod
metadata:
  name: example-pod
spec:
  containers:
  - name: example-container
    image: nginx
    env:
      - name: USERNAME
        valueFrom:
          secretKeyRef:
            name: example-secret
            key: username
      - name: PASSWORD
        valueFrom:
          secretKeyRef:
            name: example-secret
            key: password
```

Listing 2.42 Pod Uses Secret as Environment Parameter

When you roll out the two manifests using Lens and look at the pod, you can also see the environment parameters that are set in the container overview, as shown in Figure 2.29. In addition, you can see which secret the parameter comes from and can even display the decrypted value.

> **Good to Know**
>
> As with a ConfigMap, you can set all values of a secret as environment parameters. The syntax reads as follows:
>
> ```
> envFrom:
>   - secretRef:
>       name: example-secret
> ```

165

2 Basic Objects and Concepts in Kubernetes

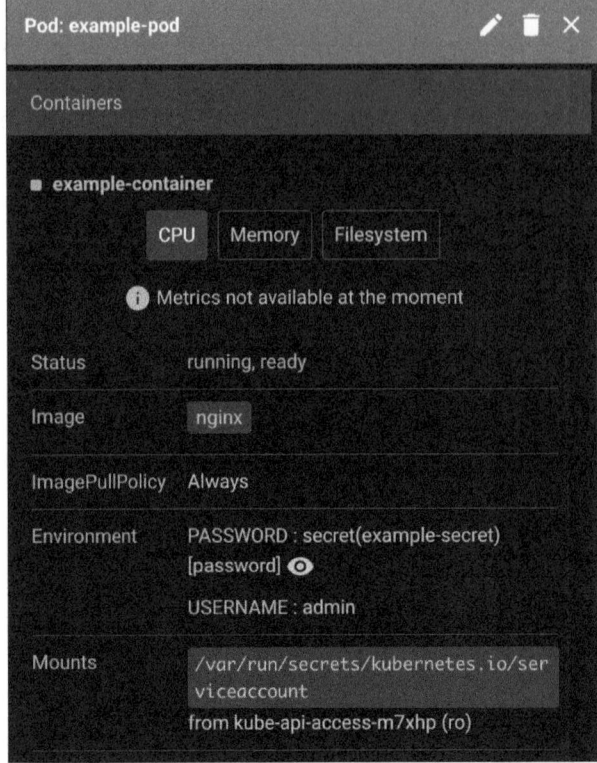

Figure 2.29 Secret Parameters in Pod Overview

Integrating a Secret as a Volume

The integration of secrets as a volume also works in the same way as with ConfigMaps. It is a useful option if you create the secrets as a dot file. The name of the file starts with a period (.) and is thus hidden. In Listing 2.43, you can see a corresponding secret manifest, and in Listing 2.44 is the matching pod manifest.

When you roll out the pod and log into it using `kubectl exec`, you should find a file named .secret-file under /etc/secret-volume. Because it is a hidden file, you need the ls -a command to be able to see the file.

```
apiVersion: v1
kind: Secret
metadata:
  name: my-secret
type: Opaque
data:
  .secret-file: SGVsbG8gV29ybGQh
```

Listing 2.43 Dot File Secret

```
apiVersion: v1
kind: Pod
metadata:
  name: dotfile-test-pod
spec:
  containers:
    - name: dotfile-test-container
      image: nginx
      volumeMounts:
        - name: secret-volume
          mountPath: /etc/secret-volume
  volumes:
    - name: secret-volume
      secret:
        secretName: my-secret
```

Listing 2.44 Pod Manifest with Secret Volume

> **Good to Know**
>
> You can also select specific keys to be integrated when you integrate the secrets as a volume. To do this, you must simply adapt the manifest as follows:
>
> ```
> secret:
>   secretName: my-secret
>   items:
>   - key: ".secret-file"
>     path: ".secret-file"
> ```

Creating Secrets Using kubectl

Like most other Kubernetes objects, you do not necessarily have to create secrets declaratively. You can also use kubectl to create secrets. In this case, it may even make sense to execute the commands within a pipeline because you want to decrypt passwords there at runtime and create them in Kubernetes. For this reason, I am presenting this option here.

> **Note**
>
> I would always prefer an encoded YAML file to imperative commands, but there are situations where you might need the commands.

The kubectl commands vary slightly depending on the type from Table 2.8 you want to create. But the basic structure remains the same. You can create the standard type Opaque, for example, as in Listing 2.45. Simply enter the key and value directly there.

```
kubectl create secret generic my-secret \
    --from-literal=username=admin \
    --from-literal=password=secret
```

Listing 2.45 kubectl create secret from-literal

However, you can also refer to files containing the value as in Listing 2.46. The file name is then used as the key.

```
kubectl create secret generic my-secret \
    --from-file=/path/to/username \
    --from-file=/path/to/password
```

Listing 2.46 kubectl create secret from-file

Secrets for Private Container Registry

In Kubernetes, you cannot deploy containers directly from your computer. You always need a container registry that manages your images and can use Kubernetes to download images. Docker Hub is a public registry that we use in many examples. As long as your Kubernetes cluster can access it via the network, you can also use Docker Hub images in your cluster.

If you develop software in a company and want to store container images, you will not want to make them publicly accessible. With the Minikube registry, you have already gained a first impression of how a private registry works. It is perfectly tailored to Minikube, and you don't need to worry about anything else in this context. Operating a private registry in a company is a little more challenging, but necessary, because in a private registry you can manage a company's images much better and have more control. Many registries also provide additional features, such as image scans. Products I have used in recent years include the following:

- Artifactory by JFrog
- Nexus
- Amazon ECR

But there are many others that fit more or less well into a company, depending on the tech stack. The important thing about a registry is that it must be well integrated into the development process, because if it is complicated to use, passwords will be stored in text files again.

2.4 ConfigMaps and Secrets

> **Good to Know**
>
> Container registries from cloud providers such as Amazon ECR can be set up quickly and easily within an account. In some companies, this means that the containers are managed decentrally; that is, each team stores the containers separately. This is neither good nor bad in the first instance, but it should be a conscious decision as to whether the containers are stored centrally or decentrally.
>
> Personally, I think central storage is better for production images because you have much more control over them. This way, rules such as these cannot simply be ignored:
>
> - Images must not be overwritten.
> - Images must not be deleted.
> - Images must be scanned for known security issues.

In most cases, private registries are also not accessible without authentication, and Kubernetes cannot simply retrieve images. For this reason, you must teach Kubernetes to authenticate itself. With Docker, you would simply use the `docker login` command and use your user name and password. Based on that, Docker generates a configuration file in JSON where it then saves the access data according to the schema shown in Listing 2.47. For each registry, this JSON file holds a string consisting of user name and password, which is Base64 encoded.

```
{
    "auths": {
        "https://index.docker.io/v1/": {
            "auth": "g4s...3rda"
        }
    }
}
```

Listing 2.47 docker/config.json

> **Good to Know**
>
> In Docker Desktop, the *config.json* file looks slightly different. There, the `auth` login is not stored in the file, and in the JSON file you will only find the reference to `"credsStore": "desktop"`.

Kubernetes is based on this authentication, and you can store the access data from the configuration file as a Kubernetes secret and reference it in the deployment manifest. For this purpose, you can generate the secret from the `config.json` file by using the following command:

```
kubectl create secret generic regcred \
    --from-file=.dockerconfigjson=<path/to/.docker/config.json> \
    --type=kubernetes.io/dockerconfigjson
```

Alternatively, you can create a secret and enter all the necessary parameters in the command:

```
kubectl create secret docker-registry regcred \
    --docker-server=<url> \
    --docker-username=<username> \
    --docker-password=<password> \
    --docker-email=<emailadress>
```

In my opinion, the second option is the better one, because you create the secret explicitly with the values you need. You can also run this command in a deployment pipeline. In addition, you can create a manifest as in Listing 2.48 and store the `config.json` file there, Base64 encoded. Choose the option that suits you best.

```
apiVersion: v1
data:
  .dockerconfigjson: eyJh…X0=
kind: Secret
metadata:
  name: regcred
type: kubernetes.io/dockerconfigjson
```

Listing 2.48 docker/config.json as Kubernetes Secret

> **Note**
> At this point, I want to point out once again that secrets should not be stored unencrypted in version management. You can run the command to create the secret in a CI/CD pipeline and insert passwords at runtime, or you can use an additional tool to encrypt secret manifests before checking them into Git.

In a deployment manifest, you want to enter the secret as `imagePullSecret`. In Listing 2.49, you can see the option marked in bold. Set the name to the name you have given the secret, and Kubernetes then can use the secret to retrieve images from a private registry.

```
apiVersion: apps/v1
kind: Deployment
metadata:
  name: my-nginx-deployment
```

```
      labels:
        app: nginx
    spec:
    ...
      spec:
        imagePullSecrets:
        - name: regcred
        containers:
        - name: my-container
          image: localhost:5000/my-nginx
          ports:
          - containerPort: 80
```

Listing 2.49 imagePullSecret in Deployment

> **Information in the Secret**
>
> Outside of your test cluster, you should never store your private access data in a Kubernetes secret. Anyone who has access to the secret can also read your passwords.
>
> Use a technical user whose authorizations are only limited to what is necessary for the application. A technical user is specifically there to give a system access to other systems.
>
> The method to create one differs from registry to registry. It is best to read the relevant documentation to find out how the authorization concept works.

> **Note**
>
> Managing secrets is never easy. One of my clients operates their Kubernetes clusters in AWS and uses AWS Secrets Manager to store secrets. At the same time, they use *sops* to encrypt secrets in GitLab, which they then create as a Kubernetes secret.
>
> As soon as it becomes complex and confusing, you need to think about how you could simplify things, especially when there is no longer a single point of truth. At the following link, you will find a Kubernetes operator that enables you to easily tap into external secret stores: *http://s-prs.co/v596422*. This could make management a little easier for you.

2.5   Establishing a Communication with Services and an Ingress

You have now learned a lot about individual pods and how you can run your application in Kubernetes. But a pod seldom comes alone, and in a world of microservices there can quickly be several hundred of them. The challenge here is that the applications want to communicate with each other. Because the pods in Kubernetes are very volatile and you do not know on which node a pod is currently running, you need a functioning service discovery.

Kubernetes provides the service object for this purpose. This object stores the information about your pods and serves as a load balancer. For the communication from outside into the cluster, there is the ingress object, which allows you to control the data traffic. Both objects are important and work together, because what would your application be if no one could find it or reach it?

Imagine your application as a store in a city. They even have several stores with the same product range at different locations. Your customers appreciate this because your stores are packed on a Saturday afternoon and customers can spread out between the stores. The service is your smart customer guidance system. It knows all your stores and keeps a precise record of where one can find them. If a new store opens, it also directs your customers to that new store.

If the service is a customer guidance system, then you can think of the ingress as a smart parking guidance system that guides your customers from outside the city to the right parking lot, where the customer guidance system (service) then takes over and guides the customer to your store. The parking guidance system asks for the customer's destination at the beginning and can even request the A38 permit and check whether the customer is authorized to drive into the city at all.

This means that the ingress

- enables access from outside the cluster,
- checks for authorization, and
- redirects the packages to the service.

The service

- knows all pod replicas and their location in the cluster,
- takes over the load balancing on the available pods, and
- redirects the requests to a pod.

Figure 2.30 shows the connections as the communication usually takes place. A pod located in the same cluster can address the service directly to reach other pods. Communication from outside the cluster can take place in two ways:

- Via the ingress, which accepts the packages on OSI layer 7 and redirects them to the corresponding pod via the service
- Via a special NodePort service that works on layer 4 and simply redirects communication arriving on a specific port to the service

2.5 Establishing a Communication with Services and an Ingress

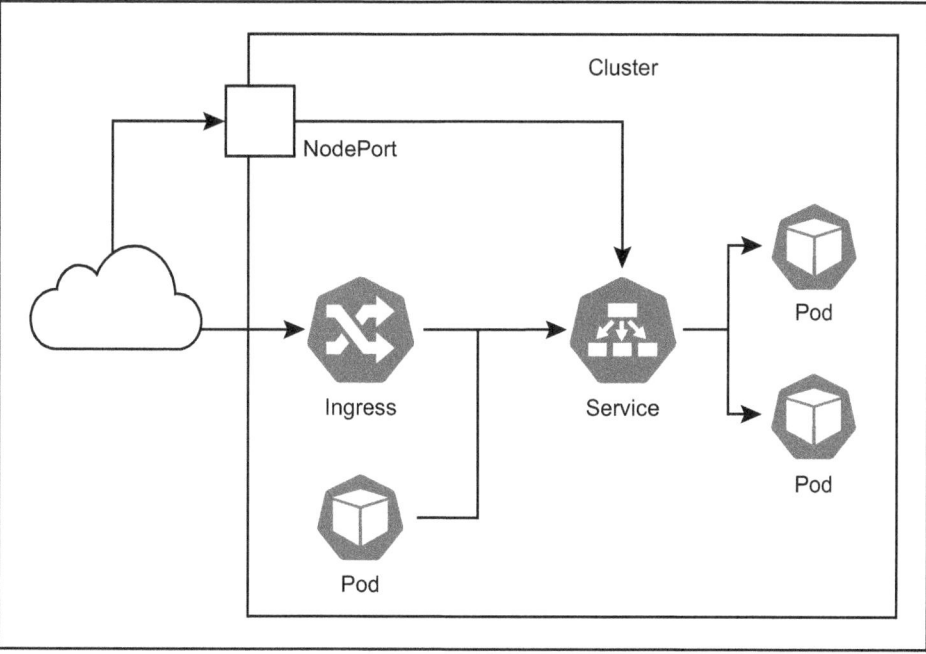

Figure 2.30 Communication with Service and Ingress

The way via NodePort is simple, but you have more control options by using an ingress. I definitely recommend using an ingress for HTTP applications, but we'll take a closer look at both.

2.5.1 Communication between Pods

A key feature of the Kubernetes design is network communication without the need for network address translation (NAT). In Kubernetes, every pod can communicate directly with every other pod, and all nodes can also communicate with all pods without the need for NAT. This is made possible by the use of real IP addresses for pods and efficient routing within the cluster. The IP address you see in a pod is also the IP address that other pods can use to reach that pod. This design reduces network complexity and ensures transparent communication within a cluster.

This means that if, for example, your frontend application is running on the same cluster as your backend application, the two can communicate with each other via the IP address. But just because something works doesn't mean it should be done that way. As you know, pods are fast-moving and transient, so normally you would never want to communicate with a single pod. You need a fixed end point that always directs you to the right pod. This is why the service object exists.

Good to Know

Interestingly, pods are also assigned a DNS address that has the following structure: [PodIP].[Namespace].pod.cluster.local.

It remains to be seen whether this DNS address is useful. If you need the IP address to create the DNS name, you can also use it directly.

2.5.2 Communication via a Service

Try to think of one of your applications in production: What is the workload like there throughout the day?

In most cases, there is no constant load. Data traffic is different in the morning than in the evening, and in the evening it is different than during the night. In Kubernetes, your application scales different numbers of pods to either match the load or to avoid tying up resources unnecessarily. In addition, it can sometimes happen that your Kubernetes deployment has to replace a nonfunctioning pod.

If the communication partner changes constantly, there needs to be a constant at one point that is the gateway for requests. To meet the requirements, you need something that monitors changes and helps you maintain communication channels. The Kubernetes service takes over these tasks for you and takes care of the following:

- **Service discovery**
  The service knows all replicas and redirects incoming data traffic to them.
- **Load balancing**
  It distributes the data traffic to all available pods.
- **A fixed end point**
  With a fixed IP address and DNS name, the service is a reliable end point.

The correct way to communicate between pods within a Kubernetes cluster is therefore always via a service. To determine which pods it should redirect the data traffic to, the service uses the concept of labels and selectors for service discovery.

One component of Kubernetes that plays a key role in this process is the *kube-proxy*, which takes care of handling the virtual cluster IPs of pods and services. For this purpose, it listens for changes on *apiserver* and enters these in the routing tables of the node. This allows traffic to be routed to the correct end points.

Good to Know

The DNS address of a service always has the same structure:
- Within a namespace, the name of the service is sufficient.
- Within the cluster, its structure is [ServiceName].[Namespace].svc.cluster.local.

2.5 Establishing a Communication with Services and an Ingress

For example, if the name of your service is `nginx-service` and it is deployed in the default namespace, you can always reach it via the following DNS address: `nginx-service.default.svc.cluster.local`.

The service object is multifaceted, and there are five different types listed in Table 2.9. Let's look at the most important of these in more detail.

Service Type	Description
ClusterIP	This is the default service type. This type makes it possible to address an application within the cluster via an internal IP address.
NodePort	You can use this type to provide a ClusterIP service that gets mapped to a port on each node in the cluster, which enables access from outside the cluster via [NodeIP]:[NodePort].
LoadBalancer	This type allows you to link the service to an external load balancer. If your cluster is integrated with a cloud provider, this can also create a load balancer.
ExternalName	You can use this service to refer to external host names via a CNAME record. For example, you could make an external database accessible via the Kubernetes service.
Headless	You should use the headless service if you do not need load balancing or ClusterIP. This allows you to connect a single pod to the service. (Note that this is a special case, and I've never come across it in the real world.)

Table 2.9 Kubernetes Service Types

The ClusterIP Service

Let's start with the most commonly used Kubernetes service. The ClusterIP service is the default service; when someone talks about a service, they usually mean this type. In the next example, you will create a service for the deployment in Listing 2.50.

```
apiVersion: apps/v1
kind: Deployment
metadata:
  name: nginx-deployment
  labels:
    app: nginx
spec:
  replicas: 2
  selector:
    matchLabels:
      app: nginx
```

```
template:
  metadata:
    labels:
      app: nginx
  spec:
    containers:
    - name: nginx
      image: nginx:1.14.2
      ports:
      - containerPort: 80
```

Listing 2.50 Example of Nginx Deployment Manifest Service

In the simplest case, you can use the imperative `kubectl expose deployment nginx-deployment` command, and Kubernetes will create a suitable service for your deployment. To do this, Kubernetes looks at the deployment manifest and defines what the service should look like based on the container ports. If you look at this in Lens, the service has the same name as the deployment, has been assigned an IP address, and can be reached via port 80. It should look like Figure 2.31.

Figure 2.31 Kubernetes Service in Lens

As a YAML manifest, a service looks like the example in Listing 2.51. Under `spec.selector`, you specify the labels that the service uses to identify the pods to which it should redirect. Under `ports`, you define the port which the service opens and the destination port to which it redirects. You can also roll out this manifest and compare the two. The two have different IP addresses and a different name. But the function is the same. Both redirect to the pods of your deployment.

```
apiVersion: v1
kind: Service
metadata:
  name: nginx-service
spec:
  selector:
    app: nginx
  ports:
    - protocol: TCP
      port: 80
      targetPort: 80
```

Listing 2.51 ClusterIP Type Service

Good to Know

You can also define multiple ports in a service and assign names to the ports. This allows you to open multiple ports in your application if you allow HTTP and HTTPS, for example. To do this, simply add another entry to your list:

```
ports:
  - name: http
    protocol: TCP
    port: 80
    targetPort: 8080
  - name: https
    protocol: TCP
    port: 443
    targetPort: 8443
```

Good to Know

It is also possible to define a fixed ClusterIP for the service. However, it must be in the CIDR range of the cluster and must not have been assigned yet. To do this, you want to set the clusterIP option as in the following example:

```
apiVersion: v1
kind: Service
metadata:
  name: nginx-service-fixed-ip
spec:
  clusterIP: 10.98.37.199
  selector:
    app: nginx
  ports:
    - protocol: TCP
      port: 80
      targetPort: 80
```

Let's now test the service briefly. As the ClusterIP service is only available within the cluster, you need a container from which you can start a query. Use the busybox container for this. In Listing 2.52, you will find a simple pod manifest to start the pod in the cluster and keep it running. Now use the kubectl exec command to connect to the pod. After that, you can send a query to Nginx via the service in the pod's command line. Run the query using wget -qO- nginx-service or wget -qO- nginx-service.default.svc.cluster.local.

```
apiVersion: v1
kind: Pod
metadata:
  name: busybox
spec:
  containers:
    - name: busybox
      image: busybox
      command: ["sh", "-c", "while true; do sleep 3600; done"]
```

Listing 2.52 Busybox Pod Manifest

You can see that you simply need the name of the service within the same namespace, but of course the full DNS name also works. Why don't you try the opposite and deploy the busybox to a different namespace?

The NodePort Service

Most of the applications you develop are not only used within the cluster. Users or applications from outside also want to reach and use your applications. A simple way to make this possible is to extend the ClusterIP service.

If you assign a NodePort to the service, every node in your cluster will forward the data traffic arriving on this port to your service. The entire thing is based on layer 4 of the OSI model and therefore gives you few options for controlling the data traffic. NodePort services are the right choice, especially for applications that do not communicate via HTTP.

Listing 2.53 shows the manifest for a NodePort service. The only difference is that you need to pass type: NodePort, and Kubernetes then will select a port and redirect it to the service. Figure 2.32 shows what the NodePort service in Lens looks like. In this case, it redirects port 30586 to port 80 of the service, which in turn redirects the traffic to port 80 of the pod.

Figure 2.32 NodePort Service in Lens

```
apiVersion: v1
kind: Service
metadata:
  name: nginx-service
spec:
  type: NodePort
  selector:
    app: nginx
```

```
ports:
  - protocol: TCP
    port: 80
    targetPort: 80
```

Listing 2.53 NodePort Type Service

> **Good to Know**
>
> If you do not want to activate a random port, you can also define a fixed port to be opened by your nodes. To do this, expand the port definition of your NodePort service as follows:
>
> ```
> ports:
>   - name: http
>     protocol: TCP
>     port: 80
>     targetPort: 8080
>     nodePort: 30586
> ```

The NodePort service is somewhat difficult to test using Minikube. As Kubernetes runs in the Docker container, it is not really accessible from the outside. As a small workaround, you can use the `docker exec -it [CONTAINER_ID] bash` command to connect to your container. There, you need to use the `curl 127.0.0.1:[NODE_PORT]` command to test whether you receive the Nginx welcome page.

A much nicer test would be to use the Raspberry Pi setup from Chapter 1, Section 1.7, because then you could also see that the NodePort is redirected from each node to your service.

The ExternalName Service

This service is a little more specific. In this case, you need to enter an external DNS name as shown in Listing 2.54. Requests that you send to the service will be redirected to the external address.

```
apiVersion: v1
kind: Service
metadata:
  name: external-dns-service
spec:
  type: ExternalName
  externalName: myservice.humanity-it.com
```

Listing 2.54 ExternalName Type Service

2   Basic Objects and Concepts in Kubernetes

But what may seem a little confusing at first makes working with external services very charming. For example, if you are running a Postgres database in production on the AWS RDS service, you can still have your pods communicate with a Kubernetes service. You just add another abstraction to the database. This allows you to turn the RDS service into a simple pod in a development environment that is sufficient for development.

Note

If you need to use external services such as databases, the ExternalName service can give you more flexibility and perhaps also save costs. Why don't you check in your company where you could use a service of the ExternalName type and simply bring databases in development environments into Kubernetes?

2.5.3 Communication via Ingress

You have gotten to know and tried out the NodePort service—a simple way to open your application to the outside world by activating a port. Unfortunately, this option leaves very little scope for analyzing and routing data traffic. The alternative for your HTTP application is an ingress.

Note

You can activate and deploy an ingress, but you can only use an ingress via a Minikube CLI tunnel. This doesn't really make testing any different than a direct tunnel to the service or pod.

However, the problem here is that Minikube itself runs in a Docker container and the cluster IP address is not accessible from your host computer. Unfortunately, I have not found an easy way around this problem.

I have used the Raspberry Pi cluster from Chapter 1, Section 1.7 for the examples here. A test cluster from your company would be even better at this point.

The ingress object works on layer 7 of the OSI model and offers to redirect incoming traffic to Kubernetes services based on rules. For example, you can

- decide to which service the traffic should be redirected based on the URL or path called;
- change the path using rewriting rules; or
- support TLS certificates to allow HTTPS traffic.

An ingress is typically a load balancer. However, the technical implementation depends on the *ingress controller*. The ingress controller provides a load balancer that takes over the task of the ingress as you define it in an ingress manifest.

2.5 Establishing a Communication with Services and an Ingress

The split between the ingress manifest and the actual implementation by the Ingress controller has a major advantage: you can decide individually which technology suits you and your cluster best. A cluster based on AWS infrastructure can rely on the AWS Load Balancer Controller, which sets up an application load balancer (ALB) in the cloud based on your manifest. For an on-premise cluster, for example, you can use the ingress Nginx controller, which builds Nginx from the manifest within your cluster. You are completely free to do this and can also use multiple ingress controllers at the same time.

Good to Know

Depending on the ingress controller, you have more or fewer options for using additional features. For example, one of my clients uses an Nginx ingress and uses the authentication feature there. An annotation sets an endpoint to the Authelia tool, the user is redirected there and authenticates themselves, and they can then use the application.

For an example of how you can customize the Nginx ingress, visit *http://s-prs.co/v596423*.

Note

To be able to use multiple ingress controllers, you must define ingress classes, which you can then reference in a manifest so that Kubernetes knows which controller is supposed to generate the ingress. You can find out more about this topic at *http://s-prs.co/v596424*.

Typically, when setting up a cluster, you need to think about which ingress controllers you want to install. In your company, the cluster admins will have already done this. You do not need to worry about the ingress controller in your test clusters.

On Minikube, you can activate the ingress controller using the `minikube addons enable ingress` command. The Raspberry Pi cluster runs the Traefik ingress controller, which is supplied directly with K3s.

Good to Know

Usually, all ingress controllers should have the same functions, but there are slight differences from controller to controller. You can get an idea of the available selection of controllers at the following link: *http://s-prs.co/v596425*.

Let's first take a look at the default backend. A default backend is often already configured in the ingress controller. This Ingress should take over all requests that cannot be assigned to any other ingress. If you do not want to rely on the implementation of the

2 Basic Objects and Concepts in Kubernetes

ingress controller or want to define your own default backend, you can do this by using the manifest from Listing 2.55.

For the example, we will again use the simple deployment with Nginx pods, which are made accessible via a service, from Listing 2.50 and Listing 2.51. The default backend extends the example and will redirect all HTTP requests that reach the cluster to the `nginx service` on port 80. The ingress does not care which Kubernetes node the request hits. Kubernetes checks the ingress rule and then redirects it accordingly.

> **Note**
> If you want to try the ingress on Minikube, you need to enable the ingress add-on via the `minikube addons enable ingress` command, then use the `minikube tunnel` command after rolling out the ingress.

> **Note**
> You can use the default backend configuration for an error page—for example, to inform your users of errors.
> Note that you should always define an ingress with rules for your application.

```
apiVersion: networking.k8s.io/v1
kind: Ingress
metadata:
  name: nginx-ingress
spec:
  defaultBackend:
    service:
      name: nginx-service
      port:
        number: 80
```

Listing 2.55 Default Backend Ingress Manifest

An ingress gives you two options for defining rules so that you can control incoming HTTP traffic:

- Path based
- Host based

A path-based rule looks at everything after the / in the calling URL. For example, you can redirect the user who calls /test to Nginx and the user who calls /test/db to a database.

> **Note**
>
> The ingress will also redirect the complete path to your application. So if /test is redirected to Nginx, then Nginx must also be able to handle the /test path.
>
> You could work around this by using rewrite rules so that the ingress controller adjusts the path before it goes to the application.

> **Good to Know**
>
> If there are multiple similar paths, the one that is the longest wins. The /test/db path could also be a subpath of Nginx, but Kubernetes will then redirect the traffic to the database.

The host can make the rules even more precise. Kubernetes looks at the URL of the host that the user calls. For example, if you call raspberry1.local/test in the browser, the ingress can redirect using the URL before the /. The rule for the /test path then only applies if the host matches as well. The host is optional, and if it is not set, then the rule applies to every HTTP request.

Let's try this out by creating a path-based ingress that points to the service. For this purpose, you should use the manifest from Listing 2.56. This ingress is very simple because it merely forwards everything from / to Nginx. This means that you do not need a rewrite rule to redirect the correct path in the URL to your web server.

```yaml
apiVersion: networking.k8s.io/v1
kind: Ingress
metadata:
  name: nginx-ingress
spec:
  rules:
  - http:
      paths:
      - path: /
        pathType: Prefix
        backend:
          service:
            name: nginx-service
            port:
              number: 80
```

Listing 2.56 Ingress Manifest for Nginx

If you create the ingress in Lens and look at it, it should look similar to Figure 2.33.

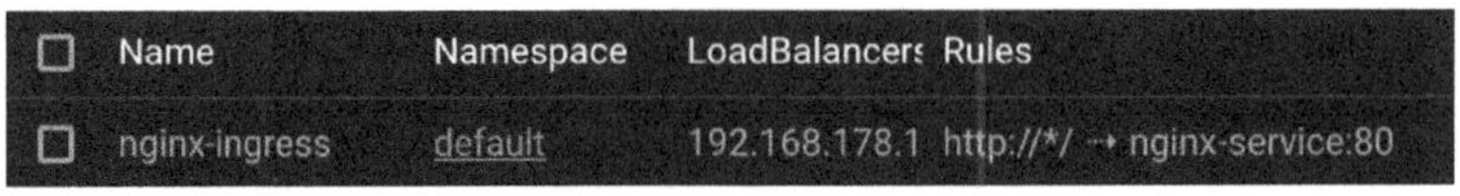

	Name	Namespace	LoadBalancers	Rules
☐	nginx-ingress	default	192.168.178.1	http://*/ → nginx-service:80

Figure 2.33 Ingress Created in Lens

In the overview, you can see the rule according to which this ingress redirects the data traffic, and you can also see the IP addresses at which the ingress is active. When you click the ingress, you should see the IPs of the Raspberry Pis there.

You can now access Nginx via your browser. You can either use the IP address of a Raspberry Pi or the host name. It does not matter which of the nodes in the cluster you call. Everyone accepts the calls and redirects them according to the ingress rules. We have used Prefix for pathType, but there is also the Exact type. You can read exactly how both work in Table 2.10.

pathType	Description
Prefix	The called path must start with the defined path. The use of uppercase and lowercase letters is important, but the final / is ignored. Examples of how the ingress would decide can be found in the documentation at the following address: *http://s-prs.co/v596426*
Exact	The complete path must be called in exactly the same way and is case sensitive. The Exact type is preferred over the Prefix type in the event of a match.

Table 2.10 Ingress pathTypes

A good way to define an ingress more specifically is to set a host name. By tightening the rule, you could, for example, use the / path more frequently and thus continue to operate multiple web servers in your cluster without a rewrite rule. The only thing you need to make sure is that the DNS points to the Kubernetes cluster, preferably via a load balancer.

You can see the change in bold in Listing 2.57. Here we use the host name of one of the Raspberry Pis, which is resolved within the local network. If you import this change, you will also see in the ingress overview that the rule for redirecting has changed. If you now try to call the IP address, you will receive a **404 page not found** error unless you have defined a default backend.

You cannot access Nginx via the URL of the second Pi either. Only if you call up *http://raspberrypi1.local/* will ingress redirect you to Nginx.

2.5 Establishing a Communication with Services and an Ingress

> **Good to Know**
>
> In the example, you can of course only access Nginx via *raspberrypi1* because the DNS points to it. Nevertheless, the second node would also redirect the traffic if you arrived there with the URL.
>
> In a real environment, the DNS would point to a load balancer in front of the Kubernetes cluster, which then redirects the traffic to any instance of the cluster. This means you do not overload a single instance.

```yaml
apiVersion: networking.k8s.io/v1
kind: Ingress
metadata:
  name: nginx-ingress
spec:
  rules:
  - host: raspberry1.local
    http:
      paths:
      - path: /
        pathType: Prefix
        backend:
          service:
            name: nginx-service
            port:
              number: 80
```

Listing 2.57 Ingress Manifest with Host Name

As you already know, you can also define multiple rules in an ingress. We have provided the preceding example as a manifest in Listing 2.58. Finally, try deploying another application and making it accessible via the ingress. It doesn't have to be a database either, but it will get you into trial and error, and the ingress concept will end up being logical and easy for you.

```yaml
apiVersion: networking.k8s.io/v1
kind: Ingress
metadata:
  name: nginx-ingress
spec:
  rules:
  - host: raspberry1.local
    http:
```

```
      paths:
      - path: /
        pathType: Prefix
        backend:
          service:
            name: nginx-service
            port:
              number: 80
      - path: /test/db
        pathType: Prefix
        backend:
          service:
            name: nginx-db
            port:
              number: 5432
```

Listing 2.58 Multiple Paths within Ingress

Chapter 3
Everything as Code: Tools and Principles for Kubernetes Operations

The best instruction is one that gets by with as few words as possible.
—Maria Montessori

The quote from Maria Montessori that opens this chapter refers to pedagogy. The fewer words we use to formulate an instruction, the easier it is to understand and the easier it is to absorb. But I also think the idea is perfect for IT. The clearer your instructions to the computer, the more likely it will do what you want.

Giving instructions to a machine is a common practice—for example:

1. You click the **Outlook** icon to open your email program.
2. You type text into the editor using your keyboard.
3. You instruct the program to send an email.

In Chapter 1, Section 1.5.5, you learned about the most common `kubectl` commands and saw how you can use them to issue instructions to Kubernetes:

- Show me my pods!
- Create a deployment for me!
- Scale up from two to three pods!

This approach is referred to as an *imperative* one: you give the machine instructions it must carry out, step by step. The challenge with instructions is to make them sustainable so that they can be repeated and executed with little effort, ideally even automatically.

Of course, you can write a script and execute it in any environment. However, imperative instructions always harbor the risk of being misinterpreted, because not every system is always in the same state. The production environment could respond very differently to the command sequence than the test environment. What if your deployment on the production environment already consists of three pods? Depending on how you formulate the last instruction, a fourth pod is then created, which does not correspond to the order; actually, nothing should have happened.

To avoid such misunderstandings, the status of the infrastructure must already be known in such a way that your instructions can access it directly. This is referred to as *everything as code*, and the approach is designed to solve precisely this type of problem: All objects are structured as text in a readable format for humans and computers. These objects are then interpreted by the computer itself, which attempts to achieve the desired state independently. You no longer work imperatively, but declaratively (Section 3.1).

This idea is not an invention of Kubernetes; it is also pursued by other configuration management tools such as Terraform or Ansible. *Infrastructure as code* (IaC) is a modern and widely used paradigm that allows you to define infrastructure components such as servers, routers, and also Kubernetes resources as code:

- Once created, you can import a manifest again and again. It does not matter whether the resources are no longer working due to an error or a new environment is to be set up. You save time because the configuration is already available and all you have to do is roll it out.
- IaC promotes the uniformity of the IT infrastructure, as you can use the same code multiple times to create identical environments. Writing it down as code always helps me to think about standardization. I inevitably ask myself which processes and setups can be simplified or improved.
- You save yourself work because you don't have to perform the same steps manually for every environment. This reduces human error, saves time and nerves, and simplifies repeatability.
- Using IaC, you can store your infrastructures in your version management system, which allows you to track changes quickly and easily and carry out rollbacks. In addition, the infrastructure is directly documented and collaboration within a team is easier. You can use your review processes for changes, and the history of infrastructure changes is worth its weight in gold when it comes to debugging.
- You can fully automate your infrastructure with CI/CD pipelines.

Everything as code is the basis for stable and simple IT operations. So let's delve deeper into the topic together in this chapter.

3.1 Declarative Configurations

Are you familiar with declarative programming using languages such as Haskell or Lisp? The difference from widely used languages such as Java or Python is the type of programming. With an imperative language such as Java, you tell the system step by step how to achieve the desired result. In Haskell, you describe what the result should be, and the way to get there remains open to the system. These are two completely different paradigms. If you are used to imperative programming, the declarative paradigm will take some getting used to.

To understand the difference between imperative and declarative programming, imagine that you want to plan a meeting with a business partner. According to the imperative paradigm, you would have to divide this task into many different steps: Call Frank. If he doesn't answer his cell phone, send him a text message. If he hasn't answered after two hours, call him again. If there is no confirmation by tonight, remove the meeting from your calendar.

These steps must be processed in the correct order, and all logical conditions must be formulated correctly; otherwise, Frank will still receive a reminder text message after his call.

For better clarity, I have shown you a simplified imperative process in Figure 3.1. You have a desired state in mind and want to get the system there by using commands. You run a command and then check whether the system has reached the desired state. If not, you must send another command to bring the state closer to the desired state. Not until your check shows that the status has been reached can you end the process.

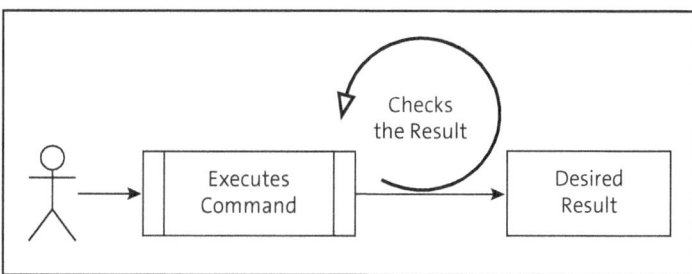

Figure 3.1 Simplified Imperative Approach

A declarative approach is much simpler: You tell your cell phone by voice control: "I'm planning a meeting with Frank this afternoon." Your cell phone then automatically tries to call Frank in the background and arrange a meeting with him. Unfortunately, he doesn't answer the phone. Your cell phone will then send a text message and a few minutes later you will receive confirmation that the appointment has been made.

You do not need to formulate and structure this process in detail each time; you are no longer involved at all. You "only" specify the result, while your cell phone independently finds a solution to receive confirmation from Frank. That sounds like a real AI assistant. Unfortunately, a voice assistant does not currently think ahead on its own and only executes one command.

The declarative paradigm is not only used in programming. It is also used in Kubernetes to provide a more abstract and flexible handling of resources and services. Take the ReplicaSet, for example. You create the ReplicaSet and define that your application should run on three pods. The ReplicaSet must then take care of achieving the desired status and creates three new pods. At the same time, it constantly monitors the current status and checks whether the desired number has been reached. If a pod fails, the ReplicaSet will start a new one. If—for whatever reason—there are already four pods, it will scale down the number without you ever explicitly requesting this.

As a developer, you do not want to worry about how the ReplicaSet fulfills its task. You leave it entirely up to the system what needs to be done to achieve the desired state.

Figure 3.2 shows a simplified representation of the declarative procedure. This time, you not only have the desired status in your head, but also define it so that the system can recognize it. In Kubernetes, you would declare this in the YAML language, which we will take a look at in Section 3.2. The system now understands which state is to be achieved and works on it independently until the state is reached. This procedure is referred to as a *reconciliation loop* as the current state is repeatedly compared with the desired state.

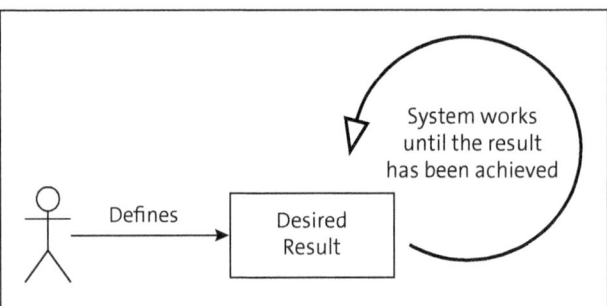

Figure 3.2 Simplified Declarative Approach

> **Note**
> It would be desirable to have Kubernetes always achieve the intended result when we declare it. Unfortunately, however, the system cannot solve all problems on its own and will of course produce error messages if a precondition is not met. For example, Kubernetes cannot deploy a pod if the declared image is not available.

The approach reminds me of the test, operate, test, and exit (TOTE) model from my natural language processing (NLP) training. This is a cognitive model that originates from psychology and is used to explain feedback and control processes in human action. You have a current state and a desired state and must do something to achieve the desired state.

Let's take brushing your teeth as an example. You get up in the morning and want to have clean teeth. Your procedure would be as follows according to the TOTE model:

1. **Test**
   You check whether your teeth are clean and realize that you need to brush them.

2. **Operate**
   You brush your teeth.

3. **Test (repetition)**
   After performing the operations, you check again whether your teeth are clean. If so, stop cleaning. If not, go back to the *operate* stage.

4. **Exit**
   You terminate the process.

In our case, this is how it works: In the imperative paradigm, you have to check the state yourself and execute commands until you have reached the desired state. In the declarative paradigm, it is always the system that works and checks until the desired state is reached. The TOTE model therefore applies in both cases. The only question is: Who has to do the work? You or the system?

If you remember Chapter 1, Section 1.5.5, there is one more thing you might be asking yourself. In that section, you tried out your first `kubectl` commands. Thus, Kubernetes also provides direct imperative commands to create or adapt resources. But why do they exist if the declarative paradigm is so much better?

In Kubernetes, the imperative approach can be used for

- simple, small, and quick changes;
- one-off operations that are rarely performed;
- debugging or troubleshooting; and
- development environments.

Let's be honest: there is no system that is perfect, and we need good developers who know what they are doing. There will always be times when you have to do it yourself, especially when something has to be done quickly. That's why it's important to me that you find your way around both `kubectl` and Lens and become confident in using them. That is your tool to be able to intervene in an emergency.

With Kubernetes, however, I recommend that you always use the declarative paradigm if possible, as this will make your life much easier. At the latest when you are working with production systems, you cannot avoid it if you want to provide reliable applications and have all changes to the system precisely logged and monitored.

> **Good to Know**
>
> One of the main advantages of the declarative paradigm is the ability to map complex systems simply and effectively *as code*. By defining the desired state, Kubernetes can automatically perform the necessary steps to achieve and maintain this state. This gives you advantages such as the following:
>
> - **Idempotence**
>   The same configuration always leads to the same end state, regardless of the current state of the system.

- **Scalability**
  You can quickly and easily make adjustments to system resources by making changes to the configuration file.
- **Repeatability**
  You can easily reproduce identical environments by using the same configuration files.
- **Self-healing**
  Kubernetes continuously monitors its status and independently takes steps to correct any deviations from the desired status.

3.2 YAML: The Language for Kubernetes

To make your intended result clear to Kubernetes, you need a language in which you can specify your resources declaratively. Kubernetes uses the YAML markup language for this purpose. In software development, you have certainly already become familiar with markup languages such as XML or JSON, and perhaps you have even used YAML in a different context. I want to take this opportunity to go into more detail about YAML, because Kubernetes uses YAML to describe all resources and states of the cluster.

YAML is a recursive acronym and stands for *YAML Ain't Markup Language*. It is currently very popular alongside JSON and impresses with its significantly better readability for humans. But what are markup languages actually used for?

> **Good to Know**
> YAML was actually only intended to be a simple markup language, which is why the acronym was originally for *Yet Another Markup Language*. However, YAML has grown considerably and is, of course, a markup language despite its name.

If you search for markup languages, you will find different types of them. The best known is HTML, which allows you to structure and format text in such a way that a machine can read and interpret it. YAML provides a format to put data into a structure that is easy to read for both machines and humans.

YAML files have the extension .yaml, and sometimes you will also see .yml. Both are fine, but according to the documentation the .yaml extension should be used.

3.2.1 Basics of YAML Syntax

If you look at YAML files, you can break each of them down to three basic elements:

- Key-value pairs
- Lists
- Nested structures

3.2 YAML: The Language for Kubernetes

Key-value pairs are the simplest form of data organization. Each pair consists of a key and an associated value. In the following example, you can map a person's data in this way:

```
name: "Kevin Welter"
company: "HumanITy GmbH."
```

You can use lists to define collections of elements. These are then grouped under a key. Each list element is indicated by a - sign. The following example shows a list of customer names.

```
customers:
  - "Kevin Welter"
  - "Sean Smith"
  - "John Doe"
```

Sometimes you have more complex structures where individual lists and key-value pairs are not enough. You now have a list of names, but there is much more information about a customer. To map this information, you can use nested structures to define entire objects. For this purpose, you use key-value pairs and lists that are arranged in hierarchies.

> **Good to Know**
>
> You can create multiple YAML documents in one YAML file. These are separated by three dashes (---):
>
> ```
> ---
> name: Kevin Welter
> ---
> name: Sean Smith
> ```
>
> In the first line of a YAML file, the dashes are optional, but they explicitly indicate that a new YAML document is starting.

In the following example, you have a list of customers who in turn have a company assigned to them. Both the customer and the company have a name, and there is further information about the company that can be entered in the substructure:

```
customers:
  - name: "Kevin Welter"
    company:
      name: "HumanITy GmbH."
      city: "Tucson"
      zip: "85706"
  - name: "Sean Smith"
    company:
```

```
    name: "Smith Inc."
    city: "Fort Worth"
    zip: "76040"
```

If you think in terms of objects, then you have the company object and you have the customer object. In this structure, the company belongs to the customer. In other cases, the customer could also be specified as a list of employees in the company:

```
company:
  name: "HumanITy GmbH."
  employees:
    - "Kevin Welter"
    - "Fabian Schaub"
```

As you can see, you have complete freedom to map your data as you or your system need it.

Indentations are of crucial importance in YAML. They define the hierarchy and structure of the data. In comparison, indentations are optional in JSON because the structure is defined by parentheses.

Indentations in YAML

- must be consistent within a document,
- define the hierarchy of an element,
- often lead to errors or misinterpretations, and
- make the file easier to read.

In the examples, I have used an indentation of two spaces in each case. Most YAML parsers and editors support an indentation depth of two or four spaces by default. There is no right or wrong here, but you should remain consistent within a document. However, this is easier said than done with large YAML files. It has often happened to me that a key-value pair was not assigned to the correct object due to an incorrect indentation and I had to debug forever to find the error. An incorrect assignment is not a syntax error and therefore your editor will not directly point out the problem.

> **Note**
> Never use tabs to structure the indentations of a YAML file! YAML requires the use of spaces instead of tabs for indentation. The interpretation of tabs between different editors and environments can vary and therefore result in conflicts.

3.2.2 Data Types in YAML

In YAML, you can find all the classic data types that you also use in other programming languages:

3.2 YAML: The Language for Kubernetes

```
string: "This is a string"
number: 123
float: 12.34
boolean: true
null value: null
```

With strings, you have several options for defining them. You can typically write a string without the quotation marks. YAML always tries to interpret the values correctly. However, if you want to use special characters such as :, ", or ' in the string, which are also used by YAML, then you absolutely need the quotation marks, as shown in Listing 3.1. It does not matter whether you use single (' ') or double (" ") quotation marks. The characters used in each case must not appear in the string itself, of course.

> **Note**
> You should follow a uniform convention within a YAML file. I always try to write a string in quotation marks because that makes it clearer for me.

YAML also provides the option of defining strings that run across several lines. By using the pipe (|) character, YAML retains the exact formatting, while > converts every line break into a space.

```
name: Kevin Welter
info: "Kevin says: 'Sometimes quotation marks are needed'"
simpleString: 'C:\Users\Kevin'
doubleString: "Line 1\nLine 2"
blockText: |
  Text in multiple lines
  Line 2
foldedText: >
  This is a long
  text broken across multiple lines for better
  legibility, but separated by spaces
```

Listing 3.1 Different String Syntax

> **Good to Know**
> Like Kubernetes, I use camel case as a convention for the keys in YAML. However, YAML does not make any specifications here. You can even use spaces in a key. I recommend that you use the programming language for which you are using YAML as a guide. For example, use camelCase for Kubernetes, snake_case for Python, and so on.

3.2.3 Anchors and Aliases

Imagine a YAML file in which you define data that is repeated frequently. Suddenly your file has more than 1,000 lines. No matter how well structured YAML is, the file becomes more unreadable the larger it gets. For this purpose, YAML provides anchors and aliases that allow you to define objects or parameters once and use them again within the file according to the *don't repeat yourself* (DRY) principle.

An anchor is set using an &anchorName, and an alias references the anchor using *anchor-Name. You can see a simple example of this in Listing 3.2.

```
favoriteNumber: &number 42
myFavoriteNumber: *number
```

Listing 3.2 Simple Anchor and Alias

You can also anchor the key-value pairs of an entire object and include them in another object. To do this, you need to use the syntax <<: *anchor, as shown in Listing 3.3. The data from basicAuthor is transferred to specificAuthor. In this case, the subject area will remain in the specificAuthor *Kubernetes*, but the name will be transferred.

```
basicAutor: &author
  name: "Kevin Welter"
  specialty: YAML
specificAuthor:
  <<: *author
  specialty: Kubernetes
```

Listing 3.3 Anchor of Entire Object

A useful real-life example where I use anchors again and again is in the GitLab pipeline tool. GitLab CI uses YAML to define pipelines, and there are many lines that are repeated over and over again. In Listing 3.4, you can see a manifest as an example. The &script anchor is set here after the .launch key. In the devJob and prdJob objects, the anchor is referenced by <<: *script_launch references, and all key-value pairs are inserted at this point.

This has the following advantage: ykou only need to define the script once, and the environments are differentiated by parameterization.

```
.launch: &script
  stage: launch
  script:
    - ./deploy.sh $ENV
devJob:
  <<: *script
  variables:
    ENV: dev
```

```
prdJob:
  <<: *script
  variables:
    ENV: prd
  when: manual
```

Listing 3.4 Pipeline Manifest in YAML

3.2.4 Single-Line YAML Notation in Documentation

If you deal with the Kubernetes documentation, you will be confronted with a YAML notation from time to time, which I would like to briefly introduce here. Not only Kubernetes's but also other documentation uses it, and I will also use it in this book on occasion: it is the single-line YAML notation.

You have already come across several manifests. The structure with lines and indentations makes a document easy to read, but if you want to refer to a specific key-value pair and include the complete hierarchy, you need a solution that saves space. An example of this is `spec.containers[].resources.limits.cpu` to reference the CPU limit from Listing 3.5.

Each `.` separates the levels of the hierarchy. This is similar to accessing nested object properties in many programming languages. The square brackets (`[]`) after `containers` indicate that it is a list. If you want to reference a specific entry in the list, you could also add an index in the parentheses.

```
apiVersion: v1
kind: Pod
metadata:
  name: my-pod
spec:
  containers:
    - name: my-container
      image: my-image
      resources:
        limits:
          cpu: "1"
```

Listing 3.5 Counterexample for Single-Line YAML Notation

3.2.5 Weaknesses of YAML

One of the main criticisms of YAML is the extensive specification, which covers a wide range of data types. This is very convenient in some situations because you don't necessarily have to put strings in quotation marks, for example, but it can lead to incorrect interpretations.

3 Everything as Code: Tools and Principles for Kubernetes Operations

A well-known example of this is the Norway problem. The Norway problem is caused by a type inference weakness in YAML when processing character strings: the country code for Norway, NO, is incorrectly interpreted as a Boolean value. An example of this is shown in Listing 3.6. If the country code NO is written in YAML without quotation marks, YAML will interpret it as False instead of the intended string, "NO".

```
countries:
  Sweden: SE
  Norway: NO # This is interpreted as Boolean false
  Finland: FI
  Germany: DE
```

Listing 3.6 Norway Problem

In Listing 3.7, you will find further values that are interpreted by YAML as Booleans. The interesting thing is that in the latest YAML specification, 1.2, which was published in 2009, the Boolean values have been restricted to True and False. Nevertheless, the old specification remains in the libraries, and the Norway problem persists.

```
yes_value: yes    # Is interpreted as True
no_value: no      # Is interpreted as False
on_value: on      # Is interpreted as True
off_value: off    # Is interpreted as False
yes: y            # Is interpreted as True
no: n             # Is interpreted as False
```

Listing 3.7 Boolean Example

[!] **Warning**

Although the new YAML specification 1.2 only interprets the True or False values as Boolean, it can happen that libraries still use the old specification for parsing. For example, Kubernetes uses the go-yaml library to parse YAML manifests.

You will find an issue posting in which this topic has been discussed for years at the following address: *http://s-prs.co/v596427*.

To avoid this problem, you can always enclose strings in quotation marks. This means that there is no room for interpretation.

In addition to the Boolean problem, there are also other misinterpretations. You can see two examples of this in Listing 3.8. The first is about port forwarding. For example, if you use the SSH port, YAML will turn the value 22:22 into a time. Of course, it has no problem with 80:80, as there is no corresponding time.

```
port-forwarding-ssh: 22:22 # Incorrectly interpreted as time
port-forwarding-nginx: 80:80 # Correctly interpreted as a character string
software-version: 1.1.0 # Correctly interpreted as a character string
database-version: 2.1 # Incorrectly interpreted as a floating point number
```
Listing 3.8 Other YAML Misinterpretations

Version numbers can also cause problems. If you stick to semantic versioning and use three numbers in each case, you won't have a problem. However, it's different if you only use two numbers, as with the database version in Listing 3.8. In that case, the number is interpreted as a float.

In general, it is best to write strings in quotation marks. This way you avoid any problems of misinterpretation. However, such a conflict occurs very rarely, and I have not yet had any critical issues because of it. If you do not import your YAML manifests directly in production, then in the worst case it could cost you some time in debugging. But as you have read this section, you will certainly remember the problem at this point.

3.2.6 Tips for Practical Use

I now want to give you a few useful tips that will hopefully make it easier for you to work with Kubernetes resources. In real life, you will be using YAML files all the time, so a good IDE or an editor with an appropriate add-on will save you a lot of headaches and time-consuming troubleshooting:

- **Comments**
  The best thing about YAML compared to JSON is that you have the option to write comments. Especially with complex manifests, commentary is worth its weight in gold. You mark a comment using #, as in the following example:
  ```
  name: "Kevin Welter" # Name of the author
  ```
- **Linting tools**
  In addition to comments, I recommend using a linting tool that checks the syntax of the YAML manifest. It is best to check which one is recommended for your development environment, as there are several on the market, but in the end they all do what they are supposed to. The most important thing is that you don't have to search forever for an incorrect indentation as the linter points it out to you.
- **Splitting files**
  As you know, you can integrate multiple documents into one YAML file. The recommendation is that you create one file per resource. For small applications, I sometimes use a single file. For larger applications, I always create a separate file for each Kubernetes object. There is no right or wrong here. Just see how it works best for you and how you can best keep an overview.

When developing, you should always make sure that you use a uniform indentation, a consistent naming convention, and anchors, because then even larger manifests will remain readable and you will enjoy writing them.

3.3 Version Management of Kubernetes Manifests

Now you have already written and seen a whole lot of different YAML files, and these need to be managed somehow. Managing Kubernetes manifests is a fundamental challenge, especially as your team grows or projects become more complex. At some point, the following question arises: How can the manifests be managed efficiently and effectively? This is where version management comes into play, an essential practice to promote order, traceability, and collaboration, which you are no doubt also familiar with from the development of your software.

The YAML manifests are at the heart of the Kubernetes architecture as they define how applications run, which resources they require, how they communicate with each other, and what status they have. In a dynamic environment where changes are made frequently and by different team members, version control is essential to maintain an overview. Without version management, you and your team could easily lose track of changes. Versioning the manifests helps you because you can

- track changes,
- quickly recognize and correct errors in new changes, and
- more easily perform rollbacks.

In software development, you have hopefully been working with version management for a long time. Nevertheless, I would like to go into this briefly, because practical knowledge is almost more important than being good at programming languages. We want to clarify the following questions:

- What is Git?
- What is the best way to manage many Kubernetes manifests?
- What branching strategies are there, and what are your experiences with them?

Note

If you are already familiar with Git and version control, you can read the section crosswise anyway. Sometimes a keyword is enough to give you a new impetus for your own work.

Before we get started, I would like to talk about a universal law that influences the processes and their structure in every company. It does not matter whether it is the structure for repositories, the structure of the CI/CD pipelines, or the structure of entire IT systems.

The law is referred to as *Conway's law*. It is a fundamental principle in software development and organizational structure that was first formulated by Melvin E. Conway in the 1960s. It states that the architecture of a software system reflects the communication structures of the organization that develops this system. In other words, the way teams communicate and interact is directly reflected in the structure and design of the software they create.

In the simplest case, this means:

- If you work in a DevOps team, you are more likely to use a monorepo and manage your application and infrastructure as code there.
- If dev and ops are separated in your company, then there is certainly also a separation in the repositories.

If you keep Conway's law in mind, you will better understand the influence of the organizational structure on the design of software systems. Try to observe the law in your company. This will help you to structure your projects.

> **Good to Know**
>
> This kind of separation also has something to do with responsibility and associated authorizations. For a while, I worked for a team that was only responsible for running the Kubernetes clusters. I supported the development of the team and the clusters. The team wanted to work according to DevOps principles. They wanted to grant the developers the greatest possible freedom.
>
> Unfortunately, this only worked as long as the developers took their responsibilities seriously—but in the end it was the members of the ops team who were on call at night and had to correct some of the developer's mistakes.
>
> This repeatedly caused trouble and ultimately led to rights being restricted and the pipeline and release structures being adapted.

3.3.1 Using Git

Git is a decentralized version control system that allows developers and teams to track every change to files and directories in a Git project. It was initially developed by Linus Torvalds in 2005 and has since become the standard for version management.

I can still remember the centralized version control system SVN. The biggest difference is that with Git, all team members have a complete copy of the repository locally and edit it there. You are therefore not dependent on the central server but can also use the Git repository locally. I haven't seen a company that still uses SVN in a long time, because the advantages of Git simply make SVN obsolete.

The most important advantages of Git are as follows:

- **Flexibility**
  Git can support you in a variety of nonlinear development workflows, allowing you to map projects of any size.
- **High performance**
  You can quickly switch between different code versions and commit new changes. Git is designed to manage code efficiently.
- **Security**
  The integrity of the source code is guaranteed in Git by means of the cryptographic hashing algorithm SHA1, which protects your code and your change history against unintentional or malicious changes.

One of the disadvantages of Git is learning how to use it. Although you can get into it very quickly, even I sometimes have a knot in my head when I have to carry out workflows that I rarely use.

There are basic terms in Git that you should know:

- **Repository (repo)**
  A *repository* is a storage location that contains the complete history of all file changes and the associated metadata. Each developer has a local copy of the repo, develops changes there, and can synchronize them with the remote repo.
- **Commit**
  A *commit* is a summary of a series of changes in the repository. Each commit contains a unique ID (the commit hash), the author of the changes, a commit message describing the changes, and a reference to the previous commit(s).
- **Branch**
  A *branch* in Git makes it possible to branch off from the main development line and work in a separate environment without affecting the main line. This is useful, for example, for developing new features or mapping specific environments. More on this will follow in Section 3.3.3.
- **Merge**
  The *merge* is the process of merging changes. Typically, this is the merging of two branches. Git provides various merging strategies to simplify the integration of changes.
- **Tag**
  Tags are references that are used to mark certain points in the version history of a repo, typically to mark release versions.

> **[+] Good to Know**
>
> Git distinguishes between remote and local repositories. The *remote* repository is located on a central server and can be used by any developer in your team.

> The development process looks as follows:
> - You first clone a remote repository to your local computer. This is then a complete *local* repository with a copy of the entire project history.
> - After you have made local commits, you can push these changes to the remote repository to share them with the team.
> - Conversely, you can retrieve changes made by others from the remote repository to keep your local copy up to date.
>
> Only the remote repository allows you to collaborate with others.

Git is designed to facilitate collaboration among developers, especially when it comes to working on the same project or even the same branch at the same time. However, despite its sophisticated merging and branching mechanisms, conflicts can arise during the merging of changes.

A *merge conflict* occurs when two developers have made changes to the same parts of one or more files and Git cannot automatically decide which version is the correct one.

Git clearly marks the areas in the files that contain conflicts. You need to open these files, find the conflict areas, and manually decide which changes you want to keep, change, or combine.

> **Note**
>
> There are many graphical tools for resolving merge conflicts that make your life easier. I use my IDE, which shows me the differences. I can then simply choose what I want to transfer.

At the following address, you will find a good tutorial for getting started, in which you will learn how to use Git locally, what commands are available, and how to use a remote repository, using GitHub as an example: *http://s-prs.co/v596428*.

But that's enough theory for now. Next, let's take a concrete look at managing Kubernetes manifests.

3.3.2 Managing Numerous Kubernetes Manifests

If you use Kubernetes more intensively and write manifests for multiple environments, then you may be asking yourself the question: What is the best way to store manifests in Git?

A good directory structure is important. Without it, projects will end up in chaos. Important elements of the structure include the following:

- **Clarity and consistency**
  The structure should be intuitive and easy to understand so that new team members can quickly get to grips with it.
- **Scalability**
  The structure must be flexible enough to scale with the growth of applications and services.
- **Separation of concerns**
  Different environments and applications should be clearly separated to avoid overlaps and conflicts.

A major challenge is that you cannot easily parameterize manifests. This leads to problems at the latest when you want to use different manifests in the production environment than in the development environment.

In Git, there are two options that you can use in this case. Both have their advantages and disadvantages:

- Folder structure within a branch
- One branch per environment

Let's take a look at two examples of a folder structure.

In an application-oriented structure, you can create a separate directory for each application or service. Within each application directory, subfolders are created for the various environments (e.g., *dev*, *staging*, *prod*). Thus, the structure could look like the one shown in Listing 3.9.

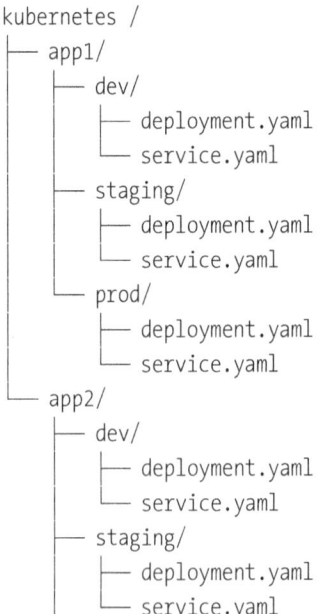

```
└── prod/
    ├── deployment.yaml
    └── service.yaml
```

Listing 3.9 Application-Oriented Structure

I recommend this structure in the following cases:

- Each application stands alone.
- Different developers are working on the applications.
- You want to emphasize the modularity of the individual applications.

This structure also makes it easier to write the CI/CD pipelines if you want to release the applications individually.

> **Note**
>
> You should take inspiration from the examples in this section, but use the structure that works best for you and modify it if necessary.
>
> If you use Helm (see Chapter 8) or Kustomize (Section 3.5), then you can also select other structures thanks to the parameterization of your templates. It is important to me that you get an idea of what is possible.

The environment-oriented structuring of your Kubernetes projects focuses on organizing your directories by environment. Within these environment directories, you create subfolders for each of your applications or services. Listing 3.10 shows an example of this.

```
kubernetes /
├── dev/
│   ├── app1/
│   │   ├── deployment.yaml
│   │   └── service.yaml
│   └── app2/
│       ├── deployment.yaml
│       └── service.yaml
├── staging/
│   ├── app1/
│   │   ├── deployment.yaml
│   │   └── service.yaml
└── prod/
    ├── app1/
    │   ├── deployment.yaml
    │   └── service.yaml
```

```
└── app2/
    ├── deployment.yaml
    └── service.yaml
```

Listing 3.10 Environment-Oriented Structure

This structure is perfect in the following cases:

- The applications belong together, such as the backend and database.
- You always roll out the applications in a package.
- The clear separation of environments makes it easier for you to move applications through the various stages of the development cycle.

This also facilitates the implementation of the CI/CD pipelines for rolling out a package. For example, you can go to the environment folder and roll out all applications.

> **Note**
>
> In these structures, you must also transfer changes made to a manifest to other environments. So you have more typing work to do, and the whole thing is more error-prone because you are copying and pasting.
>
> Thus, before you decide on a structure, you should read this entire section, in particular Section 3.3.3.

The organization and naming of your manifests in Git play a decisive role in the clarity and understanding of your configurations. A consistent naming convention helps you and your team to quickly understand the content and purpose of each file. This is particularly important in complex projects with a large number of resources.

You should define the convention depending on the repository structure. An example of this is `<application>-<resourcetype>.yaml`. For a frontend, this would be `frontend-deployment.yaml`. If you imagine this in the environment-oriented structure, then `frontend` would be found in the folder and in the name, as shown in Listing 3.11; in this case, there is no gain in information.

```
kubernetes /
├── dev/
│   ├── frontend/
│   │   ├── frontend-deployment.yaml
│   │   └── frontend-service.yaml
```

Listing 3.11 Naming Convention in Environment-Oriented Structure

3.3 Version Management of Kubernetes Manifests

> **Good to Know**
> Using the naming convention with the application in the name, you can also simply do without the subfolders for applications that should always be deployed together. Simply check your use case to see what gives you a good overview without too much overhead.

Another option is to add a version number—for example, if you want to develop the manifests further. This allows you to see at a glance which version is active in which environment.

What I have seen more often and what is important in some constellations is numbering manifests so that they are rolled out in the desired order. If you use `kubectl apply -f .` to roll out all files in a folder, `kubectl` will work its way from top to bottom. With the following naming, you can be sure that the correct sequence is adhered to during deployment:

```
01-namespace.yaml
02-deployment.yaml
...
```

In most cases, it is not important, because a deployment simply waits for missing resources such as ConfigMaps. However, if you want to deploy to a namespace, you should create that first.

> **Note**
> You can write multiple resources in a YAML file and separate them using `---`, but I recommend that you separate the resources. This allows you to create an overview of the available resources in the folder structure.

Choosing the right naming convention and folder structure for your Kubernetes manifests depends on the specific requirements of your project. Whether you choose one or the other is secondary. It is important that you choose a well thought-out convention, as this contributes significantly to the clarity, manageability, and efficiency of your project. I recommend that you take inspiration from the examples, choose a convention that meets your team's needs, and apply and optimize it consistently over time.

3.3.3 Branching Strategies

The management of manifests and the definition of naming conventions is also affected by the selected branching strategy. What you should definitely avoid is duplicating too much code and copying it back and forth. A good branching strategy can help you in this regard. I want to introduce you to approaches that I have already used in projects, whereby these approaches are also related to the division of your repositories, which we will look at in Section 3.3.4.

Good to Know

The way you organize your manifests, choose your branching strategy, and structure your repositories has a profound impact on your development process, your team collaboration, and ultimately the efficiency and scalability of your projects.

If you have been using Git for some time, you will certainly be familiar with the classic branching concepts and development workflows. Let me shed some light on the topic from the perspective of Kubernetes. Over the years, various branching flows have emerged, some of which you will no doubt already be using.

Git Flow

Git flow is the best-known workflow and was developed by Vincent Driessen in 2010. It is based on two main branches with an unlimited lifetime: *main* (or *master*) for production code and *develop* for preproduction code. Additional branches such as *feature-**, *hotfix-**, and *release-** support the development cycle. Sometimes there is still a release branch in some projects that can be used in preparation for a new release. An example of this workflow is illustrated in Figure 3.3.

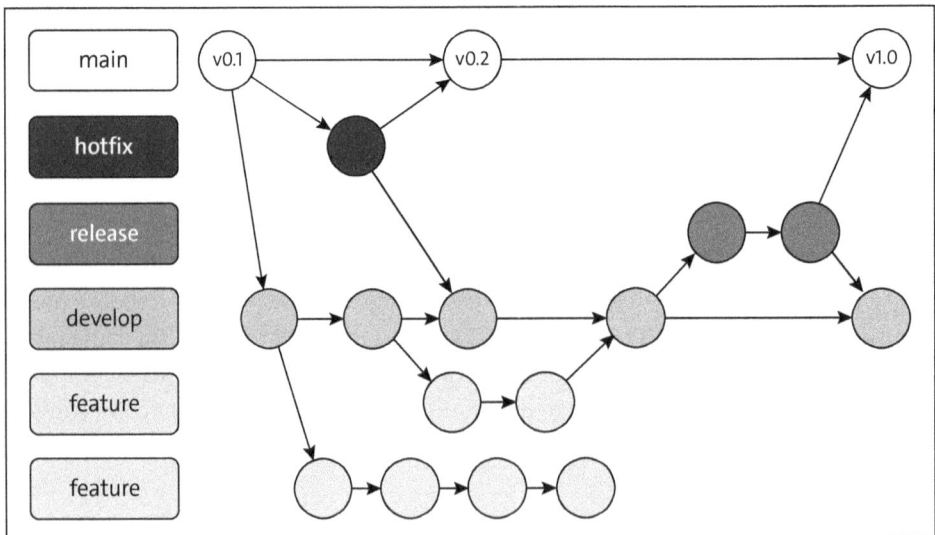

Figure 3.3 Git Flow

Git flow keeps branches clean at every stage of the project, follows a systematic naming scheme, provides extensions and support in most Git tools, and is ideal for projects

3.3 Version Management of Kubernetes Manifests

that need to manage complex software versions or have long release cycles. Disadvantages include the fact that Git histories are often difficult to read and that the CI/CD pipeline is pretty complex due to the separation of the main and develop branches.

> **Good to Know**
>
> I've seen a few variations on this branching strategy, but in the end the principles were always the same. The more heterogeneous the repository is and the more developers are working on a repository, the more likely it is that this complex strategy will be used.

GitHub Flow

GitHub flow is a simpler workflow introduced by GitHub in 2011. It has six principles, including the permanent deployability of the main branch and the creation of feature branches directly from the main branch. GitHub flow promotes CI/CD, is simpler than Git flow, and is ideal for projects that are not tied to release cycles.

Figure 3.4 illustrates a representation of the flow. A branch of the main branch is opened for a change and should find its way back into the main branch very quickly.

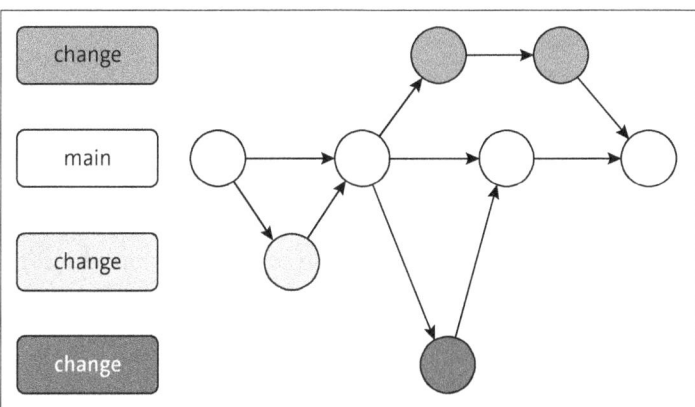

Figure 3.4 GitHub Flow

Disadvantages include potentially unstable production code and unsuitability for release planning. The development teams also need a certain amount of discipline during development, but this strategy is the best, especially if you want to implement changes directly with CI/CD.

> **Note**
>
> You can find the GitHub flow principles at the following address: *http://s-prs.co/v596429*.

209

3   Everything as Code: Tools and Principles for Kubernetes Operations

GitLab Flow

GitLab flow, which was created by GitLab in 2014, differs from GitHub flow in its environment branches, such as *preproduction* and *production*. It is based on eleven rules that help you implement CI/CD and leads to a cleaner, less messy Git history.

GitLab flow is ideal for projects that need to adapt to release cycles or that require more control over deployment. Before you deploy to production, you have to go through the intermediate step again and merge into the production branch. An example of this is shown in Figure 3.5.

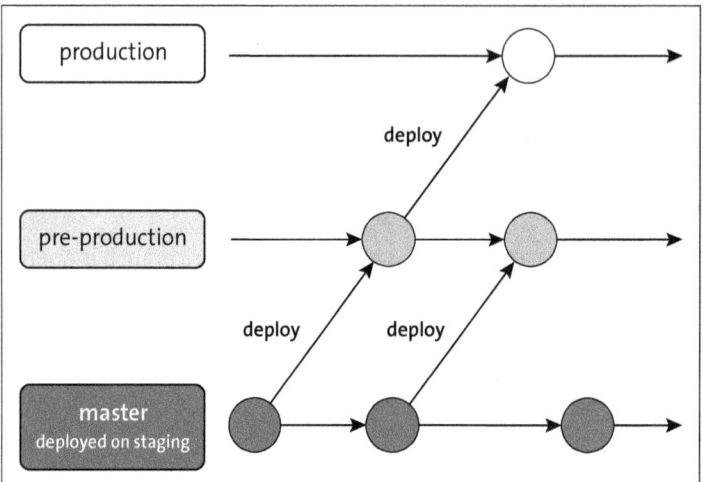

Figure 3.5 GitLab Flow

A major disadvantage is the overhead of merging into the surrounding branches. This can often lead to merge conflicts in projects because some developers have developed directly on the environment branches. This also requires discipline within the team and enforcing the rules.

> **Note**
>
> The rules of GitLab flow can be found at the following address: *http://s-prs.co/v596430*.

Depending on the setup of your repository, I would advise you to use a particular workflow. I often preferred GitLab flow in ops teams because the release cycles were of different lengths. Sometimes it took weeks before a deployment was brought from preproduction to production. The surrounding branches give you a good overview.

If you develop your manifests in a repository together with your code, then you should also use this branching strategy. I have often used Git flow for larger applications with long release cycles.

In my opinion, the ideal solution is GitHub flow. However, due to the direct integration of the automatic deployment into the main branch, it is the most difficult to master and is therefore feared by many.

> **Note**
>
> Even if GitHub flow is advertised, you should not simply switch to it. A few preliminary steps are necessary to get there. Don't forget that every merge into the main branch must be of high quality because it is rolled out directly to production.
>
> GitLab flow is much more user-friendly and more controlled.

3.3.4 Division of the Repositories

Structuring your repository is a fundamental decision that has far-reaching effects on team collaboration, development efficiency, and code maintainability. This decision is closely linked to Conway's law, which I presented at the beginning of this section.

There are two classic models for structuring repositories and, as always, many gray areas in between. There is no right or wrong here either. The important thing is that you consciously decide on a division and that you are aware that it has an influence on the branching strategy and the CI/CD pipelines.

Monorepo: One Repository for Everything

In a monorepo, the entire code is stored in a single repository. This supports close collaboration and simplifies the process of code sharing within the team. This approach makes it easier for all developers to track changes throughout the project and efficiently manage dependencies between different parts of the project. You would also integrate the Kubernetes manifests into the repository using this approach. A monorepo approach has advantages and disadvantages:

- Pros
    - Improved transparency and traceability of changes throughout the entire project
    - Simplified dependency management through central administration
    - Supports uniform development culture and practices
    - Changes to the software and manifests can be mapped in a commit or pull request
- Drawbacks
    - Can be a challenge in terms of performance and manageability for very large projects
    - Difficult with shared responsibility between teams

> **Good to Know**
> A monorepo is particularly suitable for DevOps teams that have full responsibility for the entire lifecycle of the application.

Multiple Smaller Repositories: Modularity and Independence

The use of separate repositories for different modules or services provides a clear separation of responsibilities and can increase the clarity of the project. This approach makes it easier to develop, test, and deploy independent parts of the system separately. In addition, different teams can develop the individual modules without affecting the other development processes. This approach too has advantages and disadvantages:

- **Pros**
  - Increased modularity and independence of the various project parts
  - Lower risk of merge conflicts due to isolated work areas
  - Enables specific access rights and security policies for different repositories
- **Drawbacks**
  - Can increase the complexity of integration and version management
  - Requires additional effort for coordination and communication between the teams
  - Changes that belong together must be synchronized in several pull requests and repositories

> **Good to Know**
> Multiple repositories are very suitable for teams with shared responsibility or when developers from different teams are working on specific modules.

The ideal lies somewhere in between the two approaches. I've seen excellent monorepos where the development teams had a very fast and clear development process thanks to good processes. On the other hand, I am used to splitting the repositories, as it is often clearer and responsibility often has to be shared between teams, especially in large companies.

You should therefore choose the repository structure that is best for your project and be open to adapting the structures at a later date. My experience shows me that there is always some movement in everything, which keeps things flexible.

3.4 Continuous Integration and Continuous Delivery

Continuous integration and continuous delivery (CI/CD) have become indispensable strategies not only for accelerating development and deployment processes, but also for improving quality. CI/CD pipelines play a central role in the automation of the build, test, and deployment processes and are indispensable in many companies today.

This section focuses on how you can set up and optimize CI/CD pipelines specifically for handling Kubernetes manifests. We deliberately disregard the pipeline for pure application development and focus on the deployment aspect.

By implementing CI/CD for your Kubernetes manifests, you benefit from several major advantages:

- **Faster deployment cycles**
  Automation minimizes delays and enables you to transfer changes to production more quickly. The biggest advantage is that you always run through the same pipeline and avoid manual errors.

- **Improved code quality**
  Regular integration, tests, and quality scans reveal problems at an early stage. You are forced to improve code quality at an early stage. This can sometimes be annoying, but it makes a lot of sense in the long term.

- **Improved collaboration**
  A centralized repository and automated workflows help you to work better together as a team. In a code review, a successfully completed pipeline is the first indication of a good level of maturity of the code.

- **Increased reliability**
  Automated tests and deployments reduce the risk of human error and lead to more stable releases. The release process is also faster, more reliable, and less error-prone. In addition, your stress level is lower during a rollout in production than if you have to carry out manual steps.

3.4.1 Pipeline Steps for Kubernetes

If you imagine a pipeline for your application, the steps to be carried out prior to deployment are often pretty clear. This usually includes steps such as the following:

- Static code analysis
- Vulnerability checks of the libraries
- Unit tests including a verification of the test coverage

But what are the steps for Kubernetes manifests? When can you be sure that a deployment will not lead to errors?

The answer lies in the integration of specific testing and validation steps that are tailored to the special features of Kubernetes. I will introduce two possible pipeline steps you can use.

Linting

When you develop and deploy applications in Kubernetes environments, the question often arises as to how the quality and consistency of manifests can be efficiently ensured. The linting process makes a decisive contribution to this. Linting tools analyze your YAML files to ensure syntactical correctness and check for compliance with best practices.

One tool you can use for this is Kubeconform. Such static code analyses or linting tools are particularly useful in a pull request pipeline. They give a direct indication of the quality of the new code.

Let's take a look at how Kubeconform works. To begin, follow the installation instructions at the following address: *http://s-prs.co/v596431*.

I have prepared a deployment from Chapter 2, Section 2.3 with a syntax error for you in Listing 3.12.

```yaml
apiVersion: apps/v1
kind: Deployment
metadata:
  name: my-nginx-deployment
  labels:
    app: nginx
spec:
  replicas: "2"
  selector:
    matchLabels:
      app: nginx
  template:
    metadata:
      labels:
        app: nginx
    spec:
      containers:
        - name: my-container
          image: localhost:5000/my-nginx
          ports:
            - containerPort: 80
```

Listing 3.12 invalid-deployment.yaml

3.4 Continuous Integration and Continuous Delivery

To make Kubeconform analyze this deployment, simply execute the following command:

```
kubeconform invalid-deployment.yaml
```

The result will then indicate the errors—in this case:

```
invalid-deployment.yaml - Deployment my-nginx-deployment is invalid: problem
validating schema. Check JSON formatting: jsonschema: '/spec/replicas' does not
validate … expected integer or null, but got string
```

In this deployment, the value for replicas is a string, but it should be an integer.

Of course, you can analyze not only individual files, but also entire folders. You should implement a linter in each of your pipelines. In GitLab CI, a simple step for linting could look as shown in Listing 3.13. This can be installed quickly, and the linter hardly needs any time for scanning.

```
lint_kubernetes_manifests:
  stage: validate
  image: docker.io/yannh/kubeconform:latest
  script:
    - kubeconform /pfad/zu/manifest
```

Listing 3.13 Example of Pipeline Linting Step in GitLab CI

> **Good to Know**
> What looks so simple now has already helped me a lot in some cases. Especially when things have to be done quickly, I sometimes make careless mistakes that are detected by the linter.
> I find a linter very important after running a templating engine like Helm (see Chapter 8).

Test or Validation after Deployment

Imagine you have now written Kubernetes manifests for your application and have painstakingly checked whether everything works as it should. You will continue to develop the manifests in the near future and want to avoid the manual effort of future testing. You also want to be sure that all possible contingencies have been checked. You can only achieve this through test automation, and there are also tools for Kubernetes that support you in this respect.

This is where the Kubernetes Test Tool (KUTTL) comes into play. KUTTL is a powerful framework that was developed specifically for testing Kubernetes clusters. It allows you to perform end-to-end tests that check not only the configuration, but also the actual behavior of your applications in the cluster. KUTTL helps you with the following actions:

- **Ensuring functionality**
  Postdeployment tests validate that the application works as intended and that all services communicate correctly with each other.
- **Early error detection**
  By detecting issues immediately after deployment, errors can be troubleshot quickly before they affect operations.
- **Automation and reliability**
  Automated tests increase the reliability of deployments by reducing manual checks and ensuring consistent test procedures.

KUTTL uses the strengths of Kubernetes to run tests in the same environment in which your applications run. It allows you to define test cases as Kubernetes manifests. KUTTL takes care of the setup, running the tests, and cleaning up afterward. This simplifies the testing process. Let's start with a small example. For this purpose, you need to install KUTTL for your system using the instructions provided at *https://kuttl.dev/docs/cli.html*.

Now set up a simple test suite with the folder structure from Listing 3.14.

```
.
├── kuttl-test.yaml
└── tests
    └── test-nginx
        ├── 00-assert.yaml
        └── 00-install-nginx.yaml
```

Listing 3.14 KUTTL Folder Structure

If you run this test suite, the following will happen:

1. KUTTL will roll out a deployment with Nginx in Minikube.
2. As a test, KUTTL will check whether the status of the pod is Running.
3. KUTTL will delete the deployment again.
4. KUTTL will issue a report on the tests.

The *kuttl-test.yaml* file in Listing 3.15 is located in the root directory. It defines the starting point of KUTTL and references the folder with the tests. The *tests* folder then contains the tests, which you can bundle into folders. We have a *test-nginx* folder containing two files.

3.4 Continuous Integration and Continuous Delivery

```yaml
apiVersion: kuttl.dev/v1beta1
kind: TestSuite
testDirs:
  - ./tests
```

Listing 3.15 kuttl-test.yaml

`00-assert.yaml` in Listing 3.16 describes the actual test, which will run after the deployment. `00-install-nginx.yaml` is the usual Nginx deployment in Listing 3.17. KUTTL will use it later to deploy it in Minikube and then run the test. KUTTL will check whether the status of the pod labeled `app: nginx` is `Running`.

```yaml
apiVersion: v1
kind: Pod
metadata:
  labels:
    app: nginx
status:
  phase: Running
```

Listing 3.16 00-assert.yaml

```yaml
apiVersion: apps/v1
kind: Deployment
metadata:
  name: nginx-deployment
  labels:
    app: nginx
spec:
  replicas: 1
  selector:
    matchLabels:
      app: nginx
  template:
    metadata:
      labels:
        app: nginx
    spec:
      containers:
      - name: nginx
        image: nginx:latest
        ports:
        - containerPort: 80
```

Listing 3.17 00-install-nginx.yaml

Now run the test suite using the following command:

```
kubectl kuttl test
```

At the end, KUTTL should provide a report on the tests.

> [+] **Good to Know**
> In this way, you can test and verify various setups of your application using KUTTL.

As already mentioned, KUTTL can also carry out various verifications in addition to the tests we have just tried out. This is ingenious because it allows you to check whether your rollout was successful in the CI/CD pipeline after deployment.

The options for verifications are unlimited as you can run Kubernetes commands and check the results, as shown in Listing 3.18.

In this example, the command checks whether the deployment from Listing 3.17 has been rolled out in the `default` namespace and whether a pod that has the `Running` status can be found.

```
apiVersion: kuttl.dev/v1beta1
kind: TestStep
commands:
  - command: kubectl get pod -l app=nginx -o jsonpath=
            "{.items[0].status.phase}" -n default
    expect:
      stdout: Running
```

Listing 3.18 00-check-nginx.yaml

As you can see, KUTTL is extremely powerful for verifying rollouts or even writing tests for your manifests. I can only recommend that you take a closer look at the tool and implement it in your CI/CD pipeline.

3.4.2 Pipeline Architectures

In this section, we'll take a look at the different architectures of CI/CD pipelines that can be used in Kubernetes environments. Choosing the right architecture depends on several factors, including the complexity of the application, the team size, security requirements, the desired speed of deployment, and, most importantly, the Git repository structure, which we discussed in Section 3.3.4.

There are different approaches to developing pipelines. You always have to decide between a monolithic pipeline or functional pipelines and whether you want to define the pipelines in a centralized or decentralized way:

- **Monolithic pipelines**
  In a monolithic architecture, all steps such as build, test, and deployment are handled as a single, comprehensive process. This can be useful for simple projects or small teams. However, it can slow you down in larger projects if you have to carry out each previous step every time before a deployment.

- **Functional pipelines**
  With this architecture, the CI/CD process is divided into smaller, independent parts. This promotes modularity and enables faster pipelines that can still build on each other, which allows you to separate the deployment from the build pipeline or use your own test pipelines and run them independently of each other.

- **Centralized pipelines**
  A central pipeline manages all aspects of CI/CD for multiple projects or services. This can simplify the administration, but also restrict flexibility.

- **Decentralized pipelines**
  Here, each project or repository has its own CI/CD pipeline, which provides more flexibility and allows for the adaptation to specific requirements.

> **Good to Know**
> As always, the world is not just black and white; the truth lies somewhere in the middle. Instead of centralizing everything, using pipeline templates that are integrated into other projects could also make sense. This gives you standardized pipeline steps that you can import. A good CI/CD pipeline is always found in several iterations.

I now want to introduce two pipeline concepts that I have developed for clients. Both have their charms and fulfill their purpose in the project context. But I will also show you the weaknesses of the pipelines at the end so that you can find some inspiration for your own pipelines.

Build and Deployment Separated

In the introduction to Section 3.3, I presented Conway's law.

In this example, exactly what is stated in the law applies: the build and deployment pipelines were separated because the development team should not have direct access to deployments on Kubernetes clusters that are managed by an ops team.

Take a look at the pipeline in Figure 3.6. I have simplified the pipeline steps there:

- **Build pipeline**
  - *Maven build*
    Starts the build process with Maven, compiles the code, and executes tests.

3 Everything as Code: Tools and Principles for Kubernetes Operations

- *Docker build and push*
  Creates a Docker image from the Maven build. Uploads the finished Docker image to a Docker registry.
- *Trigger deployment*
  Triggers the deployment of the new Docker image and transfers the new version number.

- **Deployment pipeline**
  - *Change of the image tag in the Helm chart*
    Updates the Helm chart with the new version number.
  - *Helm deploy*
    Runs the deployment process using Helm.

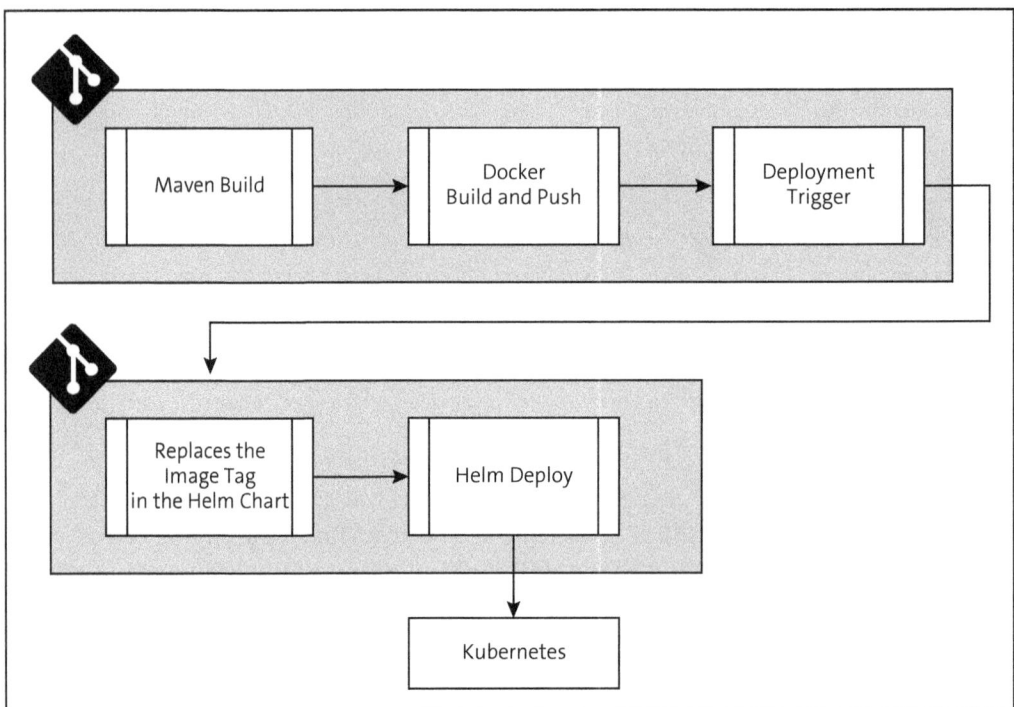

Figure 3.6 CI/CD Pipeline: Build and Deployment Separated

What you cannot see here is that the two pipelines could run independently of each other, with the deployment pipeline caching the last version number. As a result, the deployment pipeline was always able to redeploy the latest version. Upon the triggering of the deployment pipeline through the build pipeline, the process had to be approved for the production environment by an ops employee. Otherwise, the execution was blocked.

220

3.4 Continuous Integration and Continuous Delivery

> **Good to Know**
> The pipeline had many more steps, such as a static code analysis by SonarQube, but here we will only deal with the general structure of a pipeline.

The strengths of this approach are as follows:

- Not only does the division into build and deployment pipelines reflect the organizational structures between development and operations, but it also increases flexibility and control over the release process. By separating the functions into independent pipelines, the two teams can make changes faster and more securely without impacting each other.
- The independence of the deployment pipeline, which makes it possible to redeploy the latest stable version if required, underlines the reliability of this approach. In addition, the required manual approval by the ops team for production deployments provides an additional level of security that fulfills compliance requirements in this case.

Although the separation of build and deployment pipelines offers some advantages, there are also aspects to view critically:

- The strict separation can lead to silos that make communication and collaboration between development and operations teams difficult.
- Managing multiple pipelines can increase the complexity of the overall system. For my clients, I was always involved in the coordination with the two teams in order to have a common view.
- Triggering and transferring information from one pipeline to another must be carefully considered. In this case, I implemented a web hook and passed parameters.

Monolithic Pipeline with Central Templates

In this case too, the pipeline has adapted to Conway's law, as the application is developed and operated by a DevOps team. There is a single pipeline for the Java application that processes all steps in sequence. It uses central templates, which means that I didn't have to develop every step from scratch, but instead adopted the company standard. Such templates define the steps that your pipeline goes through and ensure consistency and reusability within your CI/CD processes. The templates are then customized using parameters.

You can find a representation of the pipeline in Figure 3.7.

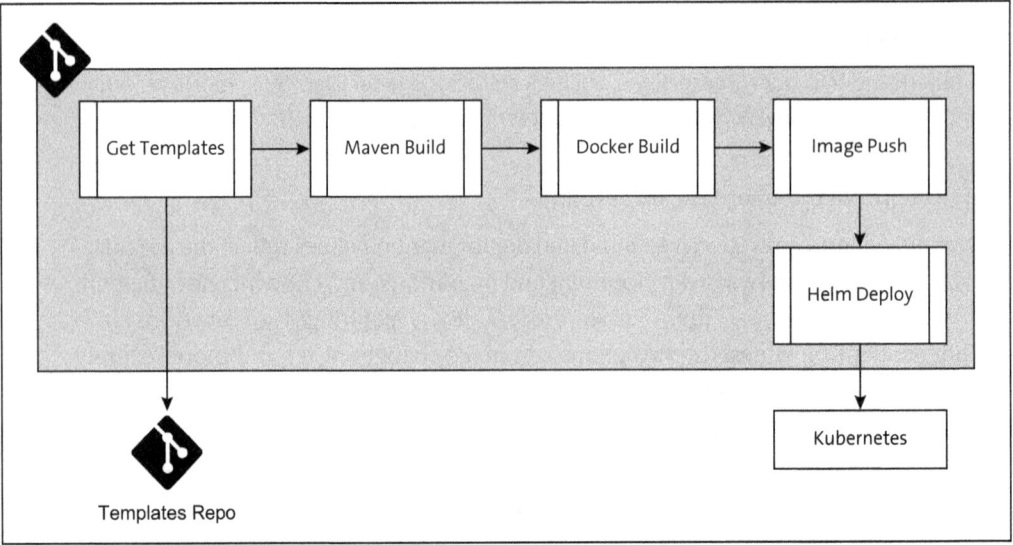

Figure 3.7 CI/CD Pipeline: Monolithic Pipeline with Central Templates

The steps of the pipeline are as follows:

- **Get templates**
  Templates and configuration files are retrieved from a central template repository.
- **Java build**
  The Java code is compiled, and tests and quality checks are also carried out.
- **Docker build**
  A Docker image is created based on the compiled Java code.
- **Image push**
  The newly created Docker image is uploaded to a Docker registry.
- **Helm deploy**
  This runs the deployment process using Helm.

Good to Know

I developed this pipeline with Azure DevOps. In my opinion, this is a very special pipeline tool, but very suitable for modularization. However, you can also implement central templates with other tools such as GitLab CI.

The strengths of this approach are as follows:

- Using a dedicated template repository as a source for pipeline templates takes some getting used to, but it's very powerful. This makes it easier for you to maintain and update your pipelines as changes can be made in one central location. The templates also ensure that all projects benefit from proven standards and do not have to keep reinventing the wheel.

- By consolidating the entire process into one pipeline, you minimize the need to synchronize multiple pipelines and develop handover points. With a shared pipeline, everyone on the team also feels responsible for the entire process. This also prevents the creation of small silos within a DevOps team.

The weaknesses of this approach are as follows:

- As all steps are carried out in a single pipeline, this quickly leads to long runtimes. If an error occurs in one of the final steps, such as deployment, this can be very annoying and slow down the development process.
- Dependence on central templates does not only bring advantages. A central template has to be maintained, and if another team is involved, this can quickly lead to conflicts. Updating to a newer template version must always be cross-checked to ensure that the pipeline continues to do what it is supposed to do. These are similar challenges in updating Java libraries.

All in all, there are many great ways in which you can structure and build your pipelines. Each structure has advantages and disadvantages, but if you follow Conway's law, you will find the right pipeline for the existing company structure.

3.4.3 GitOps

GitOps is a modern practice of software development and deployment in which Git serves as a single source of truth for the entire infrastructure and application configuration. GitOps relies on the principle of declarative configuration and allows you to deploy in Kubernetes in a different way.

I have seen with many customers that deployment is typically carried out via the CI pipeline. This means that imperative commands are executed by Jenkins, GitLab CI, or a similar tool to roll out Kubernetes manifests. The biggest difference is that you no longer use a CI tool to import new changes into Kubernetes.

The cluster itself regularly checks for changes in order to adjust the status if necessary. The process is referred to as the *reconciliation loop* and is also used by Kubernetes resources themselves to compare the current state with the expected state. The ReplicaSet pays attention to the number of pods, for example, and makes adjustments if necessary. For example, if you delete a pod, the ReplicaSet will immediately start a new one to maintain the desired state. GitOps maximizes the self-healing capabilities of Kubernetes by ensuring that the cluster state matches the desired state defined in the Git repository at all times. Whenever there is a discrepancy, the GitOps tool automatically makes corrections to restore the target state.

Good to Know

Compared to ordinary CI/CD pipelines, I see the following advantages with GitOps:

- **Declarative instead of imperative**
  GitOps uses a declarative approach in which the desired state of the infrastructure and applications is defined in Git. The tools ensure that this state is achieved and maintained. CI/CD is more imperative and focuses on the steps required to achieve a certain state.
- **Easier compliance**
  You can rely entirely on the strengths of Git. Code reviews and the dual control principle before merging to main are usually well-established processes.
- **Increased level of security**
  There is no additional tool that requires a technical user as the cluster itself retrieves the changes from Git.
- **Improved auditability**
  You only have one place to look for changes. Pipeline logs, for example, are no longer required.
- **Drift detection**
  As the reconciliation loop is run through regularly, GitOps quickly detects deviations and reports or reverses them.

In a pipeline, you would typically execute kubectl commands to deploy your manifests. With GitOps, the process looks more like the one shown in Figure 3.8. Instead of using imperative commands to adjust the cluster state, a GitOps tool continuously monitors the Git repository for changes and automatically applies them to the Kubernetes cluster. This approach enables a seamless integration of the reconciliation loop with the version control and collaboration features of Git.

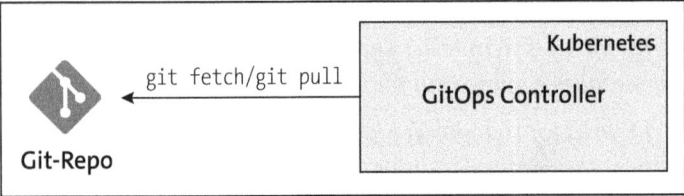

Figure 3.8 GitOps Controller

Note

There are many different GitOps tools that you should take a look at. I have often read about Flux in connection with GitLab, and I have also come across Argo CD several times. But make up your own mind, because it always depends on the tech stack you use.

You can find the repos at the following URLs:
- *https://github.com/argoproj/argo-cd*
- *https://github.com/fluxcd/flux2*

If you have the opportunity in your company, then you should talk about and evaluate GitOps tools. Unfortunately, the biggest advantage is also a disadvantage: you are more or less forced to make all changes via your Git repository. This sometimes requires a lot of getting used to, but you will see that you and your team will achieve a higher quality in your rollout through self-discipline and the methods of Git.

3.5 Templating Using Kustomize

Up to this point, you have written Kubernetes manifests and installed them on Minikube, which you have hard-coded with values. But what can you do if you want the manifest to look different for the production environment than for the development environment? This could be the case, for example, if you

- have different resource requirements;
- want to use a different image in the production environment;
- want to have fewer replicas in the development environment; or
- want to assign other labels or annotations.

The simplest approach is this: You copy the manifest and change the values. Then you would have a *deployment-dev.yaml* file and a *deployment-prd.yaml* file. I hope your alarm bells are already ringing. Unfortunately, I have experienced this all too often in companies that have made life difficult for themselves as duplication means that a developer has to make the same changes in multiple files. Sooner or later, this will lead to errors or incongruities.

But how can you proceed? The most important thing is that a manifest should be parameterized. In the simplest case, you could work in a pipeline using the sed or yq Linux tool to use the manifest as a template and replace placeholders. That's much better, but it's best not to build anything yourself and instead to rely on existing tools. In my view, there are currently two tools on the market that are highly relevant:

1. **Helm**
   - Helm is a package manager for Kubernetes that can provide Kubernetes manifests as a chart (package).
   - It offers functions for versioning and managing dependencies.
   - It also uses a templating syntax to dynamically fill a manifest with values.

2. **Kustomize**
   – Kustomize is a tool for customizing Kubernetes manifests and thus works more directly and easily.
   – Instead of templating, it uses an overlay structure to apply changes to basic YAML files.
   – Thanks to the integration in `kubectl`, you do not need any additional syntax.

Helm goes one step further as a package manager, but we will go into this in more detail in Chapter 8. For now, I would like to introduce Kustomize and the advantages that it can bring to your project:

- Kustomize allows you to make configuration adjustments without changing the original files. This means you have the flexibility to customize your configurations as needed without worrying about the integrity of your basic manifests.
- A big advantage you enjoy with Kustomize is the use of pure YAML without having to resort to template parameters. In contrast to other tools such as Helm, which require an additional processing step, you can validate and process manifests created with Kustomize directly as YAML. This makes handling easier and more transparent for you.
- You also benefit from the independence of templating engines. Kustomize uses simple editing of YAML files, so you don't need to learn any additional templating syntax.
- The reusability and modularity provided by Kustomize allow you to create basic configurations that you can reuse in different projects or contexts.
- The clear and comprehensible overlay structure of Kustomize enables you to clearly separate basic configurations from environment-specific adjustments. This makes managing different configurations for different environments clear and feasible.
- Finally, the seamless integration into Kubernetes is a decisive advantage. As Kustomize is integrated directly into `kubectl`, you do not need any additional software to use it.

Simply put, Kustomize offers you a good opportunity to provide your manifests for several environments with little effort, without having to spend a lot of time familiarizing yourself with a new tool.

3.5.1 Basic Principles of Kustomize

Let's start directly with a simple example in which a production and a development environment are defined and where different rules and requirements apply. The folder structure for the example should look as shown in Listing 3.19. We will now set up the files step by step.

3.5 Templating Using Kustomize

```
kustomize/
├── base/
│   ├── deployment.yaml
│   ├── service.yaml
│   └── kustomization.yaml
└── overlays/
    ├── dev/
    │   ├── patch-dev.yaml
    │   └── kustomization.yaml
    └── prod/
        ├── patch-prod.yaml
        ├── service-patch-prod.yaml
        └── kustomization.yaml
```

Listing 3.19 Folder Structure for Kustomize Example

First, define the basic configuration of the application. This includes a deployment with Nginx as in Listing 3.20 and a service as in Listing 3.21. Place both in the *base* folder.

```yaml
apiVersion: apps/v1
kind: Deployment
metadata:
  name: my-nginx
spec:
  replicas: 1
  selector:
    matchLabels:
      app: my-nginx
  template:
    metadata:
      labels:
        app: my-nginx
    spec:
      containers:
        - name: my-nginx-container
          image: nginx:latest
          ports:
            - containerPort: 80
```

Listing 3.20 base/deployment.yaml

```yaml
apiVersion: v1
kind: Service
metadata:
  name: my-nginx-service
```

```
spec:
  type: ClusterIP
  selector:
    app: my-nginx
  ports:
    - port: 80
      targetPort: 80
```

Listing 3.21 base/service.yaml

In addition, there is a *kustomization.yaml* file, provided in Listing 3.22, that contains a reference to all basic manifests.

```
resources:
  - deployment.yaml
  - service.yaml
```

Listing 3.22 base/kustomization.yaml

Now let's continue with the overlays. For the development environment, you want to increase the number of replicas and use a different image, which is a version that has not yet been released for production. There is a separate folder for the overlays and a subfolder for each environment. For the development environment, that is *overlay/dev*. Use the patch from Listing 3.23, and integrate it with *kustomization.yaml* from Listing 3.24.

```
apiVersion: apps/v1
kind: Deployment
metadata:
  name: my-nginx-app
spec:
  replicas: 2
  template:
    spec:
      containers:
        - name: my-nginx-container
          image: nginx:dev
```

Listing 3.23 overlay/dev/patch-dev.yaml

```
resources:
  - ../../base
patches:
  - path: patch-dev.yaml
```

```
    target:
      kind: Deployment
      name: my-nginx
```

Listing 3.24 overlay/dev/kustomization.yaml

As you can see, you only include the fields in the patch file that are supposed to be adapted in the basic file. For Kustomize to display the final result, you need to run the following command in the folder:

```
kubectl kustomize overlays/dev
```

In the output, you should see the unchanged service and the complete deployment from Listing 3.20 with the changes from Listing 3.23.

> **Good to Know**
>
> The *kustomization.yaml* file in Listing 3.24 describes which resources should receive the patch. For this purpose, you can specify the following selectors under `target`, which must all match:
>
> - group
> - version
> - kind
> - name
> - labelSelector
> - annotationSelector
>
> You do not need to specify every one, and can therefore control well what the change should be applied to.

For production, you can also implement a change for the service, which you can see in Listing 3.25.

```
apiVersion: v1
kind: Service
metadata:
  name: my-nginx-service
spec:
  type: LoadBalancer
```

Listing 3.25 overlay/prod/service-patch-prod.yaml

Also change the deployment to three replicas and to a stable image, as you can see in Listing 3.26.

```
apiVersion: apps/v1
kind: Deployment
metadata:
  name: my-nginx-app
spec:
  replicas: 3
  template:
    spec:
      containers:
        - name: my-nginx-container
          image: nginx:stable
```

Listing 3.26 overlay/prod/patch-prod.yaml

In *kustomization.yaml* in Listing 3.27, you will now find two patches with the respective selectors. Try outputting this manifest as well and check whether the changes are applied as you want.

If you are satisfied with the adjustments, you can roll out the cluster using the following command: `kubectl apply -k overlays/dev`.

```
resources:
  - ../../base
patches:
  - path: patch-prod.yaml
    target:
      group: apps
      version: v1
      kind: Deployment
      name: my-nginx
  - path: service-patch-prod.yaml
    target:
      version: v1
      kind: Service
      name: my-nginx-service
```

Listing 3.27 overlay/prod/kustomization.yaml

Congratulations! You have written your first manifests, which you can roll out to different environments using Kustomize patches. From my point of view, the syntax for the basic process is simple and easy to understand. In the following sections, we will look at a few more features that will make life easier for you.

3.5 Templating Using Kustomize

3.5.2 Resource Generator

Kustomize provides powerful generators that allow you to dynamically create resources such as ConfigMaps and secrets. These generators extract and process information directly from files, literal values, or other sources to automatically generate Kubernetes resources.

> **Good to Know**
>
> The generators are particularly useful because you do not need to maintain configuration files in a ConfigMap manifest, as you learned in Chapter 2, Section 2.4. This allows you to simply reference existing configuration files, which are then inserted into ConfigMap.

Suppose you have two configuration files that you want to include in your ConfigMap:

1. `app.properties`
   Configurations for your application
2. `logger.conf`
   A configuration file for logging your application

These files are located in the *config/dev* directory and are intended for your development environment. The structure should look as shown in Listing 3.28. The contents of the sample configurations can be found in Listing 3.29 and Listing 3.30.

```
.
├── config
│   └── dev
│       ├── app.properties
│       └── logger.conf
└── kustomization.yaml
```

Listing 3.28 Folder Structure of configMapGenerator

> **Note**
>
> The folder structure is not necessarily predefined by Kustomize; you have the freedom to design it as you see fit.

```
app.name=MyApp
app.version=1.0.0
app.environment=dev
```

Listing 3.29 app.properties

```
log.level=INFO
log.pattern=%d{yyyy-MM-dd HH:mm:ss} %-5level %logger{36} - %msg%n
log.directory=/var/log/myApp
```

Listing 3.30 logger.conf

In the root directory, you have the *kustomization.yaml* file for the generator, which references both files and also sets an additional TEST=true parameter, as shown in Listing 3.31.

```
configMapGenerator:
  - name: my-config
    files:
      - config/dev/app.properties
      - config/dev/logger.conf
    literals:
      - TEST=true
```

Listing 3.31 kustomization.yaml - configMapGenerator

If you now run Kustomize using `kubectl kustomize .`, your result should look like Listing 3.32.

```
apiVersion: v1
data:
  TEST: "true"
  app.properties: |
    app.name=MyApp
    app.version=1.0.0
    app.environment=dev
  logger.conf: |
    log.level=INFO
    log.pattern=%d{yyyy-MM-dd HH:mm:ss} %-5level %logger{36} - %msg%n
    log.directory=/var/log/myApp
kind: ConfigMap
metadata:
  name: my-config-9htd2ck66g
```

Listing 3.32 Generated ConfigMap

As you can see, the generator is very appealing and easy to use. With a generator, you can simply reference existing configuration files and save yourself double maintenance. In addition to configMapGenerator, there is also secretGenerator. You can find information on this topic at *http://s-prs.co/v596432*.

3.5.3 More Kustomize Built-Ins

The generators are referred to as *built-ins* for Kustomize. In addition to the generators, there are a few others that allow you to make in-depth customizations to your Kubernetes manifests. I now want to introduce a few selected ones here; these will help you standardize your resources and prevent unnecessary paperwork.

> **Note**
> You can find the entire selection of built-ins at *http://s-prs.co/v596433*.

One useful built-in is the *AnnotationTransformer* which you can use to define standard annotations that are attached to each resource by Kustomize. Imagine, for example, that you have the compliance guideline that certain annotations must be attached to the resources. Information about the responsible project or a cost center is necessary for some business processes, but maintaining the annotations on each resource is cumbersome.

Listing 3.33 shows how the AnnotationTransformer is activated. You only need to enter the required annotations under the `commonAnnotations` object in a *kustomization.yaml* file. Just try it out using the example from Listing 3.33.

```yaml
commonAnnotations:
  owner: kevinwelter
```

Listing 3.33 AnnotationTransformer

The generated ConfigMap should then contain the annotation, as in Listing 3.34.

```yaml
...
kind: ConfigMap
metadata:
  annotations:
    owner: kevinwelter
  name: my-config-9htd2ck66g
```

Listing 3.34 Generated Annotations

The same applies to labels, as you can see in Listing 3.35. Just like the AnnotationTransformer, the *LabelTransformer* adds the labels to each resource that's generated by Kustomize.

```yaml
commonLabels:
  owner: kevinwelter
  app: nginx
```

Listing 3.35 LabelTransformer

Another useful feature is the prefix and suffix for names. This is particularly useful if you roll out the same application more frequently, such as in a development cluster. For this, you only need the lines shown in Listing 3.36.

```
namePrefix: kevin-
nameSuffix: -dev
```

Listing 3.36 PrefixSuffixTransformer

3.5.4 Conclusion on Kustomize

You can see how easy it is to make adjustments with Kustomize. It is a wonderfully simple tool that can be used quickly thanks to its integration in `kubectl`. You have also seen that you don't need a lot of new knowledge to be able to carry out transformations or use built-ins, but the gain is huge. You save typing work and can use manifests for multiple environments and only change the fields that need to be changed by using the patches.

Note

Before you start using Kustomize in your company, I recommend that you read Chapter 8 on Helm. You should then decide which tool is best for you depending on the application.

Chapter 4
Advanced Objects and Concepts in Kubernetes

The only general principle that does not hinder progress is: anything goes.
—Paul Feyerabend

Kubernetes provides a rich set of objects and concepts that allow you to efficiently manage and scale complex applications. In Chapter 2, you got to know the basic principles of Kubernetes, which are essential for getting started and which you will use over and over again. In this chapter, we will look at concepts and objects that will allow you to delve even deeper into Kubernetes. This includes concepts such as custom resources that allow you to extend the Kubernetes API and thus open up a world in which anything is possible.

In the following sections, we will look at the following:

- **DaemonSets**
  Allow you to run a pod on any node in the cluster. Particularly useful for collecting logs, monitoring, or other services required at system level.

- **Kubernetes jobs**
  Provide an easy way to perform tasks such as batch jobs. Kubernetes jobs start, run a specific task, and end.

- **Custom resource definitions**
  Allow you to create custom resources that exist alongside the standard Kubernetes objects. This allows you to extend the Kubernetes API, which enables you to create customized solutions for specific use cases.

- **Downward API**
  Provides a way for you to inject metadata from the pod or cluster into the pod so that your application can access information without the Kubernetes API.

- **Pod priority and preemption**
  This concept enables you to assign priorities to the pods that influence scheduling.

4 Advanced Objects and Concepts in Kubernetes

Note

You should use this chapter as a reference guide. Some objects may not be important or interesting to you right now, but if you come across them in the future, you can return to this chapter.

4.1 DaemonSets

A *DaemonSet* ensures that exactly one instance of a pod is executed on each node in your cluster. Even if new nodes are added to the cluster, the DaemonSet will start the pods there. DaemonSets are particularly useful for system pods that have to be executed on each node, or applications for logging and monitoring. The *kube-proxy*, for example, is a system pod that runs on every node.

Examples and use cases include the following:

- **Log and data aggregation**
  A pod could collect logs and other data on each node, aggregate it, and send it to a central server.
- **Monitoring**
  Monitoring agents that collect system and application metrics from each node can be rolled out via a DaemonSet.
- **Network services**
  Like the *kube-proxy*, you can deploy pods that add functionality to your network.
- **Security scans**
  Applications such as Falco can be deployed on any node and perform security scans there.

Note

We will look at a simple example in this chapter to help you understand the principle. On Minikube, of course, you only have a single node, which makes it somewhat more difficult to observe the principle of DaemonSets.

Chapter 7, Section 7.4.2 contains a real use case with the Pi cluster, in which a DaemonSet is also used to collect metrics.

Let's start with the simple example in Listing 4.1. If you compare the manifest with the deployment manifest from Chapter 2, Listing 2.24, you will hardly see any differences. The syntax remains the same, which makes developing and reading the manifests easy.

```
apiVersion: apps/v1
kind: DaemonSet
```

```
metadata:
  name: my-nginx-Daemon
  labels:
    app: nginx
spec:
  selector:
    matchLabels:
      app: nginx
  template:
    metadata:
      labels:
        app: nginx
    spec:
      containers:
      - name: my-container
        image: nginx
        ports:
        - containerPort: 80
```

Listing 4.1 Nginx DaemonSet

Unroll the manifest and observe what happens. In Lens, you will find the generated DaemonSet under **Workloads • DaemonSets**. In Minikube, you can see that a pod is being created. On other cluster setups like the Pi cluster from Chapter 1, Section 1.7, you would see multiple pods.

When you deploy a DaemonSet, you should ask yourself the following question: On which nodes should a pod run?

Depending on the cluster setup, there will be taints on nodes that will prevent the pods from being deployed, typically on the master nodes. You should therefore think about tolerations, which you got to know in Chapter 2, Section 2.2.6. If you look at the example of the masters, then you need the tolerations from Listing 4.2 in a real cluster setup.

> **Good to Know**
>
> The DaemonSet controller automatically adds a few tolerations when it is created. Take a look at the pod created from the example in Listing 4.1.
>
> These tolerations are also useful, for example, to prevent *kube-proxy* from being evicted if the node's load is simply too high. Without *kube-proxy*, the node would lack functions, which in turn would result in other problems. DaemonSet pods are usually important for the node, and therefore DaemonSets must be treated differently.
>
> You can find a complete overview of the automatic tolerations at *http://s-prs.co/v596434*.

```
tolerations:
  - key: node-role.kubernetes.io/control-plane
    operator: Exists
    effect: NoSchedule
  - key: node-role.kubernetes.io/master
    operator: Exists
    effect: NoSchedule
```

Listing 4.2 Tolerations to Run on Master Nodes

Good to Know

As an alternative to the DaemonSet, you could also use *systemd* or something similar directly on the nodes. However, you will then lose the ability to manage the daemons using Kubernetes. With Kubernetes, you have a better overview of the pods, can access logs and monitoring metrics just like your applications, and you can also use YAML manifests to generate them.

As DaemonSets are usually system pods, special features exist for their communication as well. Typically, it is an application that is not used by users, as we have used it in the preceding example. Depending on the use case, you would use one of the following communication patterns:

- **Push**
  Your pods in the DaemonSet are set up to send updates to another service and are not reachable from the outside at all. An example could be a log collector that forwards the logs.

- **NodeIP and port**
  Your pods in the DaemonSet can use a host port. This makes them accessible via the IP addresses of the nodes and would allow another application to send requests to the pods—for example, to retrieve metrics.

- **Service**
  You can of course also use Kubernetes services as in Chapter 2, Section 2.5 and make the pods accessible from the outside. This is not unusual, but you must note that no specific pod can be addressed by load balancing.

- **DNS**
  An alternative would be the headless service, which takes over the service discovery. You can use it to query the DNS records of the pods.

Depending on the application, it is important to consider how the pod should be accessible.

> **Good to Know**
>
> Because the pods of the DaemonSet are more important for a node than other pods, it makes sense to set `PriorityClass` to a higher level. This way, you can make sure that the other pods are displaced and the DaemonSet pods remain on the node.
>
> We will take a closer look at the priority classes in Section 4.5.

4.2 Jobs in Kubernetes

Up to this point, you have learned about Kubernetes objects that ensure that pods run permanently. But sometimes there are tasks such as batch jobs that have a defined end, and this is where Kubernetes jobs come into play. The best way to imagine this is as follows:

A *job* in Kubernetes is like an external employee who is hired for a specific task in a company. That person has a clear project assignment and remains with the company until it has been successfully completed. The contract then ends, and the employee leaves the company. An internal employee, on the other hand, is with the company for the long term. The employee carries out work continuously and there is no defined end.

This means that jobs

- work on one-off, limited tasks;
- end after the completion of their task; and
- start pods until the task has been completed.

Deployments, on the other hand, must

- work on ongoing tasks;
- monitor pods and ensure their functionality; and
- scale as required.

Table 4.1 presents three job types that you can use. Each type has specific use cases. You could use the one-off job for database migration. The parallel jobs are useful for processing large amounts of data. For example, you could convert images or retrieve data from external services. The queue worker job is perfectly suited for processing a message queue.

Type	Description
One-off job	A single pod is started and works through a task. If it fails, a new one will be started until the pod completes successfully.

Table 4.1 Job Types in Kubernetes

Type	Description
Parallel jobs	Work continues until a defined number of pods have been successfully completed. You can run multiple pods in parallel.
Queue worker	One or more pods process a queue. If one of them completes successfully because the queue is empty, for example, then the job is completed successfully, and no new pod is started.

Table 4.1 Job Types in Kubernetes (Cont.)

4.2.1 Real-Life Kubernetes Jobs

Let's get into practice now so that you can create your first job. Listing 4.3 shows a simple job manifest. As you can see in the container specification, the job starts a busybox container, executes `sleep 3`, and then terminates.

```
apiVersion: batch/v1
kind: Job
metadata:
  name: my-job
spec:
  completions: 5
  parallelism: 2
  activeDeadlineSeconds: 60
  template:
    spec:
      containers:
      - name: sleep-container
        image: busybox
        command: ["/bin/sleep"]
        args: ["3"]
      restartPolicy: Never
```

Listing 4.3 Manifest of Kubernetes Job

You can see the new `completions` and `parallelism` options. These are the parameters that allow you to choose between the job types from Table 4.1:

- `completions` defines the number of containers that must be successfully completed for the job to be successful.
- `parallelism` defines how many pods are started simultaneously.

> **[+] Good to Know**
>
> `completions` and `parallelism` can be set as follows:

4.2 Jobs in Kubernetes

- You do not need to make any additional settings for the one-off job, as both values must be set to 1, which is the default.
- Use both options for parallel jobs. You decide how many pods should be successfully completed and how many may run simultaneously.
- For a queue worker job, you should leave the default value set for completions. This is because the job should terminate when the queue is empty. However, you can set the number of jobs that should run in parallel.

Use Lens to go to the **Workloads • Jobs** menu and create your first job using the manifest. Take a look at what Kubernetes does and how it behaves. A job named my-job should be started in the job overview. Because the manifest starts two pods at the same time, you should see in the table under **Completions** how first **2/5**, then **4/5**, and then **5/5** have been successfully completed, as shown in Figure 4.1. When you click the job, you will see in the information window that five pods have completed with the **Succeeded** status. You can also see in the events when the respective pod was started. If your pods were to output logs, you could view all logs from all pods in the top-right corner of the action bar with the **Log** icon. This will be important at a later stage.

If you go to the pod overview under **Workloads • Pods**, you will also see each of the five pods there. The job itself starts the pods as you have defined them in the manifest and monitors them in a way similar to a deployment. However, the job has a different goal when it comes to monitoring. As already mentioned, it should be possible to complete the pod in a Kubernetes job and terminate it after its task.

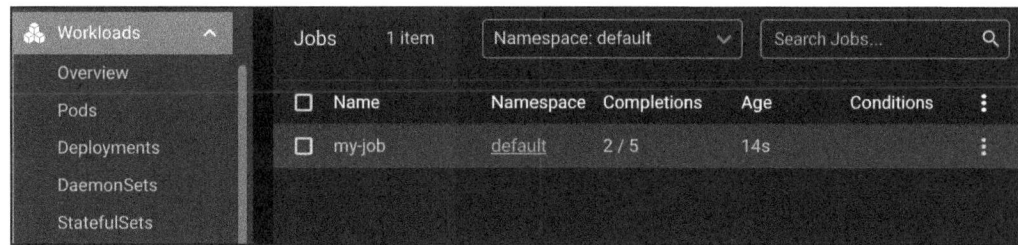

Figure 4.1 Job Overview in Lens

> **Note**
>
> Once your jobs have completed, they still remain in the overview, and the pods also remain visible. Not only does this disturb the overview, it also prevents you from importing the same job manifest again—but you can simply delete them by using Lens or kubectl. Information on how to automate the cleanup can be found at the following address: *http://s-prs.co/v596435*.

4.2.2 Queue Worker with RabbitMQ

Now let's move on to a real-life example. Let's set up RabbitMQ as a message queue in Kubernetes and publish and consume messages with Kubernetes jobs there. The architecture is deliberately kept simple. You can find an illustration of this in Figure 4.2.

For this example, we will

- deploy RabbitMQ from Bitnami as a Helm chart,
- set up a queue,
- create a container that can send and consume messages,
- create a job that sends messages, and
- create a job that consumes the messages.

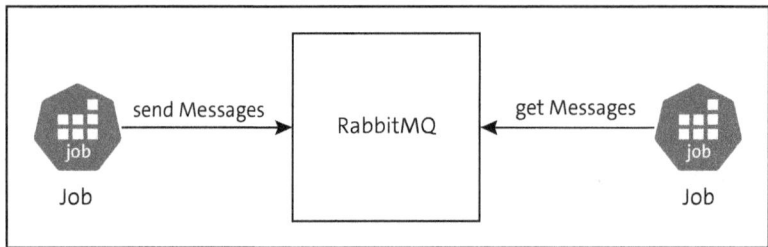

Figure 4.2 Filling and Reading RabbitMQ

Let's get started with the most important things. First, let's create a RabbitMQ instance based on the Bitnami Helm chart. Refer to Chapter 8, Section 8.1.2 to read in advance how Helm charts can be deployed via Lens. However, you do not need to have any prior experience with Helm for this exercise.

Look for the RabbitMQ chart in Lens under **Helm • Charts** (I use version 12.3.0) and install it. A StatefulSet is now created for you and a RabbitMQ pod is started. It may take a few minutes for the pod to boot up properly.

In the meantime, you can get the password for RabbitMQ from the release notes in Helm. You can also find more detailed instructions for this in Chapter 8, Section 8.1.2.

If the pod is green, go to **Network • Service**. There you will see that two services have been created. Use the "normal" service, *not* the headless service. If you look at the service, you will see that it provides multiple ports. The management UI is based on port 15672. For this purpose, you need to set up port forwarding, open the corresponding page in the browser, and log in with the data from the release notes.

I won't go into RabbitMQ in detail, as this is not necessary for the example. For the test, you'll simply create a queue that you can write to and read from.

To do this, click **Queues and Streams**. There you can create a new queue named test-queue under **Add a New Queue**. This should look similar to Figure 4.3. Once you have clicked **Add Queue**, the new queue appears directly in the overview. Then you can click the name of the queue to get more information. You will later be able to see the messages you write and read there.

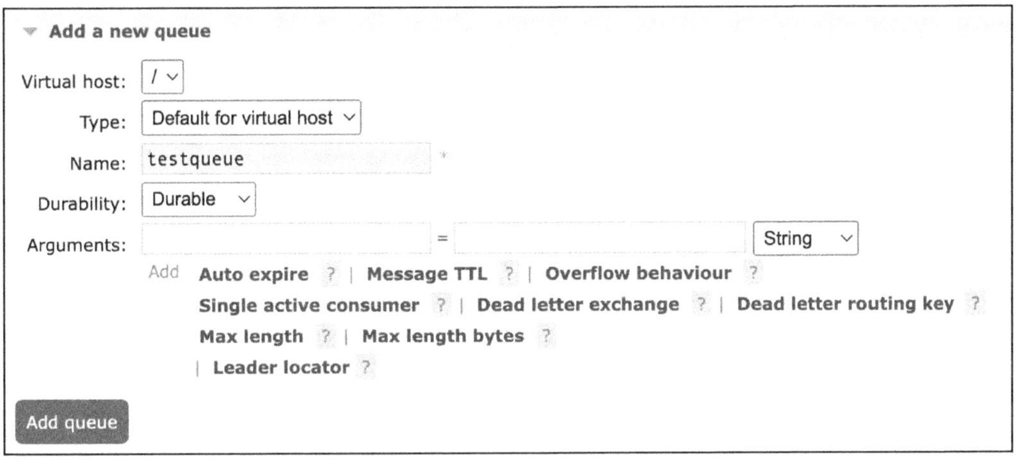

Figure 4.3 Creating Test Queue in RabbitMQ

To get messages in and out of the queue, examine Listing 4.4. There you will find a Bash script with two functions:

- The `publish` function writes a message with the content "Hello World!" to RabbitMQ.
- The `consume` function reads messages from the queue in a loop and outputs them until no more messages are available.

```bash
#!/bin/bash
RABBITMQ_HOST="rabbitmq-1698791687.default.svc.cluster.local"
RABBITMQ_PORT="15672"
RABBITMQ_USER="user"
RABBITMQ_PASSWORD="7DHPGoseHedmCUHu"
QUEUENAME="testqueue"
function publish {
    /usr/local/bin/rabbitmqadmin -H $RABBITMQ_HOST -P $RABBITMQ_PORT
    -u $RABBITMQ_USER -p $RABBITMQ_PASSWORD publish routing_key=$QUEUENAME
    payload="Hello World!"
}
function consume {
    while :
    do
        MESSAGE=$(/usr/local/bin/rabbitmqadmin -H $RABBITMQ_HOST -P
                $RABBITMQ_PORT -u $RABBITMQ_USER -p $RABBITMQ_PASSWORD
                get queue=$QUEUENAME ackmode=ack_requeue_false)
        if [[ $MESSAGE != "No items" ]]; then
            echo "Received message: $MESSAGE"
```

```
        else
            echo "No more messages in queue."
            break
        fi
    done
}
if [[ $1 == "publish" ]]; then
    publish
elif [[ $1 == "consume" ]]; then
    consume
else
    echo "Unknown command. Use 'publish' or 'consume'."
fi
```

Listing 4.4 Script for Consume and Publish Functions

The functions use parameters in lines 2 to 6, which you must replace with your values — namely, the password, the name of your created queue, and the host. The host contains the DNS name of the service. Remember the naming conventions from Chapter 2, Section 2.5.2, and use the DNS of your service.

Save the script as *rabbitmq-script.sh*, and then place it on the same level as the Dockerfile from Listing 4.5. To build the image, use the name `localhost:5000/rabbitmq-manager` to place it on the Minikube registry with `docker push`. If you encounter problems, refer back to Chapter 2, Section 2.1.2.

The Dockerfile in Listing 4.5 is kept simple. We use a special RabbitMQ tag as the base image, which is specifically designed for management. This image contains the `rabbitmqadmin` tool that you use in the script. The script is then copied into the container, made executable, and set as the entry point.

```
FROM rabbitmq:3-management
COPY rabbitmq-script.sh /rabbitmq-script.sh
RUN chmod +x /rabbitmq-script.sh
ENTRYPOINT ["/rabbitmq-script.sh"]
```

Listing 4.5 Dockerfile for Rabbit Management Container

Once you have pushed the container image to the registry, you can start the Kubernetes job that is to write messages to the queue. Use the manifest from Listing 4.6 for this purpose.

```
apiVersion: batch/v1
kind: Job
metadata:
  name: rabbitmq-publisher-job
```

```
spec:
  completions: 10
  template:
    spec:
      containers:
      - name: rabbitmq-manager
        image: localhost:5000/rabbitmq-manager
        args:
        - publish
      restartPolicy: OnFailure
```

Listing 4.6 Manifest for "Message Publisher" Job

As you can see, the job is based on the image and transfers `publish` as an argument. The container will therefore generate a message and then exit. This means that you have to increase the completions in order to write more messages, as this will generate more pods. In the example, 10 pods are started, so 10 messages are written.

Now create the job in Lens and watch as the job scales one pod at a time and writes messages to the queue.

In the RabbitMQ management interface, you can also see how the messages are written and ready to be picked up. In Figure 4.4, you can see the graphs created after the system has written and read out the messages. You should now have 10 messages set to **Ready**.

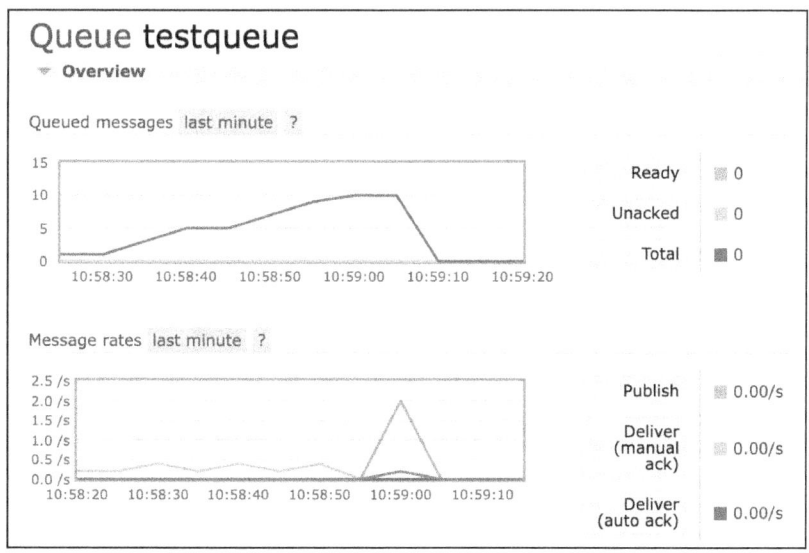

Figure 4.4 Movements on Message Graph of Queue

Now the news is ready to be consumed and processed. You will use a queue worker job to process the queue. The script is prepared accordingly, works in a loop until the queue is empty, and then terminates successfully. This is a mandatory requirement, because otherwise the job does not know when the queue is empty and cannot complete itself successfully.

Listing 4.7 shows the manifest for the queue worker job.

```
apiVersion: batch/v1
kind: Job
metadata:
  name: rabbitmq-consumer-job
spec:
  template:
    spec:
      containers:
      - name: rabbitmq-manager
        image: localhost:5000/rabbitmq-manager
        args:
        - consume
      restartPolicy: OnFailure
```

Listing 4.7 Manifest for "Message Consumer" Job

The script is given `consume` as an argument, and it leaves out completions so that it is set to the default value, 1. Now create the job via Lens and watch what happens here too. Kubernetes creates a pod that processes all messages and outputs the message in the logs. As you can see in Figure 4.4, in RabbitMQ the messages should drop to 0 again.

Congratulations! You have now created your first queue worker job and at the same time used a job to create messages. You should now use the existing manifests and play a little with the `completions` and `parallelism` values.

Try writing a total of 100 messages with two pods. Or run two queue workers that process the messages. The more you try it out, the clearer the function will become, and the better you will be able to call it when you need it.

Note

Here is a small task to complete this section: Can you find out what happens when the number of completions is reached but some pods are still working? Have fun experimenting!

4.2.3 Kubernetes CronJobs

The CronJob is perfectly suited for tasks such as the monthly billing run. It is clear that a task must be completed at a certain time, and similar to the cron jobs you know from Linux, you can use the cron format to set a time at which a Kubernetes job should be started. This means you do not need to create the job manually or via a pipeline but can leave everything to the Kubernetes CronJob, as shown in Figure 4.5. The CronJob creates a Kubernetes job at the relevant time using a template.

If you look at the manifest from Listing 4.8, you will recognize the job definition.

```
apiVersion: batch/v1
kind: CronJob
metadata:
  name: my-cronjob
spec:
  schedule: "*/5 * * * *"
  jobTemplate:
    spec:
      completions: 5
      parallelism: 2
      activeDeadlineSeconds: 60
      template:
        spec:
          containers:
          - name: sleep-container
            image: busybox
            command: ["/bin/sleep"]
            args: ["3"]
          restartPolicy: Never
```

Listing 4.8 Manifest of Kubernetes CronJob

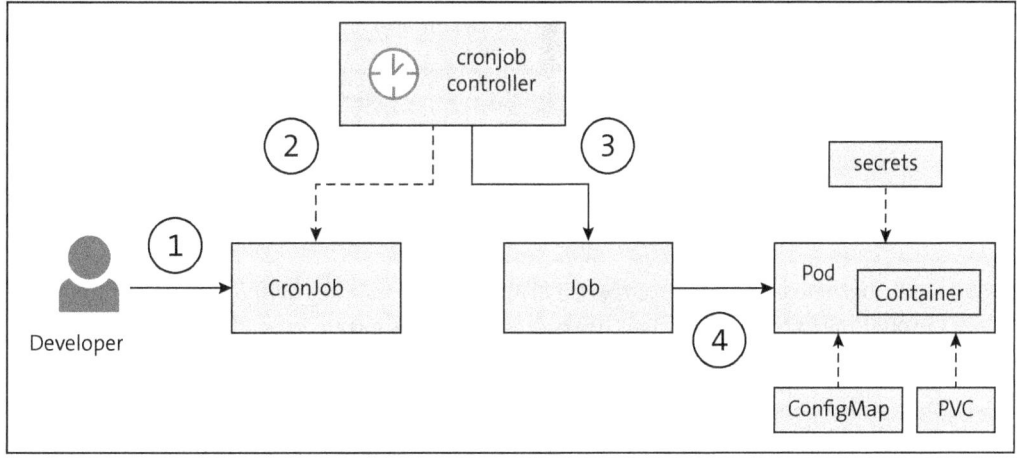

Figure 4.5 CronJob Process

You also have the schedule option, which enables you to transfer the cron expression. You can find your CronJobs in Lens under **Workloads** · **CronJobs**. If you then open the

CronJob menu by clicking on the particular CronJob, it looks like the one shown in Figure 4.6.

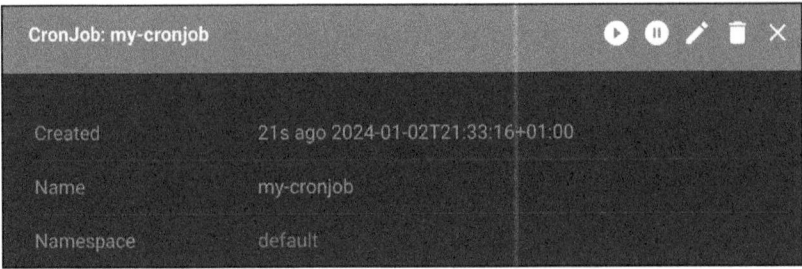

Figure 4.6 CronJob View in Lens

For CronJobs in particular, you have two additional options, which you can also see in the figure:

- You can pause CronJobs; no jobs will then be created until you unpause them.
- You can use the **Play** button to trigger a manual trigger that immediately creates a new job.

All jobs that are created by a CronJob can still be found under **Workloads • Jobs**. One advantage you have is that the CronJob names the jobs independently and you do not have to worry about name conflicts. Automatic cleanup would be advisable at this point.

> **Note**
>
> The manual **Trigger** button of the CronJob provides another advantage. You can use it to map processes that are to be controlled manually.
>
> If possible, you should avoid such manual processes, but I know that often you cannot do without them. Perhaps it will help you at some point.

4.3 Custom Resources and Custom Resource Definitions

In Kubernetes, *custom resources* (CRs) and *custom resource definitions* (CRDs) represent individual extensions of the Kubernetes API. These extensions allow you to modify the Kubernetes system according to your specific requirements.

Imagine Kubernetes as a conductor coordinating a large number of musical instruments (resources) in an orchestra (cluster). For some pieces of music, the normal instruments alone are not enough, and the orchestra needs to be enlarged. CRs are like additional, unique instruments that you add to your orchestra to create new melodies or achieve specific timbres. CRDs are like the construction and operating instructions for these instruments. You define what these new resources look like and how they should be played. Once defined, these CRDs can be used to create specific instances of the CRs—similar to a composer writing a new symphony that is then performed by the orchestra.

A CR is an elegant way to extend the Kubernetes API by defining your own resource types. It allows you to integrate your specific data structures and types into the Kubernetes system, have them validated by Kubernetes, and use them in one of your applications.

> **Good to Know**
>
> You could also map all the information you put in a CR with a ConfigMap, but then you would lose the following:
>
> - **Validation**
>   CRDs provide schema validation.
> - **Specific APIs**
>   For example, you use a CR with `kubectl` like a pod. ConfigMaps are "only" generic key-value stores.
> - **Controller integration**
>   You cannot monitor ConfigMaps with Kubernetes controllers and integrate them into automated processes.
>
> We will take a closer look at these individual topics in this section.

The CRD plays a central role in the creation of new resource types. Without a definition that describes what a CR should look like, there can be no CR. When you create a CRD, the Kubernetes API server responds by generating a new RESTful resource path for each version you specify. Once specified, the CRs behave like "regular" Kubernetes objects. You can query them using `kubectl`, for example.

4.3.1 Example: A Monitoring CR

The CR concept is sometimes a little complicated to understand—especially when there is no good use case available. Let's use an example to work out when a CR could make your life easier and how an application uses the CR.

Imagine you have developed an advanced application monitoring system specifically designed to provide detailed insight and analysis of the performance of applications within a Kubernetes cluster. Your aim now is to make this system accessible to other developers in the cluster so that they can create dashboards for their own applications. To make this possible, you want to introduce a custom resource (CR) that allows developers to define specific metrics and the application to be monitored.

4 Advanced Objects and Concepts in Kubernetes

Let's call this CR *AppMonitoringDashboard*. A manifest for it could look like the one shown in Listing 4.9. You give it the name of the application to be monitored, the desired metrics, and an update rate.

> **Note**
>
> You can find out how to view CRs and CRDs in Lens in the reference part of Chapter 1, Section 1.6.3.

```
apiVersion: monitoring.example.com/v1
kind: AppMonitoringDashboard
metadata:
  name: my-app-dashboard
spec:
  monitoredApplication: "MyApp"
  metrics:
    - "cpu_usage"
    - "memory_usage"
    - "disk_io"
  refreshRate: 60
```

Listing 4.9 "AppMonitoringDashboard" Custom Resource

But before you can create the CR, you first need a CRD. The CRD must describe the new API object. At the same time, it gives you the opportunity to validate the desired CR. The validation is necessary because you want to outsource the creation of dashboards to other developers, but your application can only create dashboards if all the important information is available.

In Listing 4.10, you can see the appropriate CRD for AppMonitoringDashboard. The structure of the CR you need can be found in the schema section. In the CRD, you also specify the names of the CRs in plural, singular, and short form. You can use the names later, such as via kubectl, to select the CRs. For example, you can reference the pod object by using the plural pods, singular pod, and short name po.

```
apiVersion: apiextensions.k8s.io/v1
kind: CustomResourceDefinition
metadata:
  name: appmonitoringdashboards.monitoring.example.com
spec:
  group: monitoring.example.com
  names:
    kind: AppMonitoringDashboard
    listKind: AppMonitoringDashboardList
```

```
  plural: appmonitoringdashboards
  singular: appmonitoringdashboard
scope: Namespaced
versions:
 - name: v1
   served: true
   storage: true
   schema:
     openAPIV3Schema:
       type: object
       properties:
         spec:
           type: object
           properties:
             monitoredApplication:
               type: string
             metrics:
               type: array
               items:
                 type: string
             refreshRate:
               type: integer
           required:
             - monitoredApplication
             - metrics
```

Listing 4.10 Custom Resource Definition for "AppMonitoringDashboard"

If you want to create your own CRD, you must make sure that the names match. Thus `metadata.name` is a composite of the plural and the group. And `names.kind` is the singular in camel case.

You will find a list of versions under the `versions` object. In the example, only v1 is available, but you could also create other versions. This allows you to develop the object further and still support older versions.

> **Note**
> More information on versioning Kubernetes objects is provided in Section 4.6.

You have now created the CRD and can create CRs as in Listing 4.9. However, the monitoring system that the CRs are supposed to use is still missing. As always, there are several possible options. To give you an idea, I will present one possible process:

1. **Monitoring the custom resources**
   You can activate a watcher in the Kubernetes API server on the CR. This watcher will inform you when changes are made or a CR is created.

2. **Extracting data from the CR**
   As soon as the monitoring app receives a notification about a new or changed CR, it extracts the relevant configuration data from the CR.

3. **Creating or updating the dashboard**
   The monitoring app then uses the extracted configuration data to create or update the corresponding dashboard.

4. **Integrating metrics data sources**
   The app configures a job to retrieve the metrics from the application.

Finally, I have provided an overview in Figure 4.7 so that you can see how the components interact.

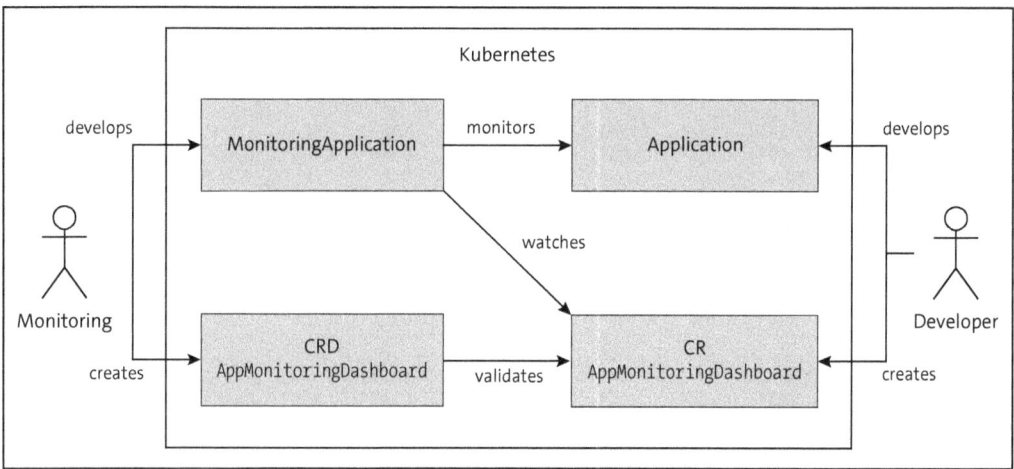

Figure 4.7 Overview of "AppMonitoringDashboard" CR and CRD

The developer responsible for monitoring develops the monitoring application and defines a CRD. The monitoring application monitors the CRs and responds to changes.

A developer who wants to create a dashboard for their application in the monitoring tool creates a CR with the necessary information. Your monitoring tool accesses the CR and will use the information from the CR to monitor the application and generate a dashboard.

4.3.2 Validation in CRD

When you create a CRD, you should have given it very clear thought and answered questions such as these:

4.3 Custom Resources and Custom Resource Definitions

- What is the object needed for?
- What should the object be able to do?
- What properties should the objects have?
- What data type does a particular property have?
- What are the thresholds a property should have?
- Where does the object's responsibility end?

You certainly don't want to create a jack-of-all-trades object, which is why it makes sense to also consider what the object is *not*. A good way to control this is to use rules for the properties that validate the values entered.

In addition to validating the data type—that is, whether an integer is actually an integer—a more detailed validation has also been possible since Kubernetes version 1.25. This allows you to set certain limits while you are still developing a CRD, which means that the developer who ultimately creates the CRs knows what is possible—and what is not.

In Listing 4.11, a simple extension of the data type validation is marked, which in this example provides an upper and lower limit for integers or defines an enumeration in our list. In the example, it protects against incorrect configurations of the metrics and against overloading due to excessively fast update cycles.

```
...
schema:
  openAPIV3Schema:
    type: object
    properties:
      spec:
        type: object
        properties:
          monitoredApplication:
            type: string
          metrics:
            type: array
            items:
              type: string
              enum: [ "cpu_usage", "memory_usage", "disk_io",
 "network_traffic", "http_requests" ]
            minItems: 1
          refreshRate:
            type: integer
            minimum: 30
            maximum: 3600
```

```
        required:
          - monitoredApplication
          - metrics
```

Listing 4.11 Validation in CRDs

> **[+] Good to Know**
>
> There are other validation options available, which you can find at the following address: *http://s-prs.co/v596436*.

Listing 4.12 contains a somewhat more complex validation. Common Expression Language (CEL) is used there to compare values with each other. This gives you even more freedom to carry out validations. You can find very detailed instructions for this at *http://s-prs.co/v596437*.

```
schema:
  openAPIV3Schema:
    type: object
    properties:
      spec:
        type: object
        properties:
          monitoredApplication:
            type: string
          metrics:
            type: array
            items:
              type: string
            x-kubernetes-validations:
              - rule: "size(self) > 0"
                message: "At least one metric must be specified."
          refreshRate:
            type: integer
            x-kubernetes-validations:
              - rule: "self >= 30 && self <= 3600"
                message: "must be between 30 and 3600 seconds."
        required:
          - monitoredApplication
          - metrics
```

Listing 4.12 Validation Using CEL

> **Note**
> During my tests, I noticed that Lens has some difficulties when it comes to deleting CRDs. For such problems, I always refer to `kubectl`. Using a simple `kubectl get crd` statement, you get your CRDs, and via a quick `kubectl delete crd [CRD-NAME]`, you can delete them.
>
> As useful as the UI of Lens is, sometimes you need the simplicity of the command line.

4.3.3 Operators

You certainly are familiar with this situation: There is a new configuration for your application, and now the application must be restarted. Or you are once again ordered by the operations team to roll out a new version of your database. Or you need to import a recovery because a problem has occurred in your database.

Even as a developer, you can't avoid taking care of repetitive tasks, especially those that require your expertise. Kubernetes can do a lot for you, but at some points even Kubernetes reaches its limits.

What if you had a little helper to support you in such tasks and make your life easier? Operators are just the thing.

In my opinion, the operator concept is one of the most important ideas in Kubernetes. Operators allow you to customize Kubernetes so that you can solve and automate any problem.

An *operator* is an extension of Kubernetes based on CRs. In the example from Section 4.3.1, the monitoring application listened to the CRs independently. If a CR is created or changed, the controller makes sure that what is specified will also be implemented. Operators can provide support for tasks such as the following:

- Rolling updates
- Updates to new versions
- Backup and recovery
- Monitoring and importing configuration changes

Where previously a person had to intervene to complete repetitive tasks, these can now be automated and simplified by an operator.

And moreover, this concept also allows you to swap out the watch functionality from your actual application, which makes it even more streamlined. You encapsulate the typical Kubernetes logic in the operator and do not need to keep it in your application.

Architecture of the Operator

An operator works by constantly monitoring the Kubernetes API for certain custom resources. As soon as a new custom resource is added or an existing one is changed, the operator will be activated. Its aim is to adjust the current state of the cluster so that it matches the desired state that you have defined in the custom resource. Figure 4.8 illustrates this process.

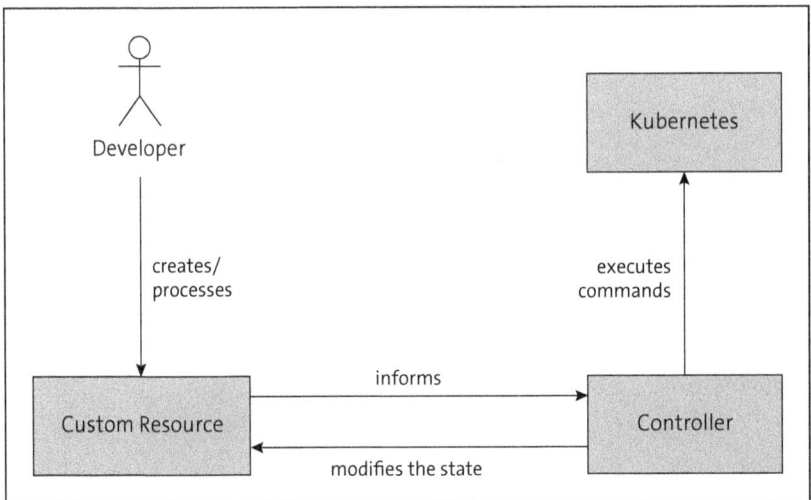

Figure 4.8 How Operators Work

Deployments and ReplicaSets work in a similar way to operators. A deployment operator monitors the definitions of deployments and, for example, adjusts the number of running pods to ensure that the desired number of replicas will be reached. Similarly, a ReplicaSet ensures that a specified number of replicas run continuously. You may recognize the pattern of the reconciliation loops we talked about in Chapter 3, Section 3.1.

If you develop your own operator, you can use it to extend the functions of Kubernetes. The operator then takes care of your own custom resource and keeps Kubernetes in the desired state.

The controller is the heart of every operator. It listens to one or more custom resources and can perform actions according to the definition in order to achieve or maintain the desired state. The controller uses the Kubernetes API to monitor the current state of the cluster, compares it with the desired state defined in the custom resource, and takes the necessary steps to resolve any discrepancies. You can imagine the procedure as follows:

- Monitoring
  The controller registers an event listener with the Kubernetes API in order to be informed about the creation, update, and deletion of custom resources.

- **Reconciliation**
  Each time a change is made, the controller retrieves the current state of the relevant resources, compares it with the desired state specified in the custom resource, and determines the actions required to correct any differences.

- **Adaptations**
  These actions can include creating, updating, or deleting Kubernetes resources to create the desired state. The controller can also call external services or adapt configurations.

> **Note**
> You can find out how to develop your own operators at the following address: *http://s-prs.co/v596438*.

Zalando's PostgreSQL Operator

Finally, I want to present a useful example that will give you an understanding of the operator. If you have ever dealt with databases, then you know that the following bullet points are important to consider in production:

- Data must not be lost.
- Data must be protected.
- Data must always be available.

This simply makes life more complicated, and a single database instance does not suffice. Remember Chapter 1, Section 1.1.4: A database is (of course) stateful, and from my point of view, operating a database in Kubernetes is the supreme discipline.

You will learn about the StatefulSet in Chapter 5, but the StatefulSet capabilities did not meet the expectations at Zalando, a company that was probably facing some operational challenges when it considered developing a PostgreSQL operator.

The operator allows you to easily deploy and manage Postgres clusters. In Listing 4.13, you can see an example of the Postgres manifest that is created as a CR. The CR has properties that the operator controller later uses to implement different things. In addition to the number of instances, you also specify the users and databases that are to be created. You can also simply pass the version of Postgres.

```
apiVersion: "acid.zalan.do/v1"
kind: postgresql
metadata:
  name: my-postgres-cluster
  namespace: default
```

```
spec:
  teamId: "myteam"
  volume:
    size: 10Gi
  numberOfInstances: 2
  users:
    admin:
      - superuser
      - createdb
    user: []
  databases:
    testdatenbank: admin
  postgresql:
    version: "13"
```

Listing 4.13 Postgres Manifest of Zalando Operator

In comparison, if you were to set up Postgres as a StatefulSet, you would have to write your own scripts to create the users, manage the rights, and take care of the database. The controller takes care of this and makes your work easier.

This is just a simple example of the Postgres manifest. The Zalando operator provides a wide range of functions, from storage extensions, backup, and recovery to a connection pool. It is definitely worth keeping an eye out for open-source operators for your use cases.

4.4 Downward API

In some cases, it is useful to access information from the pod or Kubernetes within your application. You can of course use the regular Kubernetes API to query data. However, there is another solution that provides data to your application without you having to access the regular API.

The *downward API* allows you to provide fields from pods and containers, similar to secrets and ConfigMaps. As you can see in Figure 4.9, you have two options for deployment:

- Environment parameters
- Volume

> **Good to Know**
> By making the information available via the downward API, you do not have to link your application to Kubernetes but can have it injected just like other configurations.

4.4 Downward API

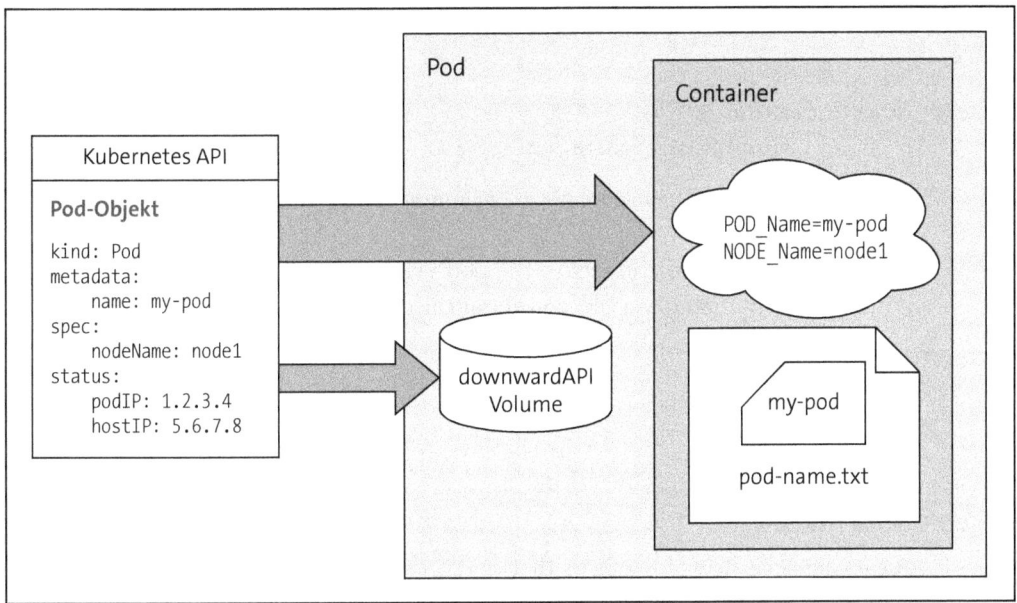

Figure 4.9 Integrating Downward API

Let's look at how you can make the fields available via environment parameters. Listing 4.14 gives an example of this, in which we set the name of the node, the pod, and the namespace as environment parameters.

```
apiVersion: v1
kind: Pod
metadata:
  name: my-nginx
  labels:
    app: nginx
spec:
  containers:
  - name: my-nginx
    image: nginx
    ports:
    - containerPort: 80
    env:
      - name: NODENAME
        valueFrom:
          fieldRef:
            fieldPath: spec.nodeName
      - name: PODNAME
        valueFrom:
          fieldRef:
```

4 Advanced Objects and Concepts in Kubernetes

```
            fieldPath: metadata.name
      - name: NAMESPACE
        valueFrom:
          fieldRef:
            fieldPath: metadata.namespace
```

Listing 4.14 Integrating Downward API as Environment Parameter

Unroll the manifest and connect to the pod via `kubectl exec`. If you now look at the environment parameters via `env`, you should also find the following entries:

```
NAMESPACE=default
PODNAME=my-nginx
NODENAME=minikube
```

Good to Know

A complete overview of the fields that you can make available via the downward API can be found at *http://s-prs.co/v596439*.

The second option is to integrate it as a volume. You may remember the syntax from the ConfigMaps and secrets in Chapter 2, Section 2.4. In Listing 4.15, you can see the manifest with the pod name and namespace. If you roll this out, you will find a separate file with the corresponding values as content for each entry under the */etc/kubeinfo* path.

Note

Not all information is available via integration as a volume. The name of the node is an example of this. For a complete overview, refer to the Kubernetes documentation.

```
apiVersion: v1
kind: Pod
metadata:
  name: my-nginx
  labels:
    app: nginx
spec:
  containers:
  - name: my-nginx
    image: nginx
    ports:
    - containerPort: 80
```

260

```
    volumeMounts:
        - name: kubeinfo
          mountPath: /etc/kubeinfo
  volumes:
    - name: kubeinfo
      downwardAPI:
        items:
          - path: "podname"
            fieldRef:
              fieldPath: metadata.name
          - path: "namespace"
            fieldRef:
              fieldPath: metadata.namespace
```

Listing 4.15 Integrating Downward API as Volume

4.5 Pod Priority and Preemption

There is another concept that allows you to control how the scheduler handles your pods and which ones it prefers. *Pod priority* and *preemption* allow you to prioritize critical workloads and thereby displace less important pods in favor of higher priority pods.

> **Good to Know**
>
> The displacement of low-priority pods is referred to as *preemption*.
>
> Imagine, for example, that you have an on-premise cluster that is currently running under full load so that a new node cannot be scaled up quickly. If you now have a job that absolutely has to go through, but there is no space on the nodes, then you have a problem. The scheduler cannot place the pod, and the pod remains in the pending status. However, you could use a higher priority to tell the scheduler to distribute other pods so that the job can start.

To assign priorities to your pods, you must use the `PriorityClass` object. Listing 4.16 shows an example with the name `high-prio`.

```
apiVersion: scheduling.k8s.io/v1
kind: PriorityClass
metadata:
  name: high-prio
```

```
value: 1000
globalDefault: false
description: "Use this Class for High Prio Pods"
```

Listing 4.16 PriorityClass Manifest

The higher the value of `PriorityClass`, the higher the priority of the pod. The Priority-Class objects can be found in Lens under **Config • Priority Classes**. After rolling out the example, you should see three priority classes, as shown in Figure 4.10.

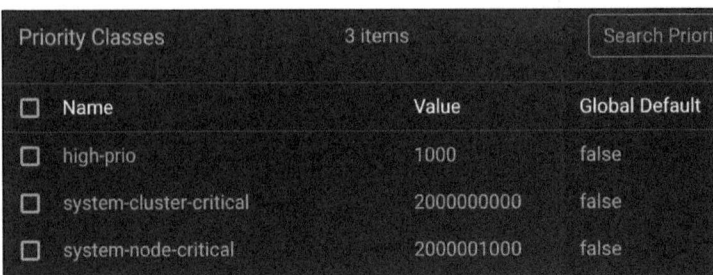

Figure 4.10 Priority Classes in Lens

In the pod manifest, you can specify `PriorityClass` via the name. An example of this is shown in Listing 4.17.

```
...
spec:
  containers:
  - name: nginx
    image: nginx
  priorityClassName: high-priority
```

Listing 4.17 Integrating PriorityClass into Pod

> **Good to Know**
>
> Since Kubernetes version 1.24, you can add a `preemptionPolicy` to the priority class. If you set the value to `never`, then the pods can have a higher priority and are only displaced later by even higher priorities, but they themselves cannot displace any other pods.
>
> This is useful if you have a pod that does not need to start immediately and therefore should not interfere with other pods. But when it runs, it should also finish its work. An example of this could be a long-running batch job.

4.6 Versioning Objects in Kubernetes

Kubernetes is also evolving, and the objects have different development cycles. You will always find these in the YAML manifests under `apiVersion`. Especially with regard to Kubernetes updates, it is important to check the API versions and update them if necessary.

>
> **Good to Know**
> All alpha and beta objects that are part of Kubernetes must be explicitly activated in `kube-apiserver` beforehand.

Kubernetes carries out versioning according to the following scheme:

- **Alpha level**
  This API version contains the designation `alpha`, as in `v1alpha1`, for example.

  The object is still in its infancy. There is no guarantee of further development or a regulated upgrade process. Its use is not recommended in production environments.

- **Beta level**
  This API version contains the designation `beta`, as in `v1beta2`, for example.

  The object is expected to become a stable object in one of the next versions. However, the definition of the object may still change. The object can also be tested with caution in a production environment.

- **Stable level**
  This API version is simply `v1`, `v2`, and so on.

  The object is fully developed, tested, and approved. The object is guaranteed to be supported in many further Kubernetes versions, and there are clear rules to ensure compatibility in further versions.

>
> **Good to Know**
> The introduction of API groups should make it easier to extend the API. The developers of Kubernetes had the following ideas:
>
> - The monolithic v1 API is organized in modular groups in order to be able to activate or deactivate entire groups. This is the cornerstone for being able to break down the monolithic API server into smaller components.
> - Groups can develop separately in the future.
> - Identically named types can be supported in different groups. Such a structure promotes innovation while ensuring stability.
> - The groups serve as the basis for extending the API with CRDs.

Despite the division into groups, it was important to the developers that interaction with tools such as `kubectl` remains simple. For example, you do not need to know the group name to display your pods.

Chapter 5
Stateful Applications and Storage

You can have data without information, but you cannot have information without data.
—Daniel Keys Moran

I can still remember how invaluable my first MP3 player with 512 MB was to me. I often had to decide which music I could keep on it and which I had to delete in order to listen to new songs again. Music streaming has completely changed our lives in that respect. Saving and backing up images has also changed. I used to make regular backups to my external hard disk. That was tedious, and I didn't want to pay for a NAS. Today, I have distributed my data in cloud storage. It's easier and cheaper than I could ever have imagined.

Storage is getting cheaper and cheaper. This sometimes means that we no longer think about what we are saving and whether it makes sense at all. The storage of data for our applications has changed as well. With modern systems that work according to the principle of software-defined storages, an upper limit is almost no longer visible. This is tempting, but data from which we do not extract information is nothing more than data waste that we should not allow to be created in the first place.

But in this chapter, I am not at all concerned with the meaningfulness of data. Instead, we are going to talk about how you can

- store data in Kubernetes in the best possible way, and
- operate applications that require and manage data.

You can't do without data, and you need applications to process it. However, the operation of a stateful application is always somewhat more complicated than a stateless application. With the StatefulSet, Kubernetes provides an object that is designed to run stateful applications. There are small but subtle differences from the other objects you have already seen, and we will take a closer look at these.

Kubernetes makes life very easy for us developers thanks to its abstractions, regardless of whether we are dealing with storage in our own data center or in the cloud. Due to abstraction, you can always allocate storage space to your application in the same way, regardless of what exactly it is. In Kubernetes, you can define a wide variety of storage classes, which are then mounted and used by the containers. In the end, the container does not care whether it is an Amazon EBS volume or an NFS volume in your own data center.

If the cluster is set up properly and the storage classes are defined, then it is easy for you to use the corresponding storage. This is a significant simplification of the development process, more freedom for developers, and a reduction in the workload for IT operations.

If you develop your application in a container, then sooner or later you will ask yourself questions such as these:

- Where do I store my data?
- What do I do if I lose data?
- How fast is the storage medium?
- What storage requirements does my application have?

This chapter is intended to get you thinking about these questions, and of course show how you can make storage available to your applications in Kubernetes.

If you have already worked with containers, then you are certainly already familiar with volumes. The concept of volumes also exists in Kubernetes. Essentially, *volumes* are directories that are accessible in one or more containers and simplify data management. The object abstracts the actual management of hard disks, and as a developer you (almost) no longer need to worry about the actual storage.

There are two main categories of volumes: ephemeral and persistent. While *ephemeral* volumes only exist during the lifetime of the associated pod and are suitable for temporary data, *persistent* volumes retain their state even after the pod has been terminated and are therefore ideal for long-term storage. With a web server such as Nginx, for example, you use an ephemeral volume to store cache data. For a database, on the other hand, you use a persistent volume to store the data securely even after a pod restart or failure.

5.1 Stateful Applications in Kubernetes through StatefulSets

Do you remember Chapter 1, Section 1.1.4, where we talked about the differences between stateful and stateless? The Kubernetes deployment from Chapter 2, Section 2.3 is intended for stateless applications.

But what do you do if you want to run a database in Kubernetes, for example? Let's take a look at the StatefulSet.

StatefulSets were specifically developed for the management of stateful applications. They differ from deployments in their ability to maintain and manage the state and identity of individual pods within the set. This makes them interesting not only for databases. You also can also use them if you

5.1 Stateful Applications in Kubernetes through StatefulSets

- have special requirements for the network and need stable identifiers, for example;
- require an orderly scaling;
- require an orderly rolling update process; or
- need stable and persistent storage.

Compared to deployment, the StatefulSet handles the pods more carefully. Order and stability are important to the StatefulSet, as this is the only way to update a database without losing data, for example. While deployments start or shut down new pods simultaneously and in no particular order, StatefulSets handle the deployment and scaling of pods in a strictly defined order and in a predictable manner.

> **Good to Know**
>
> StatefulSets work together with persistent volumes. The data is stored there for the runtime of an individual pod and can be mounted and used again by another pod. The structured naming of the pods makes it easier and more predictable to reuse the volumes.
>
> In Section 5.2, we will take a closer look at the topic of persistent volumes.

Each pod in a StatefulSet is given a unique and persistent identity. This identity is retained across pod restarts and even across recreations. The StatefulSet assigns the pods an ordinal index starting with the number 0. Contrary to a pod from a deployment, which has the structure *my-pod-hq3w8f*, the StatefulSet name has the following structure: *my-pod-0*.

The network name of the pods is also predictable and stable. For this purpose, the pod name and the index are used; in the example, that's *my-pod-0*. With a headless service, you can always refer to a specific pod in combination with the pod name.

Probably the most important difference from deployments is the behavior during rollouts and scaling. Pods in a StatefulSet are created and deleted in a strict, predictable order. This sequence is retained when scaling up or down. The first pod is not deleted until the last one has been successfully deployed, which is critical and very important in many stateful applications such as databases.

> **Good to Know**
>
> The pods in a StatefulSet are also not bound to a node. During an update or rescheduling, Kubernetes only needs to make sure that the corresponding pod also gets its volume. However, depending on the infrastructure on which the cluster is running and the volume type, the choice of nodes may be limited. For example, if you operate a cluster in AWS and use an EBS volume as a persistent volume, only nodes in the same availability zone can mount this EBS volume.

Let's start with a simple StatefulSet from Listing 5.1. Roll it out and watch the pods launch. You can observe the rollout behavior very well. The StatefulSet starts one pod after the other and waits until the pod is *ready*. In the example, this is only a few seconds, but even if an application takes minutes to initialize, the sequence will be maintained.

```
apiVersion: apps/v1
kind: StatefulSet
metadata:
  name: mysql
spec:
  serviceName: "mysql"
  replicas: 3
  selector:
    matchLabels:
      app: mysql
  template:
    metadata:
      labels:
        app: mysql
    spec:
      containers:
      - name: mysql
        image: mysql
        ports:
        - containerPort: 3306
        env:
        - name: MYSQL_ROOT_PASSWORD
          value: "secretPassword"
        volumeMounts:
        - name: mysql-persistent-storage
          mountPath: /var/lib/mysql
  volumeClaimTemplates:
    - metadata:
        name: mysql-persistent-storage
      spec:
        accessModes: ["ReadWriteOnce"]
        storageClassName: "standard"
        resources:
          requests:
            storage: 1Gi
```

Listing 5.1 Manifest for StatefulSet

To make sure you can observe the update behavior, you need to update a small thing in the template, such as the password, and roll it out again. You can see that the StatefulSet starts with the `mysql-2` pod, then updates `mysql-1`, and finally `mysql-0`. This sequence is always the same. The StatefulSet starts with the highest index and works its way to the lowest index.

> **Note**
>
> The StatefulSet relies on the readiness check to know when your application is ready. The readiness check reports the pod as ready as soon as your application is fully initialized. We will take a closer look at how to configure this check in Chapter 7, Section 7.2.

5.1.1 Pod Management Policy

As with the deployment, you also have options for the StatefulSet, which enables you to control the behavior. The pod management policy allows you to set the way the StatefulSet should handle the pods. You can configure this under `podManagementPolicy` in the manifest and choose from two options:

- `OrderedReady`

  With `OrderedReady`, your pods are managed in a strictly sequential order. This means that the next pod in the series is only started or stopped once the preceding pod in the StatefulSet has successfully transitioned to the ready state. You could already observe this process in the previous example as this is the default value.

  This policy is typically used for applications where the start sequence is important—for example, if a pod depends on the data or status of another pod. It is also useful in scenarios where a step-by-step initialization is required.

> **Good to Know**
>
> If you use `OrderedReady`, conditions may arise that require a manual intervention. If you update the pod template to a configuration that never reaches the running and ready state, the StatefulSet will stop the update and wait for your intervention.
>
> Unfortunately, it is currently not sufficient to simply update the template due to an open issue. You must then manually delete the pods the StatefulSet tried to update.
>
> You can find the open issue here: *http://s-prs.co/v596440*.

- `Parallel`

  With `Parallel`, you instruct the StatefulSet to start and stop pods in parallel. This means that, similar to deployments, no specific order is adhered to when the pods are managed. This option has no effect on the updates, which are still carried out sequentially.

The policy is suitable for applications where the start sequence of the pods is not critical. You can use it when a fast upscaling and downscaling is required, as all pods can be started or stopped at the same time.

> **Note**
>
> You cannot change the pod management policy at a later date. To activate the policy, you must delete the StatefulSet and create a new one. At the latest, you should be aware of this the moment you want to use a StatefulSets for production purposes.

Listing 5.2 show an extension of the StatefulSet manifest with `podManagementPolicy: "Parallel"`. Try out the change and observe the startup of the pods. You will see that all three pods are started at the same time. If you then check the update behavior again with another password change, you can continue to monitor the sequential updates.

```
apiVersion: apps/v1
kind: StatefulSet
metadata:
  name: mysql
spec:
  serviceName: "mysql"
  replicas: 3
  selector:
    matchLabels:
      app: mysql
  podManagementPolicy: "Parallel"
  template:
    ...
```

Listing 5.2 StatefulSet with Pod Management Policy

5.1.2 Strategies for Updates

You can also influence the way in which a StatefulSet handles updates. There are two update strategies you can choose from:

- The default option is `RollingUpdate`, which works in a similar way as deployment.
- The second option is `OnDelete`, where the StatefulSet does not automatically replace the existing pods during an update. You must then delete a pod from the StatefulSet yourself for an update so that the StatefulSet creates a new pod. This gives you much more control over the update process, but of course you have to do it yourself. Your aim should be to ensure that your application can survive an automated rolling update without any problems.

> **Note**
> Of course, your application must support rolling updates.

As you know from the rolling update of the deployment, the StatefulSet also replaces one pod after the other. This involves waiting until the new pod has fully booted up and the readiness check has been successfully completed. It is only then that the next pod will be replaced. However, there are small differences:

- Only one pod is updated at a time, and you cannot currently configure the number.
- The updates start with the pod that has the largest index and continue down to the smallest index.
- You have an additional `partition` option that allows you to define from which index the updates may be carried out.

> **Note**
> With Kubernetes version 1.24, the `maxUnavailable` option was introduced in an alpha stage. If you want to use it, you can activate it in the API server.

You can use the `partition` option to specify the index from which the updates should be carried out. The `partition` is an integer that specifies the starting point for the update within the StatefulSet. Pods in the StatefulSet with an index equal to or higher than `partition` are updated. Pods with a lower index remain unchanged even if you terminate the pod. This provides you with more granular control over the update process and more stability in your application.

Here is a small sample program. Suppose your StatefulSet has five pods, and you set the `partition` to 2 and update the container image to a new version. In this case, the pods with indexes 2, 3, and 4 will be updated. The first two pods with indexes 0 and 1 remain unchanged. This allows you to carry out the version update slowly and in a controlled manner. To update the other two pods, you want to set the `partition` back to 0 or delete the line from the manifest, as the default value is also 0.

Let's try this out to observe the behavior of the StatefulSet. To do this, you need to use the extension of the StatefulSet from Listing 5.3 and roll it out. As soon as all pods are initialized, you can make another small change to the template. You can observe how the StatefulSet sequentially updates the pods but leaves `mysql-0` unchanged. Thus, the last pod that is updated has the index that you specify in `partition`. Now you theoretically have time to test and check the new version of your application. If you also want to update the last pod to the new version, simply change `partition` to 0. The StatefulSet immediately starts replacing `mysql-0`.

5 Stateful Applications and Storage

```
apiVersion: apps/v1
kind: StatefulSet
metadata:
  name: mysql
spec:
  serviceName: "mysql"
  replicas: 3
  selector:
    matchLabels:
      app: mysql
  updateStrategy:
    type: RollingUpdate
    rollingUpdate:
      partition: 1
  template:
    ...
```

Listing 5.3 StatefulSet with Update Strategy

5.1.3 Retention Policy for Persistent Volume Claims

Kubernetes provides a new feature that allows you to control the retention of PVCs.

> **Note**
>
> Section 5.2 contains information about PVs and PVCs. However, I would like to mention a new feature that has been available in beta stage since Kubernetes v1.27. Keep in mind that the feature must be activated in the API server if you want to use it.

In the example from Listing 5.1, the pods also use PVs and PVCs. If you now scale down the StatefulSet from three to two pods or even delete it, the PVC will simply remain. If you scale it up again, the new pod will take over the PVC and thus the same PV. At first, this makes sense, as it allows the pod to continue using the same data.

However, there are situations where you may want to assign a fresh PV to a pod, such as in a test environment. For this purpose, you would currently have to delete the old PVC so that a new rollout also creates a new PVC.

Listing 5.4 shows the extension of StatefulSet with the retention policy.

```
apiVersion: apps/v1
kind: StatefulSet
metadata:
  name: mysql
spec:
  serviceName: "mysql"
```

```
replicas: 3
selector:
  matchLabels:
    app: mysql
persistentVolumeClaimRetentionPolicy:
  whenDeleted: Retain
  whenScaled: Delete
template:
...
```

Listing 5.4 StatefulSet with PVC Retention Policy

As you can see, you can define the policy for the following:

- whenDeleted
  Here you define what happens if the StatefulSet is deleted in its entirety.

- whenScaled
  This setting configures what the StatefulSet does to the PVC when you scale down the pods.

Good to Know

If a pod is replaced by the StatefulSet due to an error, the PVC will be retained, and the new pod can continue using it. It is therefore only deleted if you actively downscale or delete the StatefulSet.

You can select the following options for both cases:

- Retain
  This is the default value, and here the PVC is retained. This is how the StatefulSet would behave even without the new feature.

- Delete
  Deletes the PVC of the respective pod.

Note

If the PVC is deleted, the PV is not automatically deleted as well. However, the StatefulSet creates a new PVC and therefore a fresh PV. More on this follows in Section 5.2.

5.2 Persistent Volumes and Persistent Volume Claims

In a Kubernetes environment, persistent volumes and persistent volume claims are crucial when it comes to managing persistent storage. These two resources form the backbone for handling storage requirements in your cluster.

5 Stateful Applications and Storage

Imagine a PV as a highly flexible hard disk. The PV represents an abstraction of the actual storage medium. The data stored in the PV can be stored on an NFS, iSCSI, or a cloud hard disk such as in AWS EBS.

You can therefore think in advance about the underlying storage and how securely or quickly data needs to be accessed. In the end, however, your application does not care where the data is stored because access to the PV is always the same.

The good thing about a PV is that it outlasts the lifecycle of the pod, allowing you to store data permanently. There are of course a few pitfalls that I will go into, but if you're aware of them, then you will be able to store your data safely on PVs.

> **Good to Know**
>
> A PV does not necessarily have to be assigned to a StatefulSet. You can also use it in a deployment. However, a StatefulSet should store the data in a PV so that the data outlasts the lifecycle of the pod.

Listing 5.5 contains a simple PV manifest. As with every Kubernetes object, you define the name and labels in the metadata. In the actual manifest, you define the size of the storage, the way it can be accessed, and what type of storage it is.

```
apiVersion: v1
kind: PersistentVolume
metadata:
  name: test-pv
  labels:
    app: my-pv-app
spec:
  capacity:
    storage: 5Gi
  accessModes:
    - ReadWriteOnce
  hostPath:
    path: /tmp/test
```

Listing 5.5 Simple Manifest for Persistent Volume

> **Note**
>
> The security of your data depends on the underlying storage medium. It's like the data on your computer: if you don't have a backup and the hard disk is damaged, the data is also gone.

> In Listing 5.5, you can see a simple example with hostPath as the storage type, which places the data on a path in the file system of the worker. If this worker is terminated, the data will be lost.
>
> For this reason, you should think carefully about what data you want to store and what your availability requirements are.

The PV itself can only be requested via the PVC. An exclusive connection is created between the PV and PVC that lasts for the lifetime of the product. No other PVC can easily claim the PV. However, depending on the storage type and the access mode, multiple pods can access a PVC. We will take a closer look at this in Section 5.2.1.

In Figure 5.1, you can see a simplified illustration of how a PV and PVC interact. Assume here that a pod is supposed to use an AWS EBS volume. This is done as follows:

1. The pod uses a PVC to write to the desired volume.
2. The PVC incorporates a matching PV.
3. The PV takes over responsibility for the communication to the actual EBS volume.

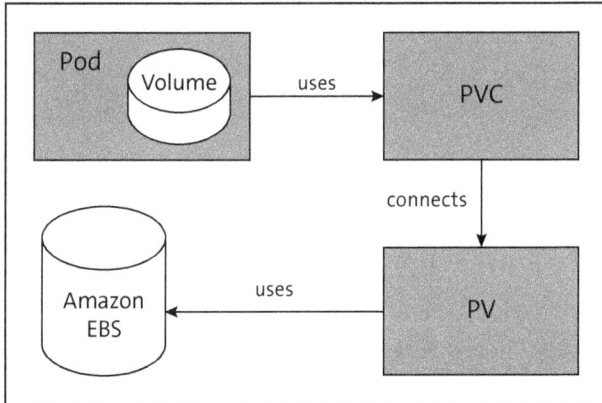

Figure 5.1 How a Pod Uses a PV

You can see the PVC as a list of requirements that a pod has for a PV. Listing 5.6 shows a simple manifest of a PVC. The requirements you describe in the PVC must be fully met by the PV; otherwise, the PV will not be claimed. The PVC waits until a suitable PV is available.

```
apiVersion: v1
kind: PersistentVolumeClaim
metadata:
  name: test-pvc
spec:
  storageClassName: ""
```

```
    accessModes:
      - ReadWriteOnce
    resources:
      requests:
        storage: 5Gi
```

Listing 5.6 Simple Persistent Volume Claim

If you compare the two manifests from Listing 5.5 and Listing 5.6, you will see that the PVC requests 5 Gi of storage space, that the same access mode is used, and that no specific storage class is required in the PVC. The configuration must be exactly the same; otherwise, the two objects will not find each other.

Just try it yourself. Copy the listings and use Lens to create a PV and then the PVC. Once you have created the PV, you can see your PV in the menu under **Storage • Persistent Volumes**, as in Figure 5.2.

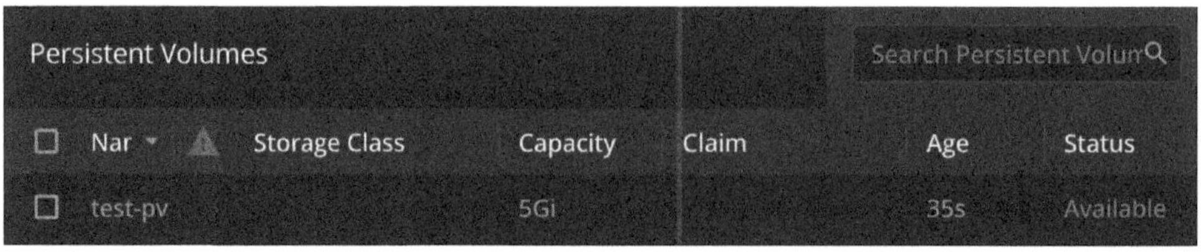

Persistent Volumes					
Nar	Storage Class	Capacity	Claim	Age	Status
test-pv		5Gi		35s	Available

Figure 5.2 New Static PV Created

The status is **Available**, which indicates that no PVC has claimed the volume yet. Next, when you create the PVC, Kubernetes will search for a matching PV and connect the two. The PV then has the status **Bound**, as in Figure 5.3.

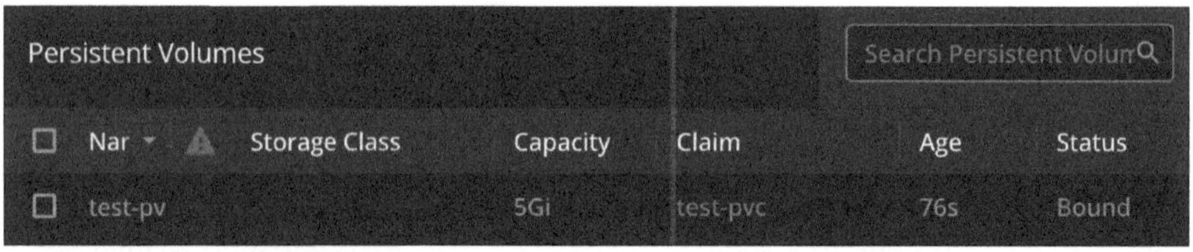

Persistent Volumes					
Nar	Storage Class	Capacity	Claim	Age	Status
test-pv		5Gi	test-pvc	76s	Bound

Figure 5.3 PV Claimed by PVC

The connection between the PV and PVC is fixed for the entire lifecycle. No other PVC may claim the volume just like that. However, if you delete the PVC, the PV will be retained, but its status changes to **Released**, as shown in Figure 5.4.

The advantage of this is that the data is not lost, and no other PVC can take over the PV and use the data. The disadvantage is that you have to intervene manually to release the PV again, back up the data, or delete it.

5.2 Persistent Volumes and Persistent Volume Claims

	Nar ▾	Storage Class	Capacity	Claim	Age	Status
	test-pv		5Gi	test-pvc	104s	Released

Figure 5.4 PV in "Released" status after PVC Has Been Deleted

Let's look at how you can release the PV again. Delete your created PVC and then click the PV that has the **Released** status. A context window opens on the right-hand side with information about the PV. There you can see under **Claim** which PVC is connected to the PV, as shown in Figure 5.5. You now want to terminate this connection, as the PVC no longer exists.

To do this, click the **Edit** (pencil) icon at the top to open the YAML manifest of the PV, and search there for claimRef, as shown in Listing 5.7.

```
spec:
  claimRef:
    kind: PersistentVolumeClaim
    namespace: default
    name: test-pvc
    uid: 180ef6fb-3703-40e3-a2ca-073f11ed89ec
    apiVersion: v1
    resourceVersion: '781131'
```

Listing 5.7 Claim Reference in YAML Manifest of PV

Here you can see in more detail which PVC is connected to the PV. Due to the unique uid, it is also not sufficient to recreate the PVC. Kubernetes recognizes that this is a new resource and will not connect the two (even if the PVC is virtually the same). Now delete claimRef from the manifest and click **Save**.

The PV will now return to the **Available** status and can be claimed again by a PVC.

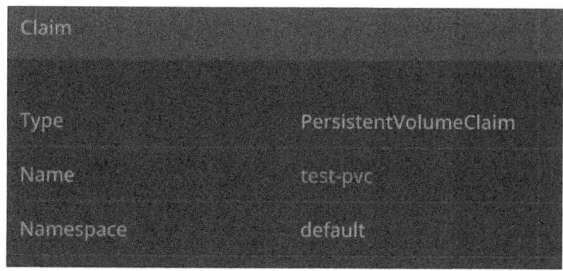

Figure 5.5 Claim Reference of PVC on PV

5.2.1 Storage Types for PVs

Kubernetes provides an abstraction to different storage media through the PVs. You learned about the hostPath in Listing 5.5. In this section, you will get to know some of the storage types Kubernetes supports out of the box. The good thing about this is that in the end, your application does not care where the data is located: regardless of whether the data is saved in a path in the host system or on a network drive, the application always saves in the same way.

Table 5.1 provides a list of the storage types Kubernetes supports out of the box in version v1.27. If you use earlier Kubernetes versions, you may also come across other types such as awsElasticBlockStore. However, these were gradually removed and switched to the CSI driver.

Type	Description
csi	The *container storage interface* supports a range of plug-ins that have been developed in accordance with the standard.
fc	The plug-in for fiber channel storage can implement storage solutions that are connected via fiber optics.
hostPath	The hostPath defines a path on the file system of a single node as storage. This approach is not recommended in production environments as there are security problems and availability is reduced.
iscsi	This plug-in is for storage that is connected via SCSI over IP.
local	The local plug-in is preferable over the hostPath. A local hard disk is mounted, so access is more secure than with hostPath. However, the availability is also reduced here because the PV is dependent on a node.
nfs	This plug-in is used to connect a network file system.

Table 5.1 Storage Plug-ins Supported by Kubernetes

One prerequisite for using a specific storage type for Kubernetes is, of course, that the corresponding storage is also connected. As a developer, you usually have little influence on which storage types can be used. However, it is important to know the options so that you can talk to the cluster admins, if necessary, and place your requirements. This enables you to understand how and where data can be stored and to communicate your storage requirements. As you can imagine, saving data on type nfs is slower than fc or scsi.

Good to Know

The clearer your application's requirements for the volume, the easier it is to select the right storage type.

5.2 Persistent Volumes and Persistent Volume Claims

Let's now take a closer look at some manifests. Listing 5.8 shows an example in which a network file system (NFS) is used. Under nfs in the manifest, you specify the IP through which the storage can be accessed and the path to which the data can be written. This is the path on the NFS. You can also specify NFS-specific mountOptions.

```
apiVersion: v1
kind: PersistentVolume
metadata:
  name: nfs-pv
spec:
  capacity:
    storage: 5Gi
  accessModes:
    - ReadWriteOnce
  mountOptions:
    - hard
    - nfsvers=4.1
  nfs:
    path: /nfs-pfad
    server: 192.168.0.5
```

Listing 5.8 Manifest of NFS Volume

I also want to show you the local volume in more detail, as you will be using simple storage solutions, especially for quick tests or during development. The local volume is preferable to hostPath. You can find a more detailed comparison in Table 5.2. You should think of the local volumes as being similar to a Docker volume that is managed by Kubernetes.

	hostPath	Local
Scheduling	The scheduler does not monitor the connection between pod and PV. The pod can be started on a different host when restarting and thus lose the data.	The scheduler takes the local PV into account and will always execute the pod on the corresponding node.
Security	The access to the host involves a certain risk as paths can be described that are not specifically intended for Kubernetes. In addition, it is difficult to control who else has access to the path.	This is a specially created and prepared volume with clear boundaries. The authorization is under the control of Kubernetes.

Table 5.2 Comparison of hostPath and Local Volumes

5 Stateful Applications and Storage

	hostPath	Local
Lifecycle	The lifecycle is linked to the node. This creates an additional risk as the `hostPath` can also be written to or deleted by other locations.	The lifecycle is linked to the lifecycle of the node.
Suitability for Production	Is only suitable for production systems in exceptional cases (e.g., a system pod within a DaemonSet).	Suitable for productive use with caution, as the data is bound to a node.

Table 5.2 Comparison of hostPath and Local Volumes (Cont.)

Listing 5.9 shows the manifest of a local volume. In addition to the path and `local`, `nodeAffinity` must also be defined. The affinity defines which node the PV is "attracted" to. In the example, we defined that the PV is only generated on a node with the host name, minikube. We explained and tested affinities in more detail in Chapter 2, Section 2.2.4.

```yaml
apiVersion: v1
kind: PersistentVolume
metadata:
  name: local-pv
spec:
  capacity:
    storage: 5Gi
  volumeMode: Filesystem
  accessModes:
  - ReadWriteOnce
  local:
    path: /tmp/test
  nodeAffinity:
    required:
      nodeSelectorTerms:
      - matchExpressions:
        - key: kubernetes.io/hostname
          operator: In
          values:
          - minikube
```

Listing 5.9 Manifest of Local Volume

> **[+] Good to Know**
>
> You have probably already noticed the `accessMode` setting in the listings. There are four access modes, but you cannot choose from all access modes for each storage, as these are dependent on the storage itself:

- `ReadWriteOnce`
  Enables reading and writing from a single node simultaneously. If you use `hostPath`, for example, you are forced to use this access mode. No other node can access the path. However, you can allow multiple pods on the same node to access the PV.
- `ReadOnlyMany`
  The PV can be mounted by several nodes with read-only access. This allows you to make data available to multiple pods, but they are not allowed to change it.
- `ReadWriteMany`
  If pods want to read and write data at the same time, they can use this access mode. One use case is a website to which customers can upload images that store your pods in an NFS.
- `ReadWriteOncePod`
  This is a beta feature in version 1.27. The access mode ensures that only one pod within the cluster can access the PV.

The documentation explains in full which storage you can use with which access mode: *http://s-prs.co/v596441*.

5.2.2 CSI Drivers for External Storage Media

Similar to the container engine, which you learned about in Chapter 2, Section 2.1.1, Kubernetes has become increasingly open in recent versions. The *container storage interface* (CSI) provides an interface that can be implemented by plug-ins. While you could still use types such as `awsElasticBlockStore`, `azureDisk`, or `gcePersistenDisk` directly in versions prior to Kubernetes v1.27, these were successively removed and reimplemented via their own CSI drivers. This makes Kubernetes more lightweight at its core.

The idea behind the CSI drivers is as simple as it is powerful: a wide variety of storage types should be supported, which must be flexibly connected. The answer to this is a standardized API that provides storage and cloud providers with an interface to provide storage for Kubernetes. As a developer, you then have the option of provisioning and using this storage simply by generating PVs and PVCs. This facilitates your work considerably because with one parameter you can decide whether the data should be stored on a single volume such as AWS EBS or in a distributed storage system such as AWS EFS.

Good to Know

Most drivers are also being further developed as open-source projects. You can find an overview of the official repositories at the following URL if you search for "Driver": *http://s-prs.co/v596442*.

5 Stateful Applications and Storage

The drivers are installed as operators in Kubernetes, which are then contacted by api-server as soon as a manifest uses the corresponding driver; after that, they provision the storage.

Listing 5.10 shows the configuration for `aws-ebs-csi-driver` under `csi.driver`. This parameter determines which storage solution is to be used.

```
apiVersion: v1
kind: PersistentVolume
metadata:
  name: test-pv
spec:
  accessModes:
  - ReadWriteOnce
  capacity:
    storage: 5Gi
  csi:
    driver: ebs.csi.aws.com
    fsType: ext4
    volumeHandle: {EBS volume ID}
  nodeAffinity:
    required:
      nodeSelectorTerms:
        - matchExpressions:
            - key: topology.ebs.csi.aws.com/zone
              operator: In
              values:
                - {availability zone}
```

Listing 5.10 Static Persistent Volume Created on Amazon EBS

> **[+] Good to Know**
>
> In production environments, the drivers are installed by cluster admins as the configurations are special in each case. However, I want to give you a better feel for the drivers and will make a brief digression here, using `aws-ebs-csi-driver` as an example.
>
> Your starting point is always the driver repository because you will find everything you need there: *http://s-prs.co/v596443*.
>
> In this case, the driver first needs certain AWS authorizations. Then the operator with Helm is installed on Kubernetes. (Helm charts are discussed in more detail in Chapter 8.) The Helm package contains everything the operator needs to recognize a configurationa as in Listing 5.10 and to provide the storage, including the operator containers as a deployment and the Kubernetes policies.

The most important thing is that you check which settings can be made for the driver prior to the installation. For example, you have the option of activating a snapshotter that allows you to make backups. These settings are also provided via Helm during installation.

In essence, that's all that needs to be done. The operator takes over the work and extends the Kubernetes API. You can use the documentation to see which specific services or configurations the driver provides and sets. Provisioning then works via the PV and PVC manifest, which the driver recognizes and executes.

5.2.3 Storage Classes and Dynamic PVs

You have now learned how to generate PVs independently. These types of PVs are also referred to as *static PVs* because you have to create and manage them yourself. However, Kubernetes can also create volumes dynamically on the basis of storage classes. As shown in Figure 5.6, this allows the system to automatically respond to requests from a PVC by providing the required PV, which means that a PVC does not wait until a suitable PV is available, but simply creates its own.

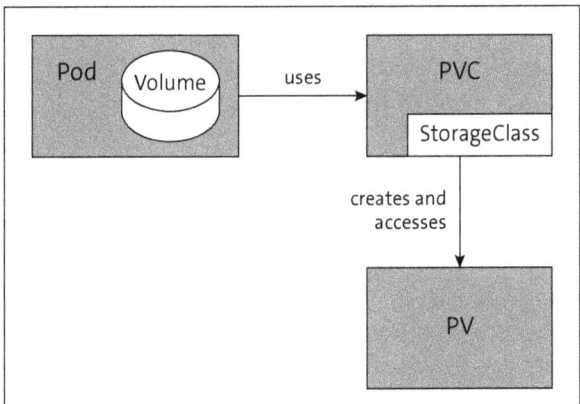

Figure 5.6 PVC Creates PV Based on StorageClass

Note

Not every storage type can be created dynamically. For example, you must create the local volume statically.

The storage classes provide an opportunity to bring more structure to the various storage offerings. You have already gotten to know some of them and know that they differ in terms of performance, availability, or manageability. In your corporate environment, you may even be given a handful of storage classes by the cluster admins as

5 Stateful Applications and Storage

these are directly dependent on the infrastructure. Clusters running on Azure require different CSI drivers than clusters in an on-premise environment.

Listing 5.11 gives an example of how `StorageClass` is defined.

```
apiVersion: storage.k8s.io/v1
kind: StorageClass
metadata:
  name: cloud-storage
provisioner: ebs.csi.aws.com
volumeBindingMode: WaitForFirstConsumer
allowVolumeExpansion: true
parameters:
  type: gp2
  encrypted: "true"
```

Listing 5.11 Example of StorageClass Manifest

In this example, Amazon EBS is used as the provisioner, as you already know from Listing 5.10.

There are two options to consider. First, by using `allowVolumeExpansion`, the storage class gives you the option of allowing the PVC to subsequently expand the memory. This is not supported by every storage type, but makes handling PVs much easier. You can simply add more storage if required and start with less at the beginning if it is still unclear how much storage is actually needed.

Another important option is `volumeBindingMode`. Table 5.3 compares the two options.

volumeBindingMode	Function	Usage
Immediate	The default mode, which ensures that a PV is generated and claimed directly when the PVC is created.	Used if a volume can be used by several nodes.
WaitForFirstConsumer	In this mode, the binding between the PVC and the PV is delayed until a pod uses the PVC as a storage request. This prevents the PV from being generated on a node that is not intended for the pod by the scheduler.	Particularly useful if the PV node is specific, as is the case with local volumes. Can be the solution if a pod does not start because the node on which the volume is running has no capacity.

Table 5.3 Comparison of Options for volumeBindingMode

5.2 Persistent Volumes and Persistent Volume Claims

> **Good to Know**
>
> In Listing 5.11, you will not see an explicitly set `reclaimPolicy`. It is important to note that the lifecycle of dynamically generated PVs corresponds to the PVC, and `reclaimPolicy: Delete` is the default. This means that the PV will be deleted as soon as the PVC is deleted.
>
> The opposite is true for static PVs. This may be useful and sufficient for some use cases, but you should make a conscious decision in favor of a reclaim policy.
>
> The reclaim policy provides an answer to the following question: What happens to the volume when the PVC's lifecycle ends?
>
> You can choose from these three options:
>
> - Retain
>   The PV remains and ends up in the released status. You can then restore the PV using manual steps.
> - Recycle
>   This is a deprecated function that empties the PV. Kubernetes recommends using dynamic PVs instead.
> - Delete
>   The PV is deleted and all data is lost.

The dynamic requirement of the PV is set by the PVC. In Listing 5.12, you can see the PVC used by the `StorageClass` from Listing 5.11. You are free to choose any name for it. For example, you could also assign the names according to speed or availability so that it is immediately clear which type of storage is to be used when the PVC is created.

```
apiVersion: v1
kind: PersistentVolumeClaim
metadata:
  name: ebs-pvc
spec:
  accessModes:
    - ReadWriteOnce
  storageClassName: cloud-storage
  resources:
    requests:
      storage: 20Gi
```

Listing 5.12 PVC that Dynamically Generates PV via StorageClass

Unfortunately, you cannot test the EBS storage class on Minikube, but Minikube itself comes with its own `hostPath` provisioner, which we will now try out to create dynamic PVs.

When you click **Storage · Storage Classes** in Lens, you will find a predefined class named `standard`. This storage class is set as the default, which also explains why you had to set the `storageClassName: ""` in Listing 5.6, as otherwise a PV is automatically generated. If you look at the details, you will also find all the settings you are familiar with from Listing 5.11.

For testing purposes, you can now adapt the code from Listing 5.6 and set `StorageClassName: standard`. The PVC generates a PV directly and is connected to it. Because the reclaim policy is set to `Delete`, you can now delete the PVC and the PV will also be deleted.

The dynamic creation of PVs on the basis of the storage classes is very simple and saves us from explicit specification of the PV. If you still have storage that can be expanded at runtime, then why and when should you use a static PV at all? There are a few reasons:

- As with local volumes, it is not always possible to generate the PV dynamically.
- As a cluster admin in particular, you have more control over the storage. This sometimes makes sense in production environments as the storage requirements are usually already clearly defined in production.
- By manually assigning the PVs to a storage, you have better options for optimizing performance.

Otherwise, you are well served with dynamic PVs, and these are more flexible, especially in development environments.

5.2.4 PostgreSQL as StatefulSet with Persistent Volume

Let's now run through a use case to put into practice what you have learned. In this context, you will

- deploy a PostgreSQL database as a StatefulSet, and then
- create a hostPath PV for the StatefulSet.

In Listing 5.13, you can see a StatefulSet manifest that already has everything you need. Try it yourself and roll it out on Minikube. I have marked the interesting parts for you.

```
apiVersion: apps/v1
kind: StatefulSet
metadata:
  name: pgsql
spec:
  serviceName: "pgsql"
  replicas: 1
  selector:
    matchLabels:
      app: pgsql
```

5.2 Persistent Volumes and Persistent Volume Claims

```yaml
  template:
    metadata:
      labels:
        app: pgsql
    spec:
      containers:
      - name: pgsql
        image: postgres:latest
        env:
        - name: POSTGRES_PASSWORD
          value: "examplepassword"
        ports:
        - containerPort: 5432
          name: pgsql
        volumeMounts:
        - name: pgsql-storage
          mountPath: /var/lib/postgresql/data
  volumeClaimTemplates:
  - metadata:
      name: pgsql-storage
    spec:
      accessModes: ["ReadWriteOnce"]
      storageClassName: "standard"
      resources:
        requests:
          storage: 10Gi
```

Listing 5.13 StatefulSet Manifest of PostgreSQL

The StatefulSet is generated, which starts a pod. This pod will create a PVC based on `volumeClaimTemplate`. This means that you do not have to create a PVC in order to transfer it to the pod. This has the advantage for you that you do not need to create a separate PVC for each new replica. The advantage of dynamic PVs also comes into play here. Due to the `standard StorageClass`, Kubernetes recognizes that the PV should be created via the `hostPath` provisioner of Minikube and executes this. So you don't have to do anything else: the PVC claims the PV and is mounted in the pod.

Under `volumeMounts`, the PVC is placed on a path within the pod, which is then used by the application.

Take a look at the resources created in Lens. There you will find the StatefulSet and the pod named `pgsql-0`, as shown in Figure 5.7. You can also see whether the PVC and PV have been generated.

5 Stateful Applications and Storage

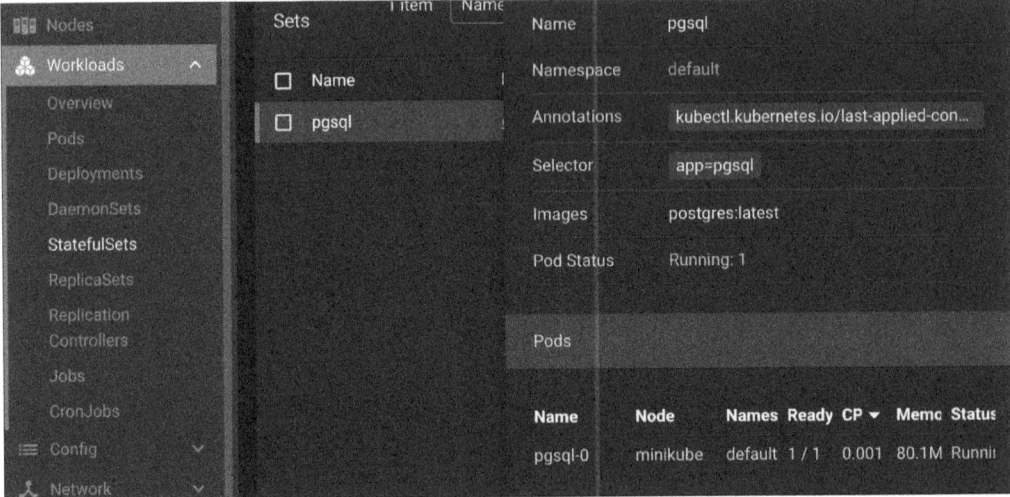

Figure 5.7 StatefulSet in Lens

Let's do a small test and see whether the data under the mount path is really stored on the PV:

1. Open the shell of the pgsql-0 pod.
2. Log onto PostgreSQL by using the psql -U postgres command.
3. Run the following commands to create a table and insert a test dataset:

    CREATE TABLE test (id SERIAL PRIMARY KEY, name VARCHAR(255));

    INSERT INTO test (name) VALUES ('data1'), ('data2');
4. Terminate the pod and wait until the StatefulSet creates a new pod. The pod will also be named pgsql-0.
5. Open the command line in this pod and log into PostgreSQL.
6. Use the SELECT * FROM test; command to check whether the dataset still exists.

The table and dataset are also available in the new pod. This means that the data is stored correctly on the PV.

> **Note**
>
> Up to this point, we have only mounted PVs as a file system in applications. But Kubernetes offers yet another option that you might come across.
>
> The volumeMode option lets you choose between two modes. The default mode is Filesystem. Kubernetes mounts the volume and will create a file system the first time. So you don't have to worry about the file system, and your application can simply read and write data.
>
> You can also set the volumeMode to Block. This mounts the volume as a block device and does not automatically install a file system. However, the pod must be able to handle a

> block device. Access to the volume should then be faster than with an additional file system layer.
>
> In practice, I have only come across the file system option so far, and this option will also be sufficient for most applications in your case.

You have successfully used a PV as storage for a database. In a production environment, it would be appropriate to turn one PostgreSQL replica into several, depending on the availability requirements. Of course, you can set your StatefulSet to deploy multiple pods, but the data between the instances will not be replicated. Depending on the application, this may require a little more development work, but it is worth it in the end.

Finally, the following question may arise: Is a PV necessarily dependent on a StatefulSet?

The answer is no. You can of course also let your pods that are managed by a deployment use PVs. This can sometimes be useful if, for example, you are building a cache that should still be available for the next pod when the pod is scheduled. But be careful! Always keep the concepts from Chapter 1, Section 1.1.4 in mind, and then consciously decide on a solution that suits you and your company.

5.3 Ephemeral Volumes

Ephemeral volumes are temporary and exist only as long as the pod that creates the volume exists. They can be used for data that does not need to persist beyond the life of the pod, such as cache data or session information. This is ideal for temporary workloads, test environments, and stateless applications.

The advantages of ephemeral volumes are as follows:

- **Performance**
  They can usually offer higher I/O performance rates as they are often stored directly on the local node and have no network latency.

- **Easier management**
  They are easier to manage because they are automatically created and deleted with the pod.

- **Lower costs**
  As the data does not need to be stored persistently, ephemeral volumes do not require complex storage solutions such as network storage, cloud storage, or backups. This significantly reduces costs and complexity.

Table 5.4 lists the types of ephemeral volumes. In general, you can keep in mind the following: if the data is no longer needed after exiting the pod, you should use an ephemeral volume.

Type	Description
emptyDir	An empty folder that is mounted at pod startup. The kubelet creates the storage locally on the root disk or even in the RAM.
configMap, secret, downwardAPI	You have already become familiar with these types in Chapter 2, Section 2.4 and Chapter 4, Section 4.4. You can use them to inject configurations as files in a volume.
CSI ephemeral volumes	This allows you to make a volume available via a CSI driver. Works in a similar way to persistent volumes.
Generic ephemeral volumes	Works like emptyDir volumes, but provides even more options, such as volumes on a network storage.

Table 5.4 Types of Ephemeral Volumes

> **Note**
> PVs of the hostPath type are of course also located directly on the node. This means that their performance hardly makes a difference compared to an ephemeral volume.

Let's look at a few examples. You have already used emptyDir in Chapter 2, Section 2.1.3 in the log collector example. In Listing 5.14, you will find the syntax you can use to create a simple emptyDir. Kubernetes will then create your volume in the default storage.

```
volumes:
- name: empty-dir
  emptyDir: {}
```

Listing 5.14 Simple emptyDir

An emptyDir provides two options. You can limit the size, and you can create the volume directly in the RAM. The latter option is particularly appealing because access to it is significantly faster than to a normal hard disk. However, you must keep in mind that the data stored on it is also limited by the pod's resource limit, which you will learn about in Chapter 7, Section 7.1. An example of this is shown in Listing 5.15.

```
volumes:
- name: empty-dir
  emptyDir:
    sizeLimit: 500Mi
    medium: Memory
```

Listing 5.15 emptyDir in RAM

As for CSI ephemeral volumes, you also need a CSI driver for generic ephemeral volumes, which will provision the volume for you. They function in a way that's similar to an `emptyDir` volume and are linked to the lifecycle of a pod. Depending on the driver, however, they have more options:

- You can also create the volume on a NAS.
- Depending on the driver, volumes may contain initial data. This is useful, for example, if your application is supposed to start with a standard set of data.
- You can set functions such as snapshots, resizing, or monitoring the storage space.

However, you are dependent on a driver that supports these functions.

> **Note**
> Aside from `emptyDir`, the other ephemeral volumes are difficult to reproduce on Minikube. However, you are already well equipped with `emptyDir` volumes; if you need more, you should talk to your cluster admins in advance anyway. Perhaps they already offer corresponding storage classes.

5.4 Other Features of Volumes

In this section, we'll introduce you to two features that are worth mentioning and that you might find useful in the future:

- Volume snapshots
- Projected volumes

Volume snapshots are particularly interesting if you have not yet developed a backup strategy for persistent volumes in your company. They can provide a simple variant that simplifies backup and recovery for you.

Projected volumes are useful if you want to use a large number of different configurations in your pod. This concept allows you to group the configurations all together and mount them under a single path.

5.4.1 Volume Snapshots

You can add a snapshot to volumes that are created using the CSI driver. In Kubernetes, a *volume snapshot* is a snapshot of the status of a storage volume; you may already be familiar with this principle from other services. AWS also offers snapshots for your EBS volumes.

Typically, you can use a snapshot much like a backup for a volume, or you can use it to copy an existing volume to use with another pod.

Use cases for snapshots include

- general backups for disaster recovery,
- backups made before updating the database, and
- creating a copy to perform an error analysis on a system other than the live system.

> **Note**
>
> Before you use volume snapshots, you should work out a backup strategy. At one of my clients, for example, we use the AWS backup, which takes automated EBS snapshots. For this reason, the company does not need additional volume snapshots.

> **Backup Is Easy: Recovery Is Tough!**
>
> You will only know whether a backup works when you want to restore it. So if you want to use volume snapshots as a backup strategy, you should test the procedure extensively.

Let's first look at the three objects you should know about:

- `VolumeSnapshotContent`
  This object represents the snapshot that was taken of a volume in the cluster. It contains the data copy of a volume at a specific point in time and serves as a basic component for data backup.
- `VolumeSnapshot`
  This object is a user's request for a snapshot. It is similar to the principle of a persistent volume claim. You can use `VolumeSnapshot` to initiate the snapshot process.
- `VolumeSnapshotClass`
  Similar to the `StorageClass`, you use this object to define the various attributes that belong to a volume snapshot.

> **Good to Know**
>
> `VolumeSnapshot`, `VolumeSnapshotContent`, and `VolumeSnapshotClass` are CRDs and not part of the core API.

5.4 Other Features of Volumes

As is usual with CRDs, there is an operator who takes care of the snapshots and listens to the CRDs (see Chapter 4, Section 4.3.3). You may need to activate the volume snapshots in your cluster before you can use them. We can simply use add-ons for the Minikube cluster. To do this, run the following commands in your command line:

```
minikube addons enable volumesnapshots
minikube addons enable csi-hostpath-driver
```

> **Note**
>
> Activating the `csi-hostpath-driver` add-on caused problems on my side. As is so often the case, stopping and restarting the Minikube cluster helped.

Activating the add-ons automatically creates a new storage class named `csi-hostpath-sc` for you, which you can find in Lens under **Storage · Storage Classes**. The three corresponding CRDs for the volume snapshot are also created and can be found under **Custom Resources · Definitions** and should look as shown in Figure 5.8.

If you click **VolumeSnapshotClass**, you will see that a `SnapshotClass` named `csi-hostpath-snapclass` has also been created.

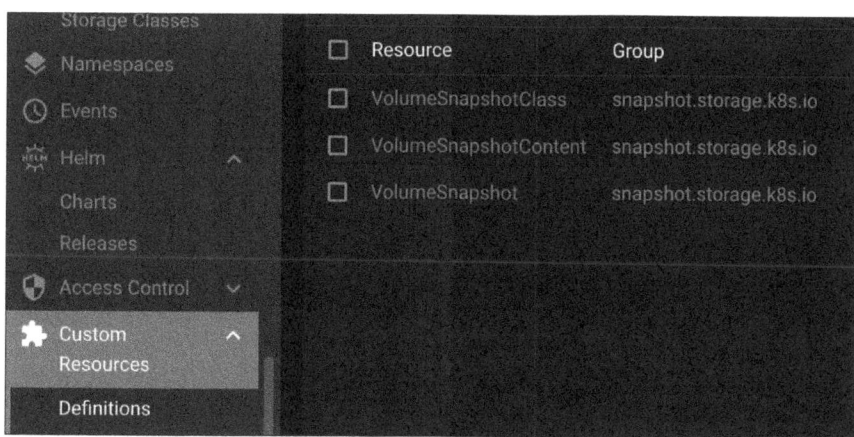

Figure 5.8 Volume Snapshot CRDs in Lens

For the example, we now need a volume that is based on the new storage class. I have prepared a PVC manifest in Listing 5.16. When you roll it out, a PV will be created automatically.

```
apiVersion: v1
kind: PersistentVolumeClaim
metadata:
  name: my-pvc
```

```
spec:
  accessModes:
  - ReadWriteOnce
  resources:
    requests:
      storage: 1Gi
  storageClassName: csi-hostpath-sc
```

Listing 5.16 PVC with CSI Storage Class

Listing 5.17 shows an example of a `VolumeSnapshot` manifest. When you import it, the snapshot controller will take a snapshot of your volume. Try it out and then take a look at the `VolumeSnapshotContent` objects. You can now find a fresh snapshot there, as shown in Figure 5.9.

```
apiVersion: snapshot.storage.k8s.io/v1
kind: VolumeSnapshot
metadata:
  name: my-pv-snapshot
spec:
  volumeSnapshotClassName: csi-hostpath-snapclass
  source:
    persistentVolumeClaimName: my-pvc
```

Listing 5.17 VolumeSnapshot Manifest

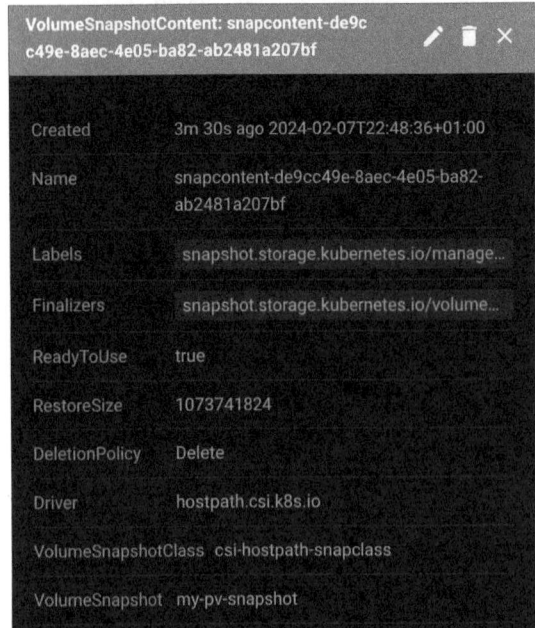

Figure 5.9 SnapshotVolumeContent

Now, of course, we also should try a restore. To do this, you must create a new PVC and link it to the backup, as shown in Listing 5.18. When you roll it out and observe the PVC, you will see that the PV generation takes a little longer than usual. The snapshot controller must restore the backup in the background.

```
apiVersion: v1
kind: PersistentVolumeClaim
metadata:
  name: my-pvc-restore
spec:
  storageClassName: csi-hostpath-sc
  dataSource:
    name: my-pv-snapshot
    kind: VolumeSnapshot
    apiGroup: snapshot.storage.k8s.io
  accessModes:
    - ReadWriteOnce
  resources:
    requests:
      storage: 1Gi
```

Listing 5.18 Restoring PVC from Backup

That was the snapshot and restore process in fast-forward mode. If you want to use the snapshot function in a cluster in your company, you must install the CRDs. You can find instructions on how to do this in the corresponding GitHub repo at *http://s-prs.co/v596444*.

> **Note**
>
> In this example, we have not even checked whether all the data has actually been restored. You can perform another test according to the following scheme:
> 1. Mount the PVC in a pod.
> 2. Place a file on the PV.
> 3. Perform the snapshot and restore processes.
> 4. Mount the new PVC in a pod.
> 5. Check whether the file is available.

5.4.2 Projected Volumes

Kubernetes provides *projected volumes* for certain volume types. These are used to combine multiple existing volume sources into a single shared volume, which is particularly useful if an application needs to access different types of configurable information, but you want to manage this information centrally. Your YAML manifest will also be a little smaller and clearer.

Supported types are as follows:

- `secrets`
- `ConfigMaps`
- `downwardAPI`
- `serviceAccountToken`

If you remember Chapter 2, Section 2.4, then you know that you can integrate ConfigMaps and secrets as volumes. A separate file is then created for each parameter under the mount path. However, as in Listing 5.19, you must perform a separate mount for each ConfigMap.

```yaml
...
spec:
  containers:
    - name: example-container
      image: nginx
      volumeMounts:
      - name: config-volume
        mountPath: /etc/config
      - name: config-volume2
        mountPath: /etc/config2
  volumes:
    - name: config-volume
      configMap:
        name: example-configmap
        items:
        - key: "config.json"
          path: "config.json"
    - name: config-volume2
      configMap:
        name: example-configmap2
        items:
        - key: "config2.json"
          path: "config2.json"
```

Listing 5.19 ConfigMaps as Normal Volumes

The folder structure then looks as follows:

```
etc
├── config
│   └── config.json
└── config2
    └── config2.json
```

The idea behind projected volumes is to bring together all types that contain configuration information so that you can mount them under a single path. Listing 5.20 shows the syntax for the projected volume. You only mount one volume and add all Config-Maps below it.

```
...
spec:
  containers:
    - name: example-container
      image: nginx
      volumeMounts:
        - name: projected-volume
          mountPath: /etc/config
  volumes:
  - name: projected-volume
    projected:
      sources:
      - configMap:
          name: example-configmap
          items:
          - key: "config.json"
            path: "config.json"
      - configMap:
          name: example-configmap2
          items:
          - key: "config2.json"
            path: "config2.json"
```

Listing 5.20 ConfigMaps in Projected Volume

After that, the folder structure in your pod will look as follows:

```
etc
└── config
    ├── config.json
    └── config2.json
```

Just try it out for yourself; use the ConfigMap and secret examples from Chapter 2, Section 2.4. The projected volume makes it easier to find the configurations, especially for your application, because there is only one folder for them.

Chapter 6
Kubernetes Governance and Security: Prepare for Production

The greatest enemy of knowledge is not ignorance, but the illusion of being knowledgeable.
—Daniel J. Boorstin

IT is characterized by constant change. Complex technologies such as Kubernetes, continuous innovation, and ever-increasing abstraction mean that it is becoming increasingly difficult to recognize any danger in advance and make appropriate preparations—which, in IT, are known as *mitigations*.

For software to be secure, many small gears have to mesh together. From your development projects, you are no doubt familiar with tools that perform vulnerability scans to alert you to weaknesses in software libraries. Or perhaps you are already using SonarQube for static code analysis, which can point out certain problems in your code. In addition, there are areas such as network security, authentication, and—if we are really precise—security in data centers. However, you don't just want to protect yourself from hackers who deliberately want to damage your company; you also want to protect your software or data from internal errors.

Assuming that a Kubernetes cluster is inherently secure or that the existing security measures are sufficient can be dangerous. I know it's hard to deal with the security concepts of Kubernetes, networks, and servers in addition to developing software. Nevertheless, a basic understanding is essential.

I first became really aware of the topic of container security in a customer project when we were presented with a catalog of specifications. It contained easy-to-apply guidelines, such as carrying out vulnerability scans of the container images or using standardized images. Then it got more complicated with rules such as the following:

- Containers must not run as root.
- Containers must not write to the root file system.
- Containers must not have any privileges.

These requirements meant that many applications had to be redesigned and rebuilt because no one had thought of these possibilities beforehand and therefore no one used them.

In this chapter, I want to introduce security and governance topics so that you are prepared for the use of Kubernetes in production and can prepare your applications for this at an early stage. I want to make the start as easy as possible for you. The aim is for you to be able to have a say with the cluster admins; fully securing and operating a cluster setup is of course much more complex.

The concepts you learn will ensure not only that the infrastructure is robust and resilient to threats, but also that you adhere to organizational policies and compliance requirements.

By *security* in Kubernetes, we mean all measures and mechanisms that serve to protect the clusters and the applications running in them from unauthorized access, misuse, and other potential security threats. This includes various aspects:

- *Authentication* and *authorization*, for ensuring that only authorized users and services have access to cluster resources
- *Data security*, for encryption of data both *at rest* and *in transit* to prevent data theft or loss
- *Vulnerability management*, for regular scans and updates to identify and close security gaps in applications and infrastructure

Governance refers to the policies, procedures, and controls used to manage and monitor the administration and operation of Kubernetes clusters. Good governance ensures that the infrastructure and applications are consistent with the business objectives, standards, and compliance requirements of the company or project. For example, you must ensure that the rules for releases are adhered to so that no code is used in production that has not been tested and checked. Or you must ensure that the personal data of customers is protected accordingly or that the company's IT resources are handled responsibly.

The requirements are as complex as they are multilayered and look slightly different in every company and often also in every project. Accordingly, it is important that you check and consider very carefully which guidelines and restrictions should apply. This is of course a completely separate work step and should not be done "on the side."

The core elements of governance include the following:

- **Policy management**
  The definition and enforcement of guidelines for the configuration and use of resources in the cluster
- **Compliance monitoring**
  Checking and ensuring that the cluster and the applications running on it comply with regulatory requirements and internal standards

- **Resource management**
  Control over the allocation and utilization of resources within the cluster to ensure efficiency and cost control
- **Audit and logging**
  The recording and analysis of activities in the cluster to investigate security incidents and check compliance with guidelines

In real life, security and governance in Kubernetes are closely linked and complement each other to create a secure, efficient, and compliant Kubernetes environment. The implementation of security measures protects the infrastructure from external and internal threats, while governance practices ensure that the use of this infrastructure complies with organizational guidelines. Both are essential for the secure and responsible operation of Kubernetes clusters in an enterprise environment.

6.1 Pod Security

For pods, there are several adjusting screws that influence their safety. The rights of a pod are also referred to as *privileges*. Privileges affect what a pod or container can do within a Kubernetes cluster.

> **Access to Secrets**
>
> If an attacker takes control of a privileged pod, they have far-reaching access to the node.
>
> Do you remember Chapter 2, Section 2.4, where we talked about secrets? A pod that has privileged access can, for example, read all the secrets that are on the node.

Pods and the containers they contain can be given specific security restrictions to limit what processes are allowed to do in the containers. For example, the user ID (UID) of the process, the Linux capabilities, and the file system group can be restricted.

> **Linux Capabilities**
>
> Linux capabilities are often only required during startup. You can swap this functionality out to the init container and thus prevent your application container from running with the capabilities.
>
> An overview of all Linux capabilities can be found at the following address: *http://s-prs.co/v596445*.

Without this restriction, a pod could control the node's network configuration, overwrite the root directory, and do many other things that you want to prevent. These capabilities are usually deactivated, but as always, there are exceptions. Some tools require extended access—for example, the following:

- **Network operations**
  Network tools need direct access to network interfaces.
- **Debugging and monitoring**
  Some monitoring tools must perform debugging at the kernel level and require correspondingly extensive privileges.
- **Storage and volume management**
  Some disk or file system management operations may require extended permissions, especially if they are outside the standard Kubernetes volume APIs.

The restrictions that you can set are referred to as the *security context*. They can be defined for the entire pod or per container and entered in the manifest, as in Listing 6.1. The policy at container level is only valid for this container and overrides rules you have set at the pod level. In the example, the container is prevented from starting the process as root.

```
apiVersion: v1
kind: Pod
metadata:
  name: nginx
spec:
  securityContext:
    runAsNonRoot: true
  containers:
  - image: nginx
    name: nginx
    securityContext:
      runAsNonRoot: true
```

Listing 6.1 Example of Set Security Context

> **[+] Good to Know**
> Not every rule at the pod level can also be set at the container level and vice versa.

If you set rules in the security context, this can result in your containers no longer starting. Unfortunately, simply telling the pod that it is not allowed to start as root does not suffice. The container must also behave accordingly; otherwise errors will occur, as in Figure 6.1.

The challenge is therefore to set the appropriate security context for each pod so that the pod has as many rights as it needs, but no more. This is referred to as the *least privileged principle*.

6.1 Pod Security

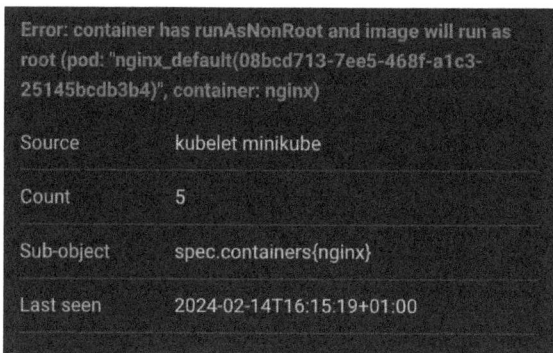

Figure 6.1 Container Must Not Start as Root

> **Good to Know**
>
> You can also set SELinux labels via the security context. However, changing the SELinux labels for a container can potentially allow the containerized process to break out of the container image and access the host file system. You should therefore handle them with care.

Your cluster admins in your company may already have given you specifications for the security contexts. We'll look at the most important ones so that you can get a feel for what you can set. You probably won't need all the rules, but you should think about what rights your pod needs.

You can set the following options at the pod level:

- runAsNonRoot
  If you set this value to true, the container must not be started as the root user.
- runAsUser
  Here you define the user ID with which the containers are executed. This means you can restrict the container to the authorizations of a specific user.
- runAsGroup
  You can use this option to set the group ID with which your container is executed.
- fsGroup
  Sets the group ID for all volumes used by the pod to set file permissions.

> **Good to Know**
>
> A complete overview of the options for the pod security context can be found in the API documentation at the following address: *http://s-prs.co/v596446*.

303

You can set the following options at the container level:

- `privileged`
  Specifies whether the container is running in privileged mode. This gives the container extensive access to the host.
- `readOnlyRootFilesystem`
  Here you specify whether the root file system of the container should be write-protected. This forces you to use volumes to store data.
- `allowPrivilegeEscalation`
  This allows you to control whether processes in the container are allowed to increase their privileges, such as by using `setuid` or `setgid`.
- `capabilities`
  You can use this information to add or remove Linux capabilities from the container. This allows you to assign necessary root user capabilities such as CHMOD to your container.

Good to Know

A complete overview of the container security context options can be found in the API documentation at *http://s-prs.co/v596447*.

6.2 Pod Security Admission

With the introduction of *pod security admission* (PSA), Kubernetes provides you with a powerful tool to define and enforce security policies for pods.

Good to Know

Previously, pod security admission was referred to as *pod security policy*. You can still find the old name in some documentation today.

PSA works on the basis of pod security standards, which are divided into three levels:

- Privileged
- Baseline
- Restricted

The pod security standards are applied to namespaces. Each of these levels specifies certain guidelines for what the security context of a pod in a particular namespace should look like.

The degree of security of the individual levels ranges from very permissive to highly restrictive. You can imagine it like an onion, as in Figure 6.2. The more layers are added, the more rights the pod has. The policies cover a wide range of security requirements so that you can control at the namespace level what the pods can and cannot do.

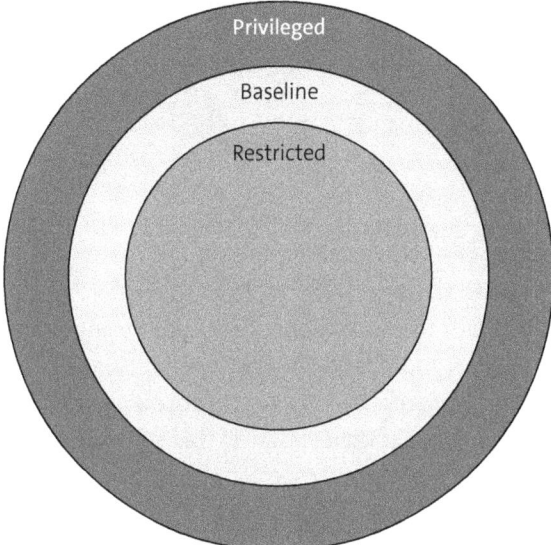

Figure 6.2 Pod Security Standards

Of course, the safest thing to do would be to give each pod restricted guidelines, but not every application can handle this.

> **Good to Know**
> Keep in mind that privileges reduce the isolation of container and host. This can lead to an application "breaking out" of the container and manipulating host resources or accessing sensitive data.

The *privileged* policy is the least restrictive policy and offers the widest possible authorizations. It is intended for system and infrastructure workloads. Under this policy, no restrictions exist, and all operations are allowed. This mode is useful for workloads that require access to system resources.

You can easily add the PSA to a namespace. An example of this is shown in Listing 6.2.

```
apiVersion: v1
kind: Namespace
metadata:
  name: example-namespace
  labels:
    pod-security.kubernetes.io/enforce: baseline
```

Listing 6.2 Namespace with PSA

> **Do Not Lock Yourself Out!**
>
> As you have also seen in Section 6.1, enforcing security context rules can cause your pods to fail to start. You should therefore ensure that all pods comply with the guidelines when introducing them into existing namespaces; otherwise, your application may fail.

6.3 Admission Controller

Say that you have a cluster that is home to many developers. Every developer knows the security requirements, but you also want to check or ensure these. You could now regularly scan all resources in the cluster and check whether the rules are being adhered to. A far more effective method is to check each resource before it even moves to the *etcd* database and thus becomes active in Kubernetes. To do this, you can simply hook into each API request and check or change the requests.

The admission controllers are an essential part of the process of an API request in Kubernetes. Their main functions include the modification (*mutate*) or validation of the content of objects. For this purpose, you can define the rules an admission controller can check.

Validating controllers look at a YAML manifest based on the rules and return the result with regard to whether the rule is complied with or whether the manifest violates it. The modifying controller will adapt the YAML manifest using a rule—for example, to set a specified CPU limit—and returns the manifest.

Every request to Kubernetes also passes through the admission controller. Figure 6.3 illustrates the process each request goes through before an object finally moves to the *etcd* database and becomes active in Kubernetes.

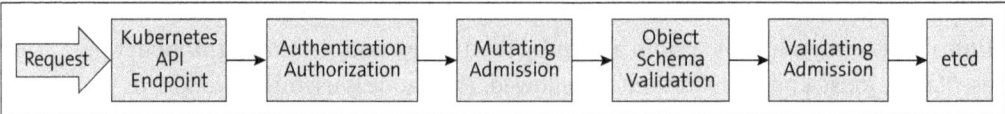

Figure 6.3 Sequence of API Request

Let's play this through with an example. When you as a user try to run a pod manifest using kubectl in your Kubernetes cluster, the corresponding API request goes through this process before the pod is actually created in the cluster:

1. **Authentication**
   First, the Kubernetes API server checks the identity of the user or service executing the request. If you remember Chapter 1, Section 1.5.3, then you know that kubectl uses the certificate in your Kubeconfig file for this purpose.

2. **Authorization**
   Once the user has been successfully authenticated, the API server checks whether the user is authorized to create a pod. This is decided, for example, on the basis of role-based access control (RBAC) guidelines.

3. **Mutating admission**
   If configured, mutating rules are applied at this point. These can manipulate the pod manifest in order to set certain labels, for example.

4. **Schema validation**
   Kubernetes checks whether the schema of the YAML manifest is correct. This step only takes place after the mutating step, as these rules could still change the manifest.

5. **Validating admission**
   In this step, you can define your own rules in addition to the schema validation. You can use them to check whether the manifests comply with your security or compliance guidelines.

6. **Persistence in etcd**
   Only now is the manifest stored in the *etcd* database and then processed further by Kubernetes, and the pod is generated.

If your request does not successfully pass one of these steps, then your pod will not launch in Kubernetes in the end. In my opinion, the greatest strength lies in the validation step. If a manifest does not comply with the rules, it will be rejected prior to the start. This allows you to enforce specific rules and thus ensure that security requirements are met, for example.

Good to Know

Kubernetes provides some standard admission controllers that you can use out of the box. For example, you can set default values by using the `DefaultIngressClass` controller or the `DefaultStorageClass` controller. Alternatively, you can use the `LimitRanger` controller to check the requests and limits of pods.

A complete list and description of the standard controllers can be found at the following address: *http://s-prs.co/v596448*.

Note

Admission controllers must be activated and deactivated in the API server. They are transferred as parameters at the start. You should involve the cluster admins for the clusters in your company. For example:

- **For activation**
  `--enable-admission-plugins=NamespaceLifecycle,LimitRanger`
- **For deactivation**
  `--disable-admission-plugins=PodNodeSelector`

The `MutatingAdmissionWebhook` and `ValidatingAdmissionWebhook` webhook controllers are particularly interesting. These allow for an unlimited expansion of the mutation and validation logic, as you can either develop code for this yourself or use one of the many available tools.

Good to Know

Tools that use the webhook controller include Gatekeeper and Kyverno.

As you can see, admission controllers, and especially webhook controllers, provide a flexible way to check and enforce governance, security, and compliance in the Kubernetes cluster. By implementing custom checks and logic, you can exercise fine-grained control over the resources in your clusters and ensure that only secure and compliant configurations are applied.

6.4 Kubernetes Policies

Policies are essential in the Kubernetes ecosystem in order to control and optimize the management of resources. These go hand in hand with the admission controllers discussed in Section 6.3, as the policies are typically validated by external tools. By using policies, you can introduce finely tuned control mechanisms to ensure that cluster usage meets your organizational requirements and best practices.

To give you a feel for the policies, I want to show you a few examples of the Kyverno tool.

Note

We will not go into the installation of Kyverno, but if you want to try out the tool, you can find more information at *https://kyverno.io*.

By using Kyverno, you have the option of writing policies that can run in either audit or enforce mode. Using *audit*, the resources are only checked, and even if they do not comply with the rules, they are deployed. However, you can view a report to get an overview of your cluster's processes. With *enforce*, the resources are blocked and are not deployed.

6.4 Kubernetes Policies

> **Good to Know**
>
> If you do not want to enforce the rules, you can use Kyverno for reporting. I set up a central reporting system for this at a client. This meant that no policies had to be rolled out in the cluster for enforcement, but there was a central overview of compliance with the security rules.

Let's now look at some examples so that you can see how you might use a policy. Listing 6.3 shows a policy that prohibits the creation of pods whose images do not originate from a permitted registry. You can use this policy, for example, to enforce that containers from the public—and therefore quite insecure—Docker Hub may not be used. As you can see, Kyverno is given a manifest snippet under `validate.pattern` that is supposed to be checked. In this case, the rule checks the image of the container. It is also defined that this policy only applies to pods.

> **Monitor First, Then Block**
>
> You should be careful when enforcing rules with Kyverno because doing so means that resources that do not comply with the policies may no longer be deployed. Even resources that are already deployed in the cluster can have problems, and in the worst case even system pods can be blocked. We had to rebuild an entire development cluster for one customer because Kyverno had blocked everything.
>
> It is best to start with audit policies and read the generated reports carefully. In the next step, you can switch to automatically enforcing the rules.

```
apiVersion: kyverno.io/v1
kind: ClusterPolicy
metadata:
  name: ensure-trusted-registry
spec:
  validationFailureAction: enforce
  rules:
    - name: trusted-registries-only
      match:
        resources:
          kinds:
            - Pod
      validate:
        message: "Only images from trusted registries are allowed."
        pattern:
```

```
    spec:
      containers:
        - image: "docker.io/trusted/*"
```

Listing 6.3 Kyverno Policy for Trusted Registries

You can find another policy in Listing 6.4. To control resource utilization within the cluster, you can use Kyverno to make sure that each pod complies with resource limits. We will take a closer look at the resources in Chapter 7, Section 7.1, but you can probably already imagine that monitoring them is important for the stability of the cluster. Pods without limits could otherwise paralyze the entire system, so it is important that no pod is started without the appropriate information.

```
apiVersion: kyverno.io/v1
kind: ClusterPolicy
metadata:
  name: require-resources-limits
spec:
  validationFailureAction: enforce
  rules:
    - name: check-resources
      match:
        resources:
          kinds:
            - Pod
      validate:
        message: "CPU and memory limits are required."
        pattern:
          spec:
            containers:
              - resources:
                  limits:
                    memory: "?*"
                    cpu: "?*"
```

Listing 6.4 Enforcing Resource Limits

We talked about the security context in Section 6.1. These rules can also be checked and enforced via Kyverno. An example of this is shown in Listing 6.5. If you do not already specify the rules via the pod security standards, an explicit rule in Kyverno can be useful.

```
apiVersion: kyverno.io/v1
kind: ClusterPolicy
```

```
metadata:
  name: disallow-root-user
spec:
  validationFailureAction: enforce
  rules:
    - name: root-user-not-allowed
      match:
        resources:
          kinds:
            - Pod
      validate:
        message: "Execution as root user is prohibited."
        pattern:
          spec:
            securityContext:
              runAsNonRoot: true
```

Listing 6.5 Enforcing RunAsNonRoot Option

6.5 Policy Objects

There are objects that act as a type of Kubernetes policy but are different from it. I want to present two of these in more detail, as they are particularly useful in larger clusters:

- Resource quotas
- Limit ranges

Resource quotas enable you to specifically control and limit the resource consumption in your Kubernetes cluster. These powerful policies allow you to set limits for resource consumption at the namespace level. By defining resource quotas, you ensure that no namespace allocates too many resources and that the availability of resources for the entire cluster remains fair and balanced. This is particularly important in larger clusters with different clients.

Good to Know

You can use resource quotas and limit ranges to protect your cluster from accidental or deliberate resource theft. What would happen if you simply started pods that block an entire node with your request requirements?

Either you have a cluster autoscaler that starts up new instances, or other pods cannot start or, in the worst case, are even displaced.

6 Kubernetes Governance and Security: Prepare for Production

You can find an example of a resource quota in Listing 6.6. It defines that

- a maximum of 10 pods may be created in this namespace;
- a maximum of four CPU requests may be made;
- a maximum of five gigabytes of memory requests may be made;
- 10 CPUs is the limit for this namespace; and
- 10 gigabytes of memory is the limit for this namespace.

```
apiVersion: v1
kind: ResourceQuota
metadata:
  name: example-quota
  namespace: my-namespace
spec:
  hard:
    pods: "10"
    requests.cpu: "4"
    requests.memory: 5Gi
    limits.cpu: "10"
    limits.memory: 10Gi
```

Listing 6.6 Sample ResourceQuota Manifest

If you define a resource quota, then it makes sense that requests and limits are also set for each pod and, if necessary, a default is enforced. This is where the `LimitRange` object comes into play. With resource quotas, you set a maximum for the namespace, while by using `LimitRange`, you can define the minimum and maximum for a single pod or container.

An example of this is shown in Listing 6.7. There you can see how to specify values for pods and containers. Limit ranges help you to ensure balanced and fair resource consumption so that all applications can run smoothly.

```
apiVersion: v1
kind: LimitRange
metadata:
  name: example-limits
  namespace: my-namespace
spec:
  limits:
  - type: Pod
    max:
      cpu: "2"
      memory: 1Gi
```

```
  - type: Container
    max:
      cpu: "1"
      memory: 500Mi
    default:
      cpu: "500m"
      memory: 256Mi
    defaultRequest:
      cpu: "250m"
      memory: 128Mi
```

Listing 6.7 Sample LimitRange Manifest

6.6 Role-Based Access Control in Kubernetes

In most cases, a role and authorization concept will already exist in a company cluster, which means there is a plan for how users log in and how they are assigned authorizations. This also depends on how the cluster is structured and managed. A large cluster will most likely have more restrictive policies than a small one as there are more users and different teams on it. Nevertheless, I would like to give an introduction in this chapter so that you can have your say and understand what is happening in the background.

RBAC in Kubernetes is a tool for defining who is allowed to do what in the cluster. It is based on a combination of roles and authorizations that determine which actions users, services, or applications are allowed to perform. RBAC enables you to precisely control access to resources in the cluster.

In the Kubernetes RBAC process, several key objects play a central role in enabling fine-grained access controls within a cluster. These objects include the following:

- `ClusterRole`
  Defines authorizations at the cluster level that can go beyond individual namespaces. One example of this is the cluster admin, who has access to all namespaces.
- `ClusterRoleBinding`
  Assigns a `ClusterRole` to users, groups, or service accounts. Only the binding enables the assigned entities to exercise the defined authorizations.
- `Role`
  Similar to a `ClusterRole`, but limited to a specific namespace. Roles define what can be done within the namespace.
- `RoleBinding`
  Binds a `role` to users, groups, or service accounts. This determines who has which authorizations in the namespace.

- **ServiceAccount**
  Special accounts used by pods to interact with the Kubernetes API server. They enable applications to access Kubernetes resources.

- **Users and groups**
  External users or groups that are not directly managed by Kubernetes but can be identified by external authentication mechanisms.

> **Good to Know**
>
> You are familiar with Kubeconfig as an authentication mechanism for the cluster. Kubernetes uses the CN field (common name), which contains the user name. This allows a role to be bound to the user via RBAC.
>
> You can find out more about user authentication at *http://s-prs.co/v596449*.

The RBAC process is relatively simple, and you may already be familiar with the principle from other tools. Figure 6.4 contains an overview of the process. You have a namespace, and the roles and RoleBindings are defined in the namespace. The RoleBindings are assigned to the users. The principle is also reminiscent of persistent volumes and persistent volume claims. The same applies to ClusterRoles and ClusterRoleBindings.

Let's now take a closer look at what exactly the individual resources do.

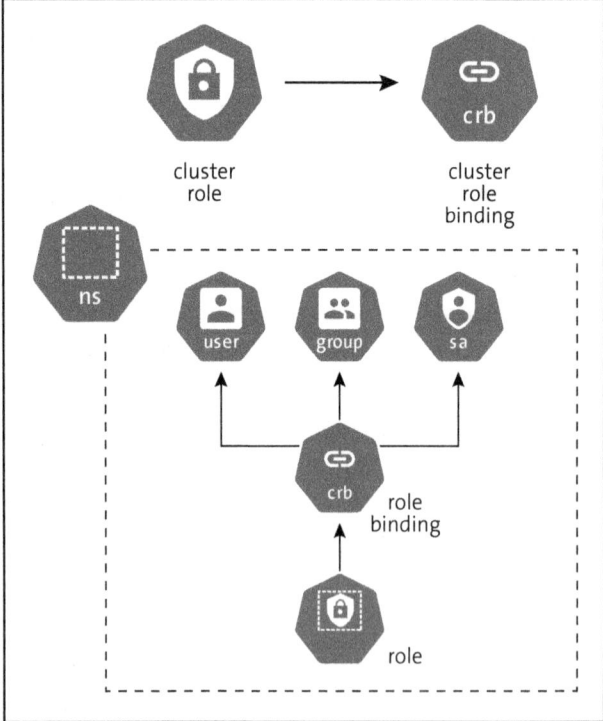

Figure 6.4 Overview of RBAC

6.6.1 Subjects: Users, Groups, and Service Accounts

Roles can be assigned to so-called subjects in Kubernetes. *Subjects* are actors that can use the Kubernetes API with the rights of the role. As mentioned previously, there are three categories of actors: users, groups, and service accounts.

In Kubernetes, users represent individual actors who require access to the system. These can be developers or administrators, for example. User authentication is a critical step to ensure that only authorized persons or processes have access to the resources and functions of the cluster.

A *group* in Kubernetes is a collection of users that are grouped together under a common name. Groups simplify the management of authorizations by making it possible to assign roles and access rights collectively. Instead of assigning specific authorizations to individual users, you can assign roles to a group, which makes the administration of authorizations in large environments more efficient.

> **Note**
>
> Typically, users and groups are not managed directly via Kubernetes. For example, if you want to establish a connection between your company's Active Directory and Kubernetes, you will need a little more configuration work, depending on the cluster structure. For one customer, for example, we used Rancher, which was also responsible for user management. OpenShift also provides a more comprehensive user management functionality than Kubernetes provides out of the box.
>
> Roughly speaking, Rancher takes over the authentication in this case and then creates an individual Kubeconfig file for your user, which you can then use to access Kubernetes.
>
> It is best to talk to your cluster admins about this. You can also read more about this at the following address: *http://s-prs.co/v596450*.

Because we want to take a more technical look at RBAC, we will leave out the users and groups and take a closer look at the service account. Service accounts are special accounts that are used for applications and services within a Kubernetes cluster. They provide an identity for processes running inside pods and allow these processes to interact with the Kubernetes API server. Service accounts are essential for automation within the cluster, as they enable applications and services to access cluster resources securely and without manual intervention.

The biggest difference from a normal user is that the service account is tied to a namespace. A default service account is automatically created when a new namespace is created. The main function of a default service account is to provide basic authentication and identity within the Kubernetes cluster. However, these accounts have no further authorizations, which means that they cannot really be used to interact with the Kubernetes API server.

Let's take a look at how you can make requests to the Kubernetes API within a pod using the service account. To do this, roll out the pod from Listing 6.8 and log into it using `kubectl exec`.

```
apiVersion: v1
kind: Pod
metadata:
  name: test-pod
spec:
  containers:
  - name: test-container
    image: curlimages/curl:latest
    command: ["sleep", "3600"]
```

Listing 6.8 Test Pod with Default Service Account

You can use the following command to read the service account token and send a query to the API server using `curl`:

```
curl -sSk -H "Authorization: Bearer \
  $(cat /var/run/secrets/kubernetes.io/serviceaccount/token)" \
  https://kubernetes.default.svc/api/v1/namespaces/default/pods
```

The token is mounted in the pod as a secret, and you will always find it on this path. If you execute the command, you will receive the following error:

```
"message": "pods is forbidden: User \"system:serviceaccount:default:default\"
cannot list resource \"pods\"
```

Using the request, you are trying to read the pods from the `default` namespace, but the service account is not authorized to do so.

Given the limited capabilities of default service accounts, it is necessary to create dedicated service accounts for most real-world applications. These dedicated accounts can be equipped with specific roles and authorizations that are precisely tailored to the needs of the respective application or service. You can find an example of a service account manifest in Listing 6.9. We will assign a role to it straight away and then use it for the query.

```
apiVersion: v1
kind: ServiceAccount
metadata:
  name: my-serviceaccount
```

Listing 6.9 Service Account Manifest

6.6.2 Roles and Role Bindings

Let's move on to the objects that enable you to define rights and assign them to an account. A *role* in Kubernetes defines a set of permissions that specify which actions a user, group, or service account can perform within a specific namespace. These authorizations include reading, writing, or deleting Kubernetes resources such as pods, deployments, and services.

You can find an example of a role in Listing 6.10. As you can see, a policy has three sub-objects:

- apiGroups
- resources
- verbs

apiGroups group the various API resources in Kubernetes. These are divided into groups to make it easier to expand the system. Each API group contains related resources. For example, the apps API group contains resources that have to do with applications, such as deployments, ReplicaSets, and StatefulSets.

resources are the specific objects to which a role has access. When you define a role or ClusterRole, you list the resources for which the role is to grant authorizations. For example, the resources could be pods, services, or deployments.

verbs define the operations that may be performed on the specified resources. You can adjust these to a very detailed extent, as there are several of them. Verbs that you can control using RBAC are as follows:

- create
- delete
- deletecollection
- get
- list
- patch
- update
- watch

> **Good to Know**
>
> You can use the following command to view all API objects and their verbs:
>
> `kubectl api-resources --sort-by name -o wide`

As you can see, these three settings allow you to define very precisely which authorizations are set for which resources.

6 Kubernetes Governance and Security: Prepare for Production

> **Note**
> Always remember the least privileged principle when designing roles. Assign users or service accounts only the minimum necessary authorizations that they need for their tasks.

Let's now continue the example from the previous section. To do this, roll out the role from Listing 6.10 in your cluster. This will allow the service account to access pods in the `default` namespace with `get`, `watch`, and `list`.

```
apiVersion: rbac.authorization.k8s.io/v1
kind: Role
metadata:
  namespace: default
  name: pod-reader
rules:
- apiGroups: [""]
  resources: ["pods"]
  verbs: ["get", "watch", "list"]
```

Listing 6.10 Manifest of Role

Now bind this role to the service account from Listing 6.9 using `RoleBinding` from Listing 6.11. If you roll this out now, this role will be assigned to the service account.

```
apiVersion: rbac.authorization.k8s.io/v1
kind: RoleBinding
metadata:
  name: read-pods
  namespace: default
subjects:
- kind: ServiceAccount
  name: my-serviceaccount
  namespace: default
roleRef:
  kind: Role
  name: pod-reader
  apiGroup: rbac.authorization.k8s.io
```

Listing 6.11 RoleBinding Manifest

Now that you have assigned the role to the service account, you still need to assign the service account to the test pod so that it can use it. Adapt the manifest as in Listing 6.12. Roll out the new pod, log back in to it using `kubectl exec`, and run the `curl` command from the previous section.

You should now receive a JSON object with all pods of the namespace.

```
apiVersion: v1
kind: Pod
metadata:
  name: test-pod
spec:
  serviceAccountName: my-serviceaccount
  containers:
  - name: test-container
    image: curlimages/curl:latest
    command: ["sleep", "3600"]
```

Listing 6.12 Test Pod with Its Own Service Account

6.6.3 Conclusion

As you have seen in the example, creating and assigning roles and authorizations with RBAC is very simple. You only need to know in advance which accesses the pod requires to the Kubernetes API. You can then create the role accordingly and assign the rights via a service account.

Of course, the simplicity of RBAC also harbors a danger. The more complex and detailed the roles become, the more likely it is that an error will creep in and pods will gain access that they should not have. This can be particularly dangerous if a pod is granted admin access. In the end, assigning authorizations is always like a balancing act: overly permissive roles can pose unintended security risks, while overly restrictive roles can limit the functionality of your applications. It is therefore important to check the authorizations regularly and ensure that they still meet the current requirements.

Chapter 7
Developing Applications for Kubernetes: Ready for Production

Failures are the norm in large-scale systems.
—Google

In Google's white paper on Borg, this is the first sentence in the chapter on availability. And even today, we can only agree with this statement: the larger the system, the higher the probability that an error will occur somewhere. However, the aim of a distributed system is not to make the individual components more fail-safe, but to build the system itself in such a way that a failure can be tolerated.

This starts with the hardware level, for example. Hard disks have long been interconnected in so-called redundant arrays of independent disks (RAIDs) to ensure data availability in the event of a single disk failure. Kubernetes is also designed and developed in such a way that the failure of an individual component, such as a master or worker, is manageable and can be compensated for. However, your applications must also be prepared accordingly.

A big mistake that I unfortunately see far too often is to believe that a traditional application can simply be packaged like a present using *lift and shift*, and with a new pink ribbon on the container, the application runs with all the benefits of a cluster system. But there are other concepts and assumptions behind old applications, some of which were developed several decades ago. Take, for example, a classic Java EE application that runs on a JBoss application server. Such applications are usually implemented monolithically, have to perform many tasks, and are scaled vertically. If more power is required, you simply add more CPU and memory.

One assumption in such an application is that stability prevails and that we are dealing with long-running processes. A few years ago, for example, I experienced a JBoss application at a customer that took about 30 minutes to start up. The application first filled its storage with data from a database in order to be able to work afterward. You can certainly imagine how unpleasant the failure of a machine is on which this application is running.

Kubernetes, on the other hand, follows the concept that *errors are the rule*. An application must be able to cope with a failure without the overall system suffering as a result, requiring a decoupled software architecture based on the separation of concerns principle from Chapter 1, Section 1.1.5. This makes it possible to replace or redevelop parts of the overall system without any problems or major dependencies.

> **Note**
>
> An important concept is the transience of components and containers. Where updates were installed on a server in traditional environments, today a new container is built to replace the old one. The components are not built to last forever, so they are transient and ephemeral. Remember the pets and cattle example from Chapter 1.

The new concepts alone do not make a good application, and they bring other challenges with them. In this chapter, I want to provide some tools that will enable you to make your application ready for production for Kubernetes.

7.1 Managing Pod Resources

If many pods share the resources on a server, then these must also be managed. This typically involves the two resources of CPU and memory. When you deploy a pod, it will consume as many resources as it needs without any further settings. This can lead to other pods not receiving sufficient CPU and becoming correspondingly slower. If there is no memory available, this even leads to an "out of memory" error and the pod terminates.

In the manifest of your pods, you can enter two pieces of information about your resources that Kubernetes should take into account: requests and limits.

Requests allow you to transfer the "normal" consumption of your pod. This value is used by Kubernetes to assign the pod to a node that can still provide sufficient resources. Limits can be used to set the maximum amount of resources the pod can receive.

> **Good to Know**
>
> Several versions of Kubernetes now also offer the management of ephemeral storage. You can find more information on this in Chapter 5, Section 5.3.

Let's assume you have an application that typically requires 512 MiB of memory and 0.5 core CPU. Kubernetes wants to run it on a node that has a total of two core CPUs and 4 GiB of memory. Kubernetes checks all pods running on this node and adds the requests together. Four pods of your application can therefore run simultaneously on this node. If your pod requests more than is available, Kubernetes will try to run it on another node.

Let's assume that four pods are now running on the node, thus filling that machine to capacity. However, the requests do not say anything about how much the pods actually consume. If your application has nothing to do, then it is possible that the node as a whole has nothing to do either. This should be avoided as far as possible in order to make optimal use of resources, which is why setting the right requests is so important. What you can decide, however, is the maximum that a pod may use.

If you set a limit of one core CPU and 1 GiB of memory for the pods, then the individual pods can double their resources. This can be useful, for example, if you have an application that has a load peak from time to time. This allows the peak to be intercepted, but the pod does not block resources unnecessarily. However, there is a risk that more resources will be allocated than are available. This then leads to throttled pods or, in the worst case, to out of memory errors.

You can find a graphical representation of this in Figure 7.1.

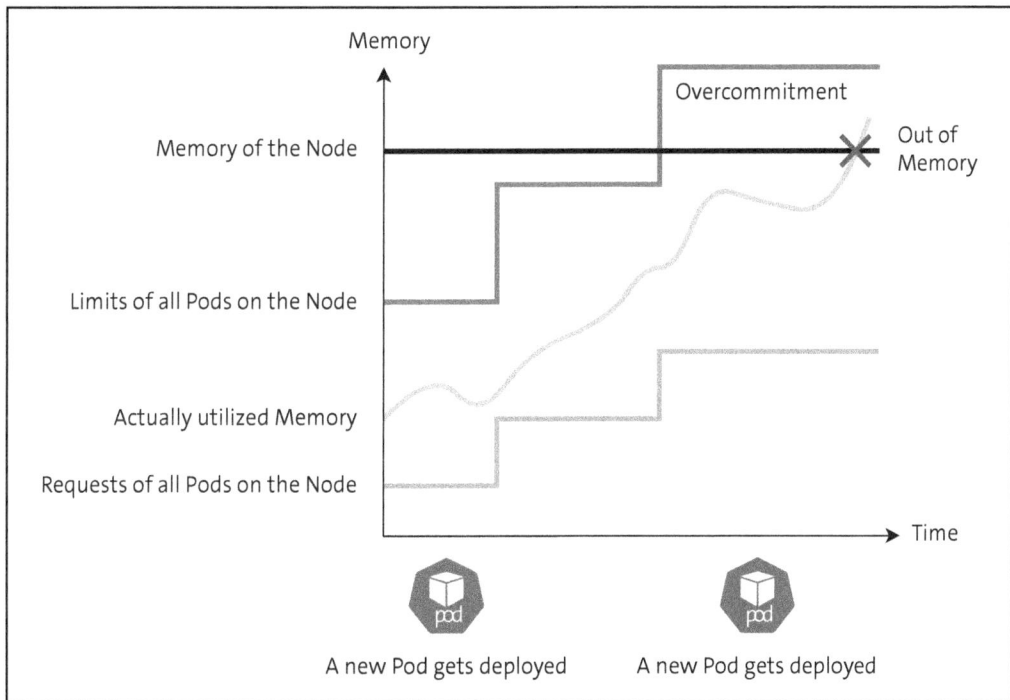

Figure 7.1 Overcommitment of Resources

You can see that the limit is above the node size at a certain time. This procedure is referred to as *overcommitment*. If the actual consumption exceeds this threshold, then out of memory errors will occur. However, Kubernetes sets a limit beforehand, which leads to the expulsion of pods in order to mitigate out of memory issues. Unfortunately, this does not always work.

> **Note**
> If your limits are higher than your requests, then you have an increased risk of running into a resource bottleneck. If you do not yet know exactly how many resources your application needs, then you should try out several iterations and observe your application. You can then slowly refine the values.

> **Be Careful when Setting Limits and Requests**
> Especially in a production cluster, you should be careful when setting requests and limits. An incorrect value can not only affect your application, but also steal resources from other containers on the same host.
>
> The best solution to start with is to set requests and limits to the same value as this prevents a node from being overcommitted.

Let's take a look at the definition of resources in the manifest. You will find different terms in the documentation, such as *millicore* or *millicpu*, which both have the same meaning. They come from the Latin word *mille*, meaning *thousand*. So if you request 500 millicore, this is equivalent to 0.5 CPU. The requirement of 1 CPU means that one CPU core of the computer is actually used.

You typically specify the size of the memory in the Mi (mebibytes) or Gi (gibibytes) unit. You could theoretically also specify this in bytes, but that would make the manifest unreadable. In Listing 7.1, you can see what a YAML manifest for an Nginx pod could look like. You can also use the resource specification in your deployment configurations. Just try it out right away!

```yaml
apiVersion: v1
kind: Pod
metadata:
  name: nginx
spec:
  containers:
    - name: nginx
      image: nginx
      resources:
        limits:
          cpu: 200m
          memory: 256Mi
        requests:
          cpu: 200m
          memory: 256Mi
```

Listing 7.1 Resource Definition in YAML

To conclude this section, I invite you to try something out again. What happens if you set the limits too low? And how can you even recognize this?

You can simply use the example from Listing 7.1 and set the memory limit and the memory request to `memory: 256Ki`. To do this, open Lens and create a new resource. It is even better if you test it on one of your deployments. After saving, you will see the new pod trying to start but running into an error. In Figure 7.2, you can see the out of memory error (`OOM-killed`), which occurred of course because 256 kilobytes of RAM is a bit meager.

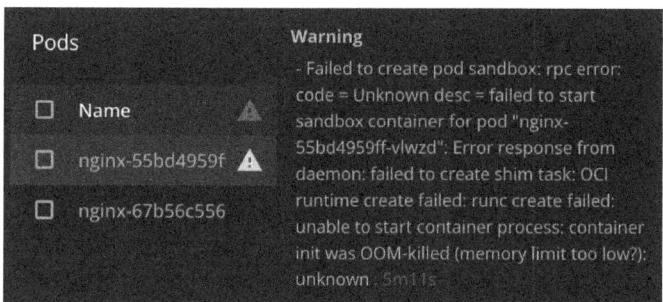

Figure 7.2 Error when Starting Pod with Insufficient Memory

In this case, you have clearly given the pod too little memory. There will be other cases where this is not quite so clear. If your application wants to allocate memory during operation but is not allowed to do so, then out of memory errors will occur as well. For this reason, it is important that you check your application with a load test and adjust the resources according to the findings.

> **Good to Know**
>
> As you now know, it is very dangerous to have no resource limits set at all. For this reason, it is common for cluster admins to set default values. These are implemented via so-called limit ranges, which we discussed in Chapter 6, Section 6.5. The administrator can set default values as well as minimum and maximum values here.
>
> There is also the option of limiting an entire namespace with the resource quotas. For example, the administrator can specify that the total of the memory limit in your namespace may not exceed 10 Gi.
>
> You should therefore pay attention to such specifications in your company as they can get in the way of your resource requirements.

7.2 Readiness, Liveness, and Startup Probes

When was the last time you had to call your internet provider because your DSL stopped working? Did the support person there also ask you first of all whether you had restarted your router? It's probably the same for you as it is for me, and you've already restarted everything before you even think of calling anywhere.

Even when operating our applications, a restart is sometimes worth its weight in gold. But neither you nor a colleague from IT operations wants to be called on the weekend or at night to restart the pod in Kubernetes. The good news is that Kubernetes can do this on its own. You just have to teach it to do so.

For the self-healing functionality to work, Kubernetes provides three different testing mechanisms to test how your application is doing. These so-called probes monitor the health of the application and derive actions from this monitoring if, for example, your application no longer works. You can define the following probes:

- **Liveness probe**
  This probe allows you to check whether your application is working properly. Kubernetes can automatically restart your application if the liveness probe fails for a defined period of time. In everyday language, the liveness probe is also simply referred to as a *health check*.

- **Readiness probe**
  This probe checks whether your application is ready to accept and process requests, which enables you to ensure that your application is fully operational before it receives requests. The Kubernetes service removes pods from load balancing if the readiness probe fails. Not only is this useful for initialization, but you can also use this function to remove pods from load balancing that are currently experiencing problems.

- **Startup probe**
  The startup probe is like a first check to see if everything in your container has started up correctly. If you define a startup probe, all other probes are deactivated until the startup probe is successful. You can use the startup probe if the initialization of one of your applications takes a very long time. If the startup probe fails, the kubelet will restart the pod.

The action performed by the liveness or startup probe depends on the restart policy, which you learned about in Chapter 2, Section 2.1.7. For example, if the policy is configured to Never, the liveness probe will not restart the container.

Good to Know

I personally have never used the startup probe as most modern applications boot up very quickly. However, if you have a legacy application that can take several minutes to complete its startup, then you should use the startup probe.

Its advantage lies in the temporary deactivation of the liveness and readiness probes. This prevents the pod from being recognized as *unhealthy* and restarting before it has even fully booted up.

The probes represent a useful tool for improving the self-healing of your applications. However, the implementation of probes is not recommended for every application. The following questions will help you decide whether you should implement one of the probes:

- **Liveness probe**
  - Can the application get into a state from which it cannot recover itself?
  - Could an automated restart solve the problem?
  - Could you define the probe in such a way that no unnecessary restarts have to be accepted?
- **Readiness probe**
  - Do conditions have to be met before the applications can process requests?
  - Could you check these conditions?
  - Does it make sense to remove the pod from the load balancing if the probe fails?
- **Startup probe**
  - Do you use one of the other two probes?
  - Does your application have a long startup process?
  - Could the other two probes ensure a restart before the application is fully booted?

In Section 7.2.1, we'll look at how probes can be defined and what options you have for implementing them. You can then answer one or two additional questions for which you may not yet have an ideal answer.

> **Good to Know** [+]
>
> If the application in your container can itself ensure that the container terminates with an error, you do not need a liveness probe. In that case, you can use the restart policy to define how the kubelet should react.
>
> With a liveness probe, however, you are more flexible and can tell the kubelet when it should take action according to the restart policy.

> **Avoid Endless Restarts** [!]
>
> A warning at this point: make sure that the liveness probe only really becomes active if
>
> - the application cannot rectify the error independently, and
> - restarting the pod can help.
>
> Incorrect implementation leads to repeated restarts of the pods, and the application will no longer be usable.
>
> Take, for example, an application that requires a database but can also respond to requests without the database via a caching mechanism. This liveness probe should

> not include the database because otherwise the self-healing attempts of Kubernetes would lead to a total failure.

7.2.1 How to Define Probes

You can design a probe in many different ways—for example, as follows:

- By executing a command in the container using `exec`
- By checking a TCP connection on a specific port
- By sending an HTTP request

> **[+] Good to Know**
>
> Since Kubernetes v1.27, a liveness probe can also be used with the gRPC protocol. You can find out more on this in the documentation at *http://s-prs.co/v596451*.

Let's start with the HTTP probes as you will probably need them most often. In Listing 7.2, you can see how the probes are defined in YAML under the `spec.template.spec.containers[]` object. You can create separate probes for each container within the pod. In Table 7.1, you can find the available options for configuring the probes.

```yaml
readinessProbe:
  httpGet:
    path: /health
    port: 8080
  initialDelaySeconds: 5
  failureThreshold: 1
  periodSeconds: 5
livenessProbe:
  httpGet:
    path: /health
    port: 8080
  initialDelaySeconds: 10
  failureThreshold: 3
  periodSeconds: 10
startupProbe:
  httpGet:
    path: /health
    port: 8080
  failureThreshold: 30
  periodSeconds: 10
```

Listing 7.2 Example of HTTP Liveness, Readiness, and Startup Probes

7.2 Readiness, Liveness, and Startup Probes

For an HTTP probe, you need to let Kubernetes know through which port and on which path an HTTP GET request should be sent. All response codes in the range >=200 && < 400 are considered a success, whereas everything else is an error.

The appeal of an HTTP probe is that you can customize the health check of your application. There are no limits, and you can implement anything from a simple check to a complicated query. You are also free to decide whether you want to implement one path for all probes or a separate path for each probe. We will run through an example of this in Section 7.2.2.

Option	Function	Default Value
initialDelaySeconds	Use this option to tell the kubelet how long it will wait before executing the first probe.	0
failureThreshold	Here you define how often a check must fail in succession before the entire probe fails.	3
periodSeconds	Defines how often the probe will be executed. With a value of 15, for example, the kubelet checks every 15 seconds.	10
successThreshold	For the readiness probe, you can define how often the check must run successfully before the pod status gets set to *ready*. This value must be set to 1 for the liveness and startup probes.	1
terminationGracePeriod-Seconds	Here you can tell the kubelet how long it should wait after scheduling the pod before forcing the deletion.	30

Table 7.1 Configuration Options

> **Good to Know**
>
> Kubernetes recommends using the same API endpoint for the readiness and liveness probes, but with different values for failureThreshold. This causes the pod to switch to the *not ready* status before it gets restarted by the kubelet. This has the advantage that the pod is removed from the load balancing of the service. However, you are free to decide how you define the API endpoints.

Another option is to check a TCP port. An example of this is shown in Listing 7.3. The kubelet attempts to open a socket on the container and the corresponding port. If that does not work, the probe will fail.

As not every application can respond to HTTP requests, this is a good option for monitoring databases or queues, for example.

```
livenessProbe:
  tcpSocket:
    port: 8080
  initialDelaySeconds: 5
  periodSeconds: 5
```

Listing 7.3 Sample TCP Liveness Probe

The third option for a probe is to run a command on the container using `exec`. Listing 7.4 shows an example in which the `cat /tmp/health` command is executed in the container. If this file does not exist, the probe will fail.

```
livenessProbe:
  exec:
    command:
    - cat
    - /tmp/health
  initialDelaySeconds: 5
  periodSeconds: 5
```

Listing 7.4 Sample Exec Liveness Probe

You can also run more complex commands. However, I only recommend such types of probes if an HTTP probe or a TCP probe is not possible or useful.

7.2.2 Testing Probes Using an Example

I have prepared an example so that you can try out the readiness and liveness probes. In this section, we will

- create a Python application that responds to HTTP probes;
- build a Docker image using the Python application;
- deploy the application as a deployment on Minikube; and
- test the probes with the application to see how Kubernetes responds.

We will make the Python application configurable using environment parameters. This makes it much easier to test later by making simple configuration changes.

Let's start with the Python application that you can find in Listing 7.5. We use the Flask web framework to provide a web server that listens for the /ready and /health paths. The application expects READY_TIME and UNHEALTHY_TIME as environment parameters, which you can set later in the deployment. The READY_TIME parameter allows you to define how many seconds the application needs to tell Kubernetes that it is ready,

7.2 Readiness, Liveness, and Startup Probes

while you can use the UNHEALTHY_TIME parameter to define after how many seconds after starting the application the *unhealthy* status will be returned.

You can immediately observe how Kubernetes will behave thanks to the way it works.

> **Note**
> You are welcome to add log messages to the application. This also allows you to check the container logs to see when and how often Kubernetes performs the checks.

```
from flask import Flask
import os
import time
app = Flask(__name__)
# Start time of the application
start_time = time.time()
# Read environment variables
ready_time = int(os.environ.get('READY_TIME', 5))
unhealthy_time = int(os.environ.get('UNHEALTHY_TIME', 15))
@app.route('/ready')
def ready():
    # Returns OK if the application runs longer
    # than 'ready_time' runs in terms of seconds
    if time.time() - start_time > ready_time:
        return 'OK', 200
    else:
        return 'Not Ready', 503
@app.route('/health')
def health():
    # Returns OK as long as the application runs less
    # seconds than 'unhealthy_time'
    if time.time() - start_time < unhealthy_time:
        return 'OK', 200
    else:
        return 'Unhealthy', 503
if __name__ == '__main__':
    app.run(host='0.0.0.0', port=8080)
```

Listing 7.5 "app.py" for Health Checker Application

In the next step, we will prepare everything so that you can package the application in a container image and store it in the Minikube registry from Chapter 1, Section 1.4.7. For this purpose, you need the (very simple and straightforward) *requirements.txt* file from Listing 7.6 and the Dockerfile from Listing 7.7.

```
flask
```

Listing 7.6 requirements.txt File for Health Checker Application

```
FROM python:3.9-slim
WORKDIR /app
COPY requirements.txt .
RUN pip install --no-cache-dir -r requirements.txt
COPY app.py .
EXPOSE 8080
CMD ["python", "./app.py"]
```

Listing 7.7 Dockerfile for Health Checker Application

The files should be located in a folder on the same level, as shown in Listing 7.8.

```
.
├── Dockerfile
├── app.py
└── requirements.txt
```

Listing 7.8 File Structure of Health Checker Application

The Dockerfile is based on the official Python image, installs the dependencies from the *requirements.txt* file in the subsequent step, and then copies the application into the image. Use the following commands to build the image and store it in the Minikube registry:

```
docker build -t localhost:5000/health-checker .
docker push localhost:5000/health-checker
```

[»] **Note**

Remember to activate the Docker host of Minikube so that you can also reach the registry. If you encounter problems, take another look at Chapter 1, Section 1.4.7.

Now everything is ready to deploy the application on Minikube. You can use the deployment manifest from Listing 7.9 for this purpose.

```
apiVersion: apps/v1
kind: Deployment
metadata:
  name: health-checker
spec:
  replicas: 1
  selector:
```

```yaml
      matchLabels:
        app: health-checker
  template:
    metadata:
      labels:
        app: health-checker
    spec:
      containers:
        - name: health-checker
          image: localhost:5000/health-checker
          ports:
            - containerPort: 8080
          env:
            - name: READY_TIME
              value: "5"
            - name: UNHEALTHY_TIME
              value: "15"
          readinessProbe:
            httpGet:
              path: /ready
              port: 8080
            initialDelaySeconds: 3
            periodSeconds: 3
          livenessProbe:
            httpGet:
              path: /health
              port: 8080
            initialDelaySeconds: 5
            periodSeconds: 5
```

Listing 7.9 Deployment.yaml for Health Checker Application

Under `spec.template.spec.containers[].env`, you give the container the environment parameters that are defined in the application. Roll out the deployment using Lens and observe the pod that is created.

You will see the pod start and the container in it go through the following statuses:

1. The container is initially in **Not Ready** status. You can recognize this by the orange box in Lens.
2. After five seconds, the container switches to **Ready**.
3. After approximately another 20 seconds, the container turns orange again because the health check has failed.
4. Kubernetes restarts the pod.

The pod will get stuck in this cycle. Kubernetes keeps trying to restart the pod if the health check fails. In Figure 7.3, you can see what the pod looks like after some time. Kubernetes counts the number of restarts, and the application becomes set to **Unhealthy** again shortly after it has reported **Ready**.

Figure 7.3 Pod in Restart Circuit

> **Note**
>
> You may notice that the container does not switch to the **Not Healthy** state immediately after the UNHEALTHY_TIME has expired. This is due to the livenessProbe settings in our deployment. The kubelet only checks whether the application is still alive every five seconds. Because we did not define a failureThreshold, Kubernetes takes the default value of 3. It therefore takes up to 19 seconds for the container to be considered *unhealthy*.

You will also find the following message in the pod's events, which will give you an indication of the problem (here you will see the status code that you have defined in the application):

```
Liveness probe failed: HTTP probe failed with statuscode: 503
```

Now you should try out the different setting options for the probes. You can also adjust the environment parameters and test the behavior of Kubernetes. If you want to go one step further, then extend the example with a startup probe.

> **Note**
>
> Try to implement the probes for one of your own applications. Find out which values are best for the probes. Try to answer the following questions:
> - How long does your application need to be ready?
> - When is your application considered unhealthy?
> - How long should Kubernetes wait before restarting your application?
>
> You will find that there is no one-size-fits-all answer. Use a few iterations to test the behavior.

7.3 Scaling and Load Balancing

In Chapter 1, Section 1.1, we talked about the concepts behind Kubernetes. Along with self-healing, horizontal scaling is one of the best features of Kubernetes that significantly simplifies IT operations. The idea behind this is to simply scale another container when the load increases, and to do so fully automatically. A metric monitors the load on the container based on the CPU or the number of messages in a queue, for example. If the metric rises above a defined threshold value, then a new container is set up and the load will be distributed to all existing containers via a load distributor. Of course, the number of containers will be reduced again when the load decreases. As shown in Figure 7.4, this principle allows you to consume only what you need. In the cloud, that means that you save money, and that's in line with the *pay-as-you-go* principle.

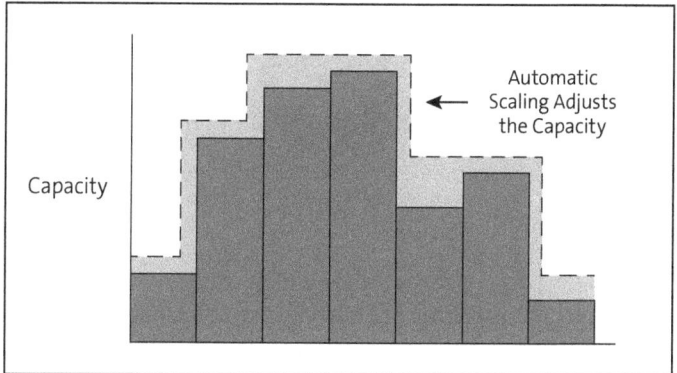

Figure 7.4 Automatic Scaling Based on Consumption

In my first job after graduating, I worked for a company that was right in the middle of a cloud migration. The old sales platform ran 24/7 on a powerful server system from HP in the company's own data center. It was a huge server rack full of computing power. This server needed plenty of capacity to cope with the rush of buyers at a sales event. But most of the time, the server was only running at 40% capacity (and that's a good utilization!) and was unnecessarily heating up the data center.

However, not every application is designed for horizontal scaling. As described in Chapter 1, Section 1.1.4, stateless applications are predestined for this. Horizontal scaling is a requirement that must be taken into account in the software architecture. But if everything fits, then you and the operations team will be able to sleep soundly.

7.3.1 Horizontal Pod Autoscaling

For horizontal pod scaling, the *horizontal pod autoscaler* (HPA) object is available. The HPA enables your applications to respond dynamically to changes in the load by automatically increasing or decreasing the number of pods, as shown in Figure 7.5. It can monitor certain metrics such as CPU utilization and scale automatically if threshold values are exceeded or not reached.

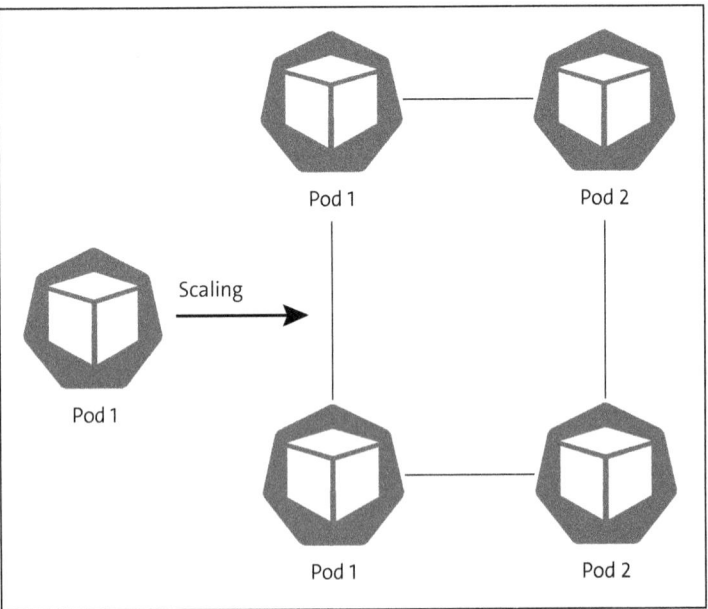

Figure 7.5 Horizontal Scaling

Horizontal in this context means that the number of pods is increased; that is, the cluster grows in width. The counterpart to this is vertical scaling, which is discussed in Section 7.3.2.

Good to Know

The HPA process is a control loop that runs and checks regularly. The standard value is 15 seconds. This means that scaling does not take effect immediately if the threshold value of a metric is exceeded.

Let's jump straight into an example. For Minikube, you want to run the `minikube addons enable metrics-server` command in preparation. The metrics server then collects the metrics from the kubelets for the pods that are needed for the HPA.

I had to stop and restart Minikube after activating the add-on so that the HPA could get the metrics.

Note

If you want to install the metrics server on an "ordinary" cluster such as the sample Raspberry Pi cluster, you can find more information at the following address: *http://s-prs.co/v596452*.

For this example, we are using the Apache pod, which Kubernetes provides specifically for this use case. You can find the manifest in Listing 7.10.

```yaml
apiVersion: apps/v1
kind: Deployment
metadata:
  name: apache-hpa
spec:
  selector:
    matchLabels:
      run: apache-hpa
  template:
    metadata:
      labels:
        run: apache-hpa
    spec:
      containers:
      - name: apache-hpa
        image: registry.k8s.io/hpa-example
        ports:
        - containerPort: 80
        resources:
          limits:
            cpu: 300m
          requests:
            cpu: 300m
---
apiVersion: v1
kind: Service
metadata:
  name: apache-hpa
  labels:
    run: apache-hpa
spec:
  ports:
  - port: 80
  selector:
    run: apache-hpa
```

Listing 7.10 HPA Example: Apache Deployment with Matching Service

Roll out the manifests for the deployment and the service. You can then roll out the HPA object from Listing 7.11. There you define that the monitored metric is the CPU and that the autoscaler can scale between a minimum of one pod and a maximum of three pods.

Note

You can find the HPA example from the Kubernetes documentation at the following address: *http://s-prs.co/v596453*.

```
apiVersion: autoscaling/v1
kind: HorizontalPodAutoscaler
metadata:
  name: apache-hpa
spec:
  scaleTargetRef:
    apiVersion: apps/v1
    kind: Deployment
    name: apache-hpa
  minReplicas: 1
  maxReplicas: 3
  targetCPUUtilizationPercentage: 50
```

Listing 7.11 Manifest of Horizontal Pod Autoscaler

Good to Know

The HPA expects the definition of requests and limits from Section 7.1. This makes sense because if the pod can simply use the entire CPU during your load test, then it is difficult to see a result.

We now need to put the application under load in order to experience the HPA in action. You can use the `kubectl` command from Listing 7.12 for this purpose. Make sure that you create the load generator pod in the same namespace as the Apache pod. This is the only way it can reach Apache with the command provided as we use the name of the service. If your load generator is in a different namespace, you must adapt the URL.

```
kubectl run -i --tty load-generator --rm --image=busybox:1.28 \
    --restart=Never -- /bin/sh -c "while sleep 0.01; do wget \
    -q -O- http://apache-hpa; done"
```

Listing 7.12 Generating Load Generator

Now observe the behavior of the HPA and the deployment. As in Figure 7.6, you can see that the load on the pods increases and the HPA scales new pods.

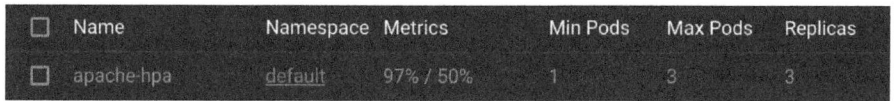

Figure 7.6 HPA during Load Phase

> **Note**
> Regarding the command from Listing 7.12, it is important that you write the command in your console in one line. Simply copying and pasting the multiline command caused problems for me.

You have now created a very simple HPA and seen it in action. The HPA becomes particularly interesting when you use custom metrics. If you have a suitable application, you will find more information on the following page: *http://s-prs.co/v596454*.

7.3.2 Vertical Pod Autoscaling

While the HPA adjusts the number of pods to handle the load, the *vertical pod autoscaler* (VPA) focuses on the resource allocation of the individual pods. The VPA optimizes the CPU and memory requirements of the pods running in your Kubernetes cluster. This enlarges or reduces the size of the pod as required, as you can see in Figure 7.7.

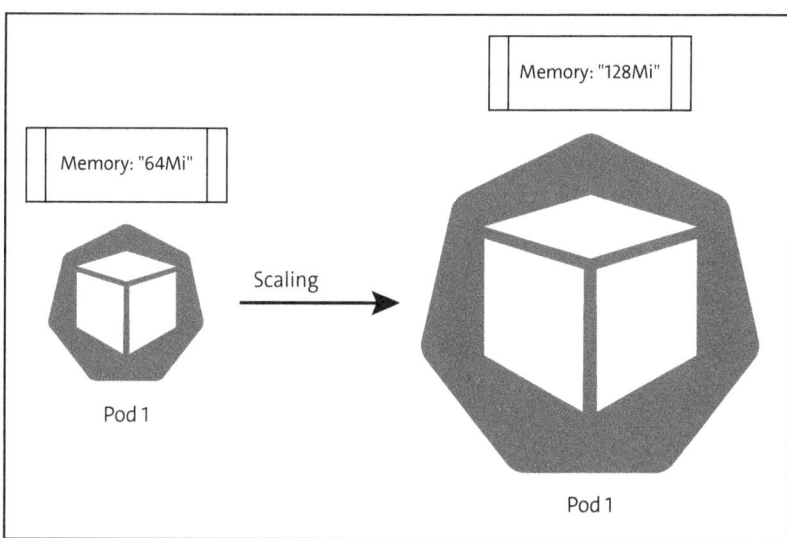

Figure 7.7 Vertical Scaling

The VPA continuously monitors the resource utilization of the pods and compares it with the defined requests and limits. If it determines that the resource requirements are not ideal, then it adjusts the requirements.

Note

I used the VPA in a project for Prometheus, which you can read about in Section 7.4. This was a good way to make Prometheus scalable without having to synchronize multiple replicas. What I found very critical about it is that the requests and limits are not recognizable at a glance. In addition, the pod behaves in a different way than the manifest in version management suggests.

For me, the VPA was an invisible magic hand that I found difficult to understand. The HPA is much easier because you can quickly see how many replicas of a pod are currently running.

If you have the option, it is best to develop your application in such a way that it can scale horizontally.

Let's briefly go through an example. We use the Apache pod from Listing 7.10 again, but now we use a VPA. For this reason, make sure to delete the HPA for this example if you have not already done so.

To install the VPA, you first need a set of CRDs. You can install them using the following commands:

```
kubectl apply -f https://raw.githubusercontent.com/kubernetes/autoscaler/vpa-release-1.0/vertical-pod-autoscaler/deploy/vpa-v1-crd-gen.yaml
kubectl apply -f https://raw.githubusercontent.com/kubernetes/autoscaler/vpa-release-1.0/vertical-pod-autoscaler/deploy/vpa-rbac.yaml
```

Then you can import the VPA object from Listing 7.13 and start the load generator again as in the previous example (see Listing 7.12). Observe the pod and the way the VPA handles it.

```
apiVersion: autoscaling.k8s.io/v1
kind: VerticalPodAutoscaler
metadata:
  name: my-vpa
spec:
  targetRef:
    apiVersion: "apps/v1"
    kind: Deployment
    name: apache-hpa
  updatePolicy:
    updateMode: "Auto"
```

Listing 7.13 VPA Manifest

7.3 Scaling and Load Balancing

You have now become familiar with both options for the automatic scaling of Kubernetes. I always prefer horizontal scaling to vertical scaling. First, it allows you to create multiple pods that run on different nodes, which ensures greater reliability. Second, the requests and limits of a single pod are set in such a way that it can still find space even on well-utilized nodes. This increases the capacity utilization and thus the efficiency of your cluster.

In addition, applications that can scale horizontally are usually more robust. But that also means that scaling is already part of the application, and in the development phase you already need to think about how the shutdown of a pod works and how the overall application can survive it. This way, you can make sure that your application survives an unintentional failure of a pod and can be scaled accordingly. Of course, scaling during operation can help, but it does not save poorly programmed apps whose architecture has a bottleneck.

In real life, you must select the scaling type that best suits your application. For example, applications that depend on a stable state are difficult to scale horizontally: databases are a prime example in this respect. With a web server like Apache, the result depends on whether the requests can be distributed well to different pods via a load balancer.

7.3.3 Cluster Autoscaler

For the sake of completeness, I also want to mention the cluster autoscaler. This tool is particularly interesting if you have a very volatile load on your applications, but it is usually the responsibility of the cluster admins. It allows you to automatically start new nodes and delete old nodes. Especially in public cloud environments, you can save money immediately. Figure 7.8 shows a graphical representation of the scaling.

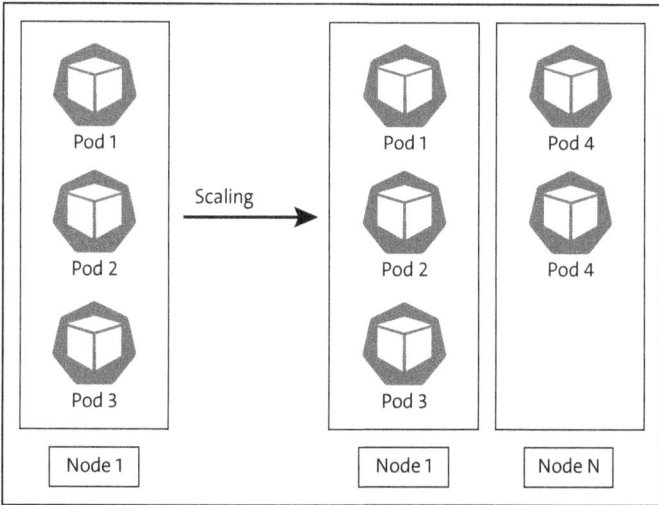

Figure 7.8 Cluster Autoscaling

> **Good to Know**
>
> In my opinion, the cluster autoscaler provides several advantages:
>
> - **Cost efficiency**
>   By adapting the cluster size to the actual load, you avoid the costs of unused resources.
> - **Scalability**
>   The cluster autoscaler allows your cluster to grow and shrink with the requirements of your applications, which is essential for scalable, cloud-native applications.
> - **Improved developer experience**
>   You need to worry very little about capacity. If you want to carry out a quick load test, the cluster can simply map that independently.

How does the cluster autoscaler work? It continuously monitors the utilization of the pods and nodes in your cluster and detects when pods cannot be started because not enough resources such as CPU or memory are available on the existing nodes. Based on this knowledge, the autoscaler then initiates the addition of new nodes to provide the required resources. At the same time, it also recognizes when nodes are underutilized and removes these nodes to save resources and costs. Remaining pods are evicted and started on other nodes. This empties the node, and then it can be switched off.

You need the cluster autoscaler in particular if you want to manage Kubernetes clusters in large, dynamic environments and have to think about geographical scaling in the cloud. If you are not yet using it and are looking for more information, you can find the GitHub repository at the following address: *http://s-prs.co/v596455*.

7.4 Monitoring

One key to stable operations is *monitoring*. Especially in the volatile world of Kubernetes, you need a good toolset to collect metrics, send alerts, and assist with debugging when necessary. Not only do these tools give you insight into the performance and health of your applications, but they also allow you to proactively respond to issues before they become critical.

The complexity and dynamics of Kubernetes clusters with their numerous pods, services, and other resources place special demands on monitoring. You need to be able to collect and analyze the right data quickly in order to make informed decisions.

> **Good to Know**
>
> I will introduce specific tools that I have frequently used in companies. However, there are also competing products for each of these tools that work according to similar

principles. I will mention these at one point or another, but you will see that if you know one, you know them all.

You can always apply this basic knowledge to other tools.

If you already feel confident with the basic principles, you can go directly to Section 7.4.2. There we will bring all the tools together in a demo.

7.4.1 Introduction: Prometheus, Grafana, and Alertmanager

The Prometheus ecosystem is used for monitoring. When it comes to monitoring applications in a Kubernetes cluster, this tool stack is widely used and has a correspondingly large community. The stack includes the following tools:

- Prometheus
- Grafana
- Alertmanager

As a powerful open-source tool for monitoring and alerting, *Prometheus* has established itself as the de facto standard for monitoring Kubernetes clusters. At its core, Prometheus is a time series database, collects metrics from endpoints, and stores them. With its powerful query language, you can create complex queries to get exactly the insights you need. In addition, Prometheus supports alert rules that notify you as soon as certain thresholds are exceeded.

Grafana is often the first choice to visualize the data collected by Prometheus. It provides a flexible platform for creating dashboards that provide a clear view of metrics. Grafana also supports data from many other sources, making it a versatile tool for monitoring.

The *Alertmanager* tool that is part of the Prometheus ecosystem is often used for alerting purposes. It allows you to send notifications via various channels such as email, Slack, or webhook, based on the alert rules you have defined in Prometheus.

Figure 7.9 shows the architecture of the components and how they interact with each other.

But why is Prometheus the first choice when it comes to Kubernetes?

Prometheus was specifically developed to work in modern, dynamic environments such as Kubernetes. It collects metrics via a pull mechanism in which it regularly queries endpoints (*targets*) to collect relevant data. In a Kubernetes environment, these targets can be pods, services, or node instances.

A key aspect of the integration is the service discovery of Prometheus within Kubernetes. Due to this feature, Prometheus automatically discovers new pods or services that provide metrics. Monitoring would not be possible otherwise as Kubernetes resources are started up and shut down dynamically. This automatic detection is therefore crucial to ensure that Prometheus always collects up-to-date data as you cannot possibly introduce new pods and nodes to your monitoring system "manually" after each automatic scaling.

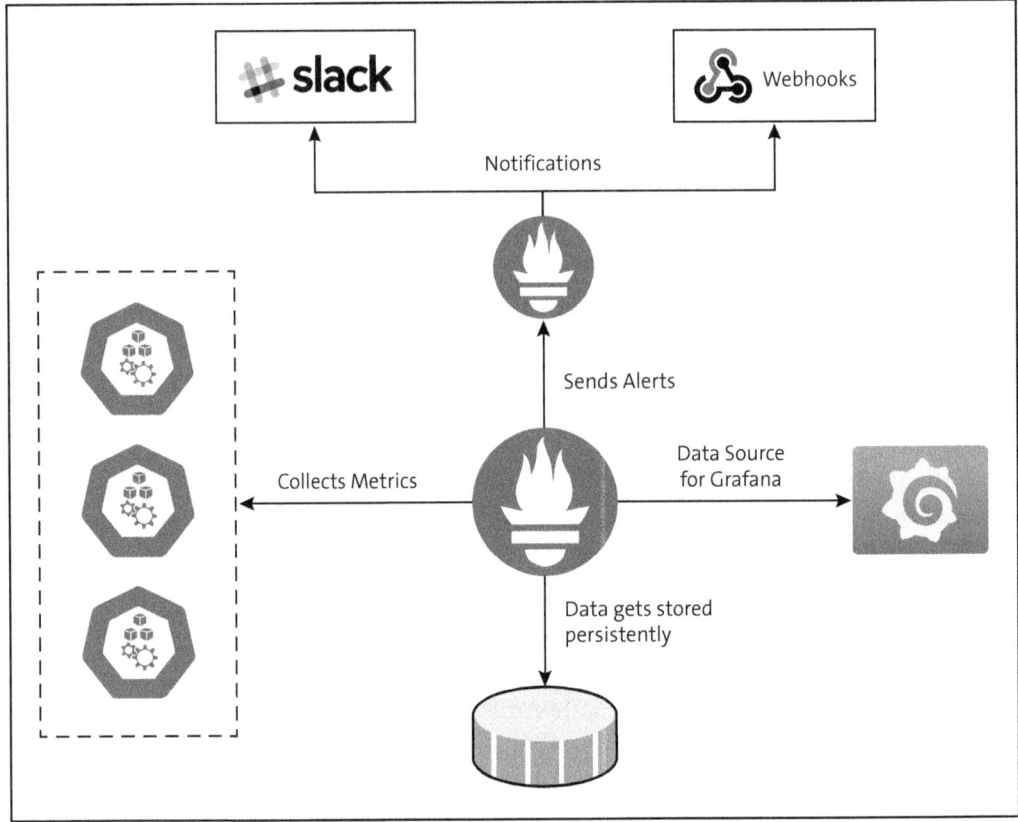

Figure 7.9 Monitoring Architecture with Prometheus, Grafana, and Alertmanager

To recognize the pods it should query, Prometheus uses the Kubernetes API and checks for certain annotations. An example of this is shown in Listing 7.14.

```
apiVersion: v1
kind: Pod
metadata:
  name: my-app-pod
  annotations:
    prometheus.io/scrape: "true"
    prometheus.io/path: "/metrics"
    prometheus.io/port: "8080"
...
```

Listing 7.14 Configuration of Prometheus Scraping

Another point in Prometheus's favor is its many custom exporters developed by the community. This gives you a direct interface for many tools to provide metrics for Prometheus. *Exporters* act as bridges between Prometheus and the systems or applications to be monitored. You collect metrics from these sources, convert them into the format expected by Prometheus, and make them available via an HTTP endpoint.

Examples of this include the following:

- **Node exporter**
  This collects hardware and operating system metrics from the host. The node exporter is essential for gaining insight into the resource utilization and performance of the physical or virtual machines running your Kubernetes cluster.
- **Kube-state-metrics**
  This extends the metrics provided by Kubernetes by collecting detailed information about the state of Kubernetes objects such as deployments, pods, and nodes.
- **Grok exporter**
  This allows you to convert log files into metrics through queries and export them for Prometheus.

The exporters are generally very easy to implement, and you can use them to have Prometheus monitor almost everything. An example of how to use the node exporter can be found in Section 7.4.2.

The extensibility and flexibility of Prometheus make it a good monitoring tool, which is why it is so widely used. It's best to try it out for yourself to get a feel for it. You can find the instructions for this in Section 7.4.2.

7.4.2 Monitoring on the Pi Cluster

Now let's get everything into the application and use Prometheus to monitor the Raspberry Pi cluster. The goal is to see at a glance how the Pis are doing. Here, I am particularly interested in the temperature, as I run the small computers without a fan. At the same time, I also want to see how much RAM is still available.

The sequence of our work steps will be as follows:

1. Install the node exporter on Kubernetes.
2. Install Prometheus via a Helm chart.
3. Configure Prometheus so that the node exporter queries metrics.
4. Install the Raspberry Pi exporter.
5. Extend the configuration of the node exporter.
6. Install Grafana via a Helm chart.
7. Create a dashboard to create the temperature display.

We will use the node exporter for this purpose. It already has a lot to offer, but for the sample use case, we'll need to add more metrics for the Pis.

The node exporter can be installed in various ways. It typically runs directly on a server and provides the metrics for Prometheus via a REST interface. For this example, let's try out the node exporter as a DaemonSet. Kubernetes then ensures that the service runs on every node.

> [!] **Access to the hostPath Volume**
> The node exporter pod accesses the host's file system through the hostPath volume. This can be a danger in production environments, and you should be aware of this. For this reason, you should clarify such a setup with the cluster admins beforehand.

Installing the Node Exporter on Kubernetes

First you need to create the `monitoring` namespace if you do not already have it. You can find the manifest for the DaemonSet in Listing 7.15.

```yaml
apiVersion: apps/v1
kind: DaemonSet
metadata:
  labels:
    app.kubernetes.io/component: exporter
    app.kubernetes.io/name: node-exporter
  name: node-exporter
  namespace: monitoring
spec:
  selector:
    matchLabels:
      app.kubernetes.io/component: exporter
      app.kubernetes.io/name: node-exporter
  template:
    metadata:
      labels:
        app.kubernetes.io/component: exporter
        app.kubernetes.io/name: node-exporter
    spec:
      containers:
      - args:
        - --path.sysfs=/host/sys
        - --path.rootfs=/host/root
        - --no-collector.wifi
        - --no-collector.hwmon
```

```
        - --collector.filesystem.ignored-mount-points=^/(dev|proc|sys|var
            /lib/docker/.+|var/lib/kubelet/pods/.+)($|/)
        - --collector.netclass.ignored-devices=^(veth.*)$
      name: node-exporter
      image: prom/node-exporter
      ports:
        - containerPort: 9100
          protocol: TCP
      resources:
        limits:
          cpu: 250m
          memory: 180Mi
        requests:
          cpu: 102m
          memory: 180Mi
      volumeMounts:
      - mountPath: /host/sys
        mountPropagation: HostToContainer
        name: sys
        readOnly: true
      - mountPath: /host/root
        mountPropagation: HostToContainer
        name: root
        readOnly: true
    volumes:
    - hostPath:
        path: /sys
      name: sys
    - hostPath:
        path: /
      name: root
```

Listing 7.15 Manifest for DaemonSet of Node Exporter

You will see that the node exporter will be accessible on port 9100. Provide the pods with hostPath volumes so that the node exporter can also access the corresponding paths in order to obtain metrics. Under `spec.template.spec.containers[].args`, you can see the configurations specific to the node exporter. You will expand these later. You can find the right service for the DaemonSet in Listing 7.16. Roll out both Listing 7.15 and Listing 7.16 in your cluster.

```
kind: Service
apiVersion: v1
metadata:
  name: node-exporter
```

7   Developing Applications for Kubernetes: Ready for Production

```
  namespace: monitoring
  annotations:
      prometheus.io/scrape: 'true'
      prometheus.io/port:   '9100'
spec:
  selector:
      app.kubernetes.io/component: exporter
      app.kubernetes.io/name: node-exporter
  ports:
  - name: node-exporter
    protocol: TCP
    port: 9100
    targetPort: 9100
```

Listing 7.16 Manifest for Node Exporter Service

Installing Prometheus via a Helm Chart

Now let's roll out Prometheus in your cluster. You'll use the Helm chart from Bitnami for this purpose. To do this, use Lens to search for Prometheus in your cluster under **Helm • Charts**. The Helm chart should look like the one shown in Figure 7.10. Now click **Install**. The Helm chart opens as a file. Select the **monitoring** namespace here and click **Install**.

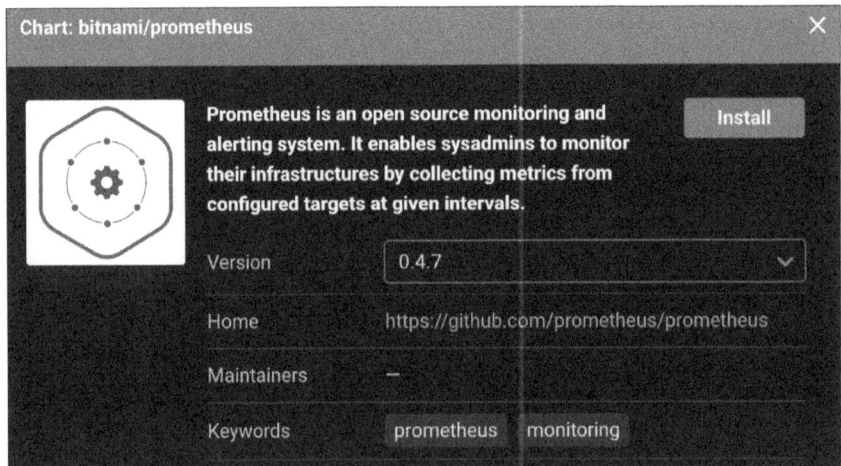

Figure 7.10 Prometheus Helm Chart in Lens

After a few minutes, Prometheus should have started successfully. The Helm chart also includes the alert manager, which is not relevant in this case. If you now set up port forwarding for the Prometheus service and open the page, you will see the fresh UI of Prometheus.

7.4 Monitoring

Next, you need to get Prometheus to fetch the data from the node exporter so that you can also query it in Prometheus.

The configuration of Prometheus is created as a ConfigMap. Let's expand this so that it knows where to fetch the data from the node exporter. To do this, go to the Prometheus ConfigMap in Lens under **Config • ConfigMaps**, which should have a name similar to `prometheus-1702161396-server`.

Click **Edit** and the YAML manifest will open. Listing 7.17 shows a section of it. The changes you need to add are marked in bold. Then save the ConfigMap, switch to the pod overview, and delete the current Prometheus pod so that the ReplicaSet builds a new pod that pulls the new ConfigMap.

```
...
data:
  prometheus.yaml: |
    global:
      external_labels:
         monitor: prometheus-1702161396
    scrape_configs:
      - job_name: 'node-exporter'
        kubernetes_sd_configs:
          - role: endpoints
        relabel_configs:
        - source_labels: [__meta_kubernetes_endpoints_name]
          regex: 'node-exporter'
          action: keep
      - job_name: prometheus
        kubernetes_sd_configs:
          - role: endpoints
            namespaces:
              names:
                - monitoring
        metrics_path: /metrics
        relabel_configs:
...
```

Listing 7.17 Extension of Prometheus Scrape Configuration

> **Good to Know**
> The better option for activating a new ConfigMap would be to use an operator. There is also an operator for Prometheus that takes over the responsibility for managing ConfigMaps. This means that you do not need to restart the pods if you change the configuration. I have skipped this for our example as it focuses on monitoring. In production

environments where you cannot simply restart services for a configuration change, you should definitely use operators.

To check whether the configuration is effective and Prometheus queries the metrics of the node exporter, you can check under **Status • Targets** in the Prometheus UI whether the metrics appear accordingly. It should look like Figure 7.11.

Figure 7.11 Node Exporter as Target of Prometheus

Extending the Node Exporter with Metrics from the Raspberry Pi

Now Prometheus can query the metrics of the node exporter, and we want to extend it in such a way that we get specific metrics for the Raspberry Pis. I found the following GitHub repository for this: *http://s-prs.co/v596456*. It may be a little older, but it is clearly programmed and works perfectly. The way it works is simple:

- You install a timer on the Pi that regularly triggers a shell script.
- The shell script runs simple commands to obtain the data, which is then provided as a metric for Prometheus.
- The data is stored in a file on the Pi's file system.

For the installation, you need to run the following command on each of your Raspberry Pis:

```
curl -fsSL "https://raw.githubusercontent.com/fahlke/raspberrypi_exporter/master/installer.sh" | sudo bash
```

This command loads the *installer.sh* file from the repository and executes it. You probably know that the concept of executing content from the internet directly into the shell is fundamentally insecure, so take a quick look at the code before you carry out the installation that provides the script and the timer.

You can then use the commands from Listing 7.18 to check whether the service is running and if the metrics are written to the file under the */var/lib/node_exporter/textfile_collector* path.

```
# Check if the service is running
systemctl status raspberrypi_exporter.timer
# Check if the metrics are written to disk
grep -E "^rpi" /var/lib/node_exporter/textfile_collector/raspberrypi-
metrics.prom
```

Listing 7.18 Commands for Checking Raspberry Pi Exporter

The next step is to extend the node exporter so that it can read the metrics in the file. For this purpose, let's extend the DaemonSet from Listing 7.15. To do this, go to the **Workloads • DaemonSets** overview in Lens and edit the YAML manifest of the node exporter. In Listing 7.19, the changes you need to make are printed in bold. For the pod to be able to access the host's file, a hostPath volume must be created that releases this exact path. The volume is then mounted in the pod so that the node exporter can access the file. As a further argument, we give the application the path to the file under the --collector.textfile.directory option so that the node exporter can read the file. As soon as you save the manifest of the DaemonSet via the **Save** option, the pods will automatically be replaced.

```
...
  spec:
    volumes:
      - name: sys
        hostPath:
          path: /sys
          type: ''
      - name: root
        hostPath:
          path: /
          type: ''
      - name: pi
        hostPath:
          path: /var/lib/node_exporter/textfile_collector
          type: ''
    containers:
      - name: node-exporter
        image: prom/node-exporter
        args:
          - '--path.sysfs=/host/sys'
          - '--path.rootfs=/host/root'
          - '--no-collector.wifi'
          - '--no-collector.hwmon'
          - '--collector.textfile.directory=/var/lib/
node_exporter/textfile_collector'
          ...
```

```
        volumeMounts:
          - name: sys
            readOnly: true
            mountPath: /host/sys
            mountPropagation: HostToContainer
          - name: root
            readOnly: true
            mountPath: /host/root
            mountPropagation: HostToContainer
          - name: pi
            readOnly: true
            mountPath: /var/lib/node_exporter/textfile_collector
            mountPropagation: HostToContainer
...
```

Listing 7.19 Extension of Node Exporter Configuration for Pi Exporter

It may now take some time for the first data to be displayed in Prometheus. You should then be able to query them as shown in Figure 7.12.

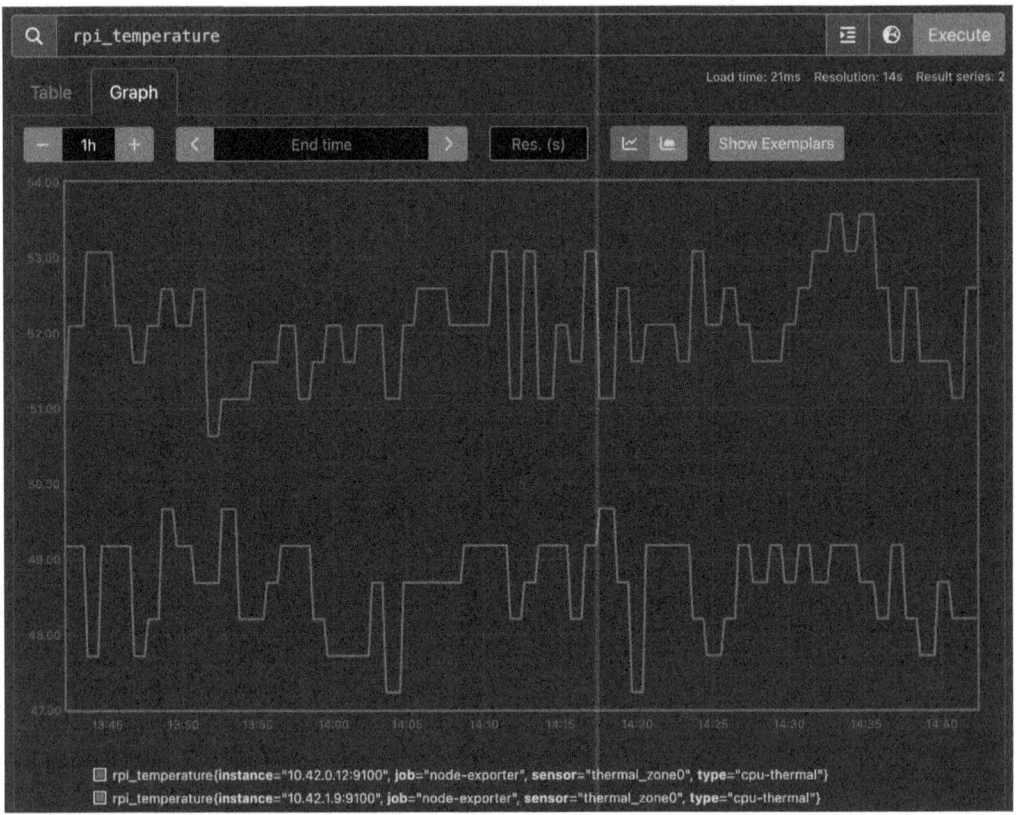

Figure 7.12 Querying Temperature Metrics in Prometheus

7.4 Monitoring

> **Note**
> If no metrics appear, you should check the logs of the node exporter. This will show you if it cannot access the file. You should also check the volume configuration again because if the path is not mounted correctly in the pod, the application within the pod will not be able to access it.

Installing and Configuring Grafana via a Helm Chart

Now let's install Grafana to display the metrics in a more appealing way and organize them in a dashboard. To do this, you want to search for Grafana in the Helm charts of Bitnami and deploy the chart via Lens as with Prometheus. Make sure that you use the same namespace. Then set up port forwarding and open the graphical user interface of Grafana. The user name and password for the login can be found in the Helm release, as described in Chapter 8, Section 8.1.2, Listing 8.4.

The first step you need to take now is to link Prometheus and Grafana. To do this, you can specify Prometheus as the data source in the menu under **Connections • Data Sources**. Click **Add Data Source** and then **Prometheus**. If you have deployed the two applications in the same namespace, you can enter the service as a URL, as shown in Figure 7.13. If they are deployed in different namespaces, you should look again for the DNS naming convention described in Chapter 2, Section 2.5.2 and enter the URL accordingly. Otherwise, you do not need to change anything in the configuration. Scroll down and click **Save & Test**.

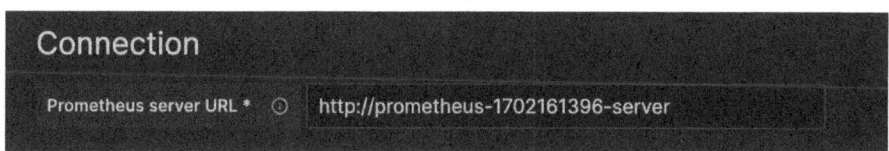

Figure 7.13 Kubernetes Service as Prometheus URL

Once the data source has been set up, you can start creating a dashboard. We'll load a standard dashboard for the node exporter so that you don't have to start from scratch. Grafana provides several standards that you can load very easily. The node exporter dashboard can be found at the following address: *http://s-prs.co/v59457*.

You only need the ID of the dashboard to load it; in this case, that's 1860. Then you can click **Create Dashboard** under **Home • Dashboards** and then **Import a Dashboard**. Enter the ID there, as in Figure 7.14; select Prometheus as the data source in the subsequent step; and then click **Import**.

7 Developing Applications for Kubernetes: Ready for Production

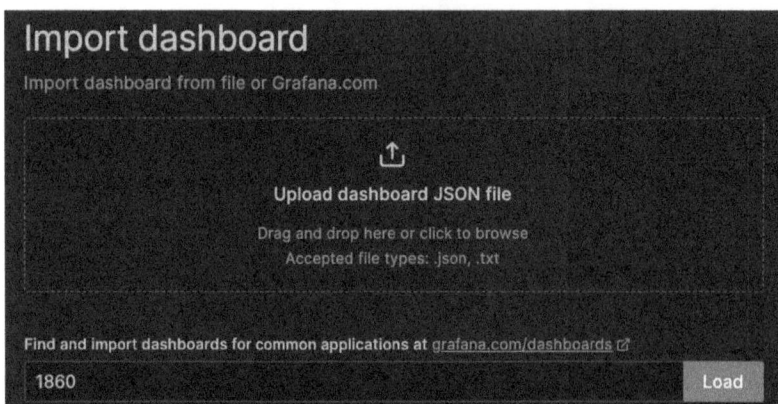

Figure 7.14 Entering Dashboard ID for Import

Next, let's expand the Grafana board to include the new Raspberry Pi metrics. To do this, you can simply add a new visualization via **Add**. The new window should then look like Figure 7.15.

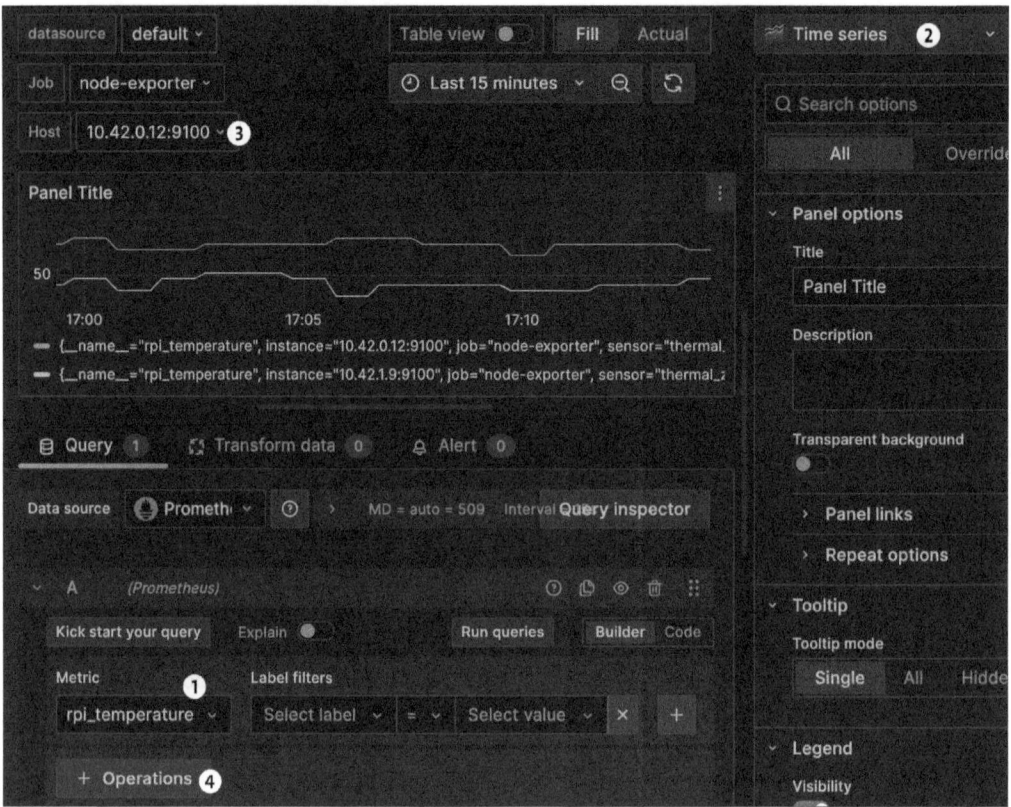

Figure 7.15 Creating Your Own Visualization in Grafana

354

> **Note**
>
> Now you can experiment with the dashboard and view the various metrics. It is important that you also select your correct Raspberry Pi as the host in the dropdown menu as in Figure 7.15 ❸; otherwise, Grafana will not display any data.

Select the metric you want to display ❶. In the dropdown menu, you will find all metrics known to Prometheus. There you should also find the new Raspberry Pi metrics starting with rpi. In this example, use the temperature to see how high it currently is.

Now you can select the type of visualization, depending on how you want to use the data ❷. A time series is a nice tool to see the progression, but I use the gauge type for this display because I'm only interested in the current temperature. Under **Operations** ❹, you can manipulate the data even further. For example, aggregation makes sense for the gauge. Try a max aggregation, and select the instance as the label. You can save the new metric to the dashboard via **Apply**.

As you can see, there are many ways to visualize your data. Feel free to play around with it some more.

Chapter 8
Orchestrating Kubernetes Using Helm

Modularity in programming is like putting together a jigsaw puzzle: every part has its place, and when everything comes together, a masterpiece is created.

Using Kubernetes comes with a number of challenges, especially when it comes to efficiently managing, deploying, and scaling applications. As you have learned throughout this book, each application entails a variety of manifests for deployments, services, volumes, and more. Managing these manifests can quickly become confusing, especially if you are trying to keep your applications consistent across multiple environments. In addition, there are questions such as the following to consider:

- How can you ensure that your Kubernetes manifests are reusable and easy to update?
- How can you efficiently manage complex applications with many dependent resources?
- How can you ensure a uniform configuration across development, staging, and production?

These and similar questions lead us to the search for a solution that not only simplifies the provision and management of applications, but also emphasizes modularity and reusability. This is where Helm comes into play.

Helm is based on the philosophies of *modularity* and *don't repeat yourself* (DRY). Tasks should therefore be divided into the simplest possible substeps and be reusable. Not only is Helm a templating tool, but it also serves as a package manager for Kubernetes manifests and supports you in deploying, managing, and orchestrating applications. It is therefore obvious why it has become an indispensable tool in the Kubernetes world.

The answer lies in the numerous advantages Helm provides to you:

- Reusability of manifests through parameterization
- Deployment of complex Kubernetes applications as a simple and configurable package
- Simple centralization and standardization within your company
- Extensive ecosystem of predefined charts for numerous applications

In the course of this chapter, we will delve deep into the world of Helm. You will learn how Helm overcomes the challenges of managing Kubernetes applications and how you can use it to make your deployments more efficient and modular. Prepare yourself to discover the possibilities Helm opens up for your Kubernetes management.

Note

The examples in this chapter are all based on Helm version 3.

8.1 Helm: The Kubernetes Package Manager

Helm is much more than a tool for parameterizing Kubernetes manifests. It is similar to the Homebrew package manager for macOS or the corresponding Linux tools. You select your software, specify where the package comes from, and can then install it on Kubernetes. A Helm package includes all the manifests the application needs to run.

In Helm, there are three key terms you need to know:

- A Helm package is referred to as a *chart* and contains all the manifests and configurations you need to deploy an application in Kubernetes.
- If you deploy a Helm chart in Kubernetes, it is referred to as a *release*. You can usually deploy a chart multiple times in a cluster, and each instance is a separate release.
- As with other package managers, charts can be stored and distributed in a *repository*.

The parallels to containers and management by images, which are stored in repositories and instantiated to running containers, are obvious. However, there is a significant difference in terms of flexibility: while a container image is unchangeable, you can deeply customize Helm charts with specific configurations for each release. This gives you the freedom to adapt and optimize applications exactly as required.

You can imagine the development process as shown in Figure 8.1. You work locally on a chart and develop the Kubernetes manifests. You can either deploy these directly to Kubernetes or store them in a Helm repository. From there, you can also access other Helm charts, configure them, and then transfer them to Kubernetes.

But let's do one thing at a time.

Let's first take a look at what is contained in a Helm chart. As you will see later, a Helm chart usually contains multiple Kubernetes manifests. Let's take a web application that uses a database as an example. The Helm chart contains the deployment, the service, and the ingress manifest of the web application as well as the StatefulSet and service manifest of the database. So you get everything Kubernetes needs to start the application in one chart.

8.1 Helm: The Kubernetes Package Manager

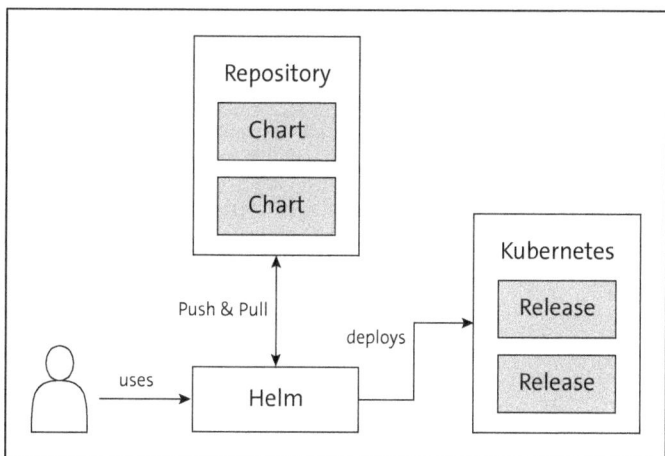

Figure 8.1 Architecture of Helm

The difference from ordinary Kubernetes manifests is the templating of Helm. Helm comes with a templating engine based on Go so that you can adapt the manifests to different environments or requirements using configurations.

Take a look at the release process in Figure 8.2. You can see that you need a *values.yaml* file in addition to the Helm chart, which contains the parameters needed by your chart. The magic of Helm lies in these values.

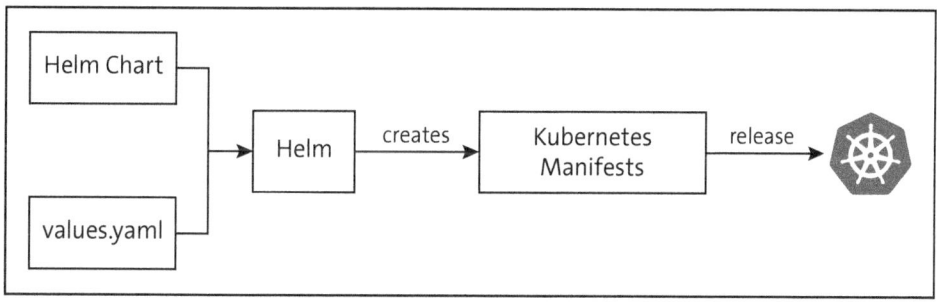

Figure 8.2 Process for Releasing Helm Chart in K8s

Each finished chart comes with a *values.yaml* file that sets all the necessary parameters. You can view this as a kind of default value that enables you to easily roll out the chart. Things start getting magical when you start to overwrite the values in different situations, such as to replace the default values with environment-specific values. You do not need to adapt the existing *values.yaml* file for that, but simply give Helm another *values.yaml* during the templating process.

Let's take the *values.yaml* file from Listing 8.1, which is provided in a Helm chart, as an example.

```
name: humanity-backend
version: 1.0.0
```

Listing 8.1 Default "values.yaml" File

The name of the application and the version to be rolled out are set there. You now want to customize the chart for your development environment and roll out a more recent version. To do this, simply give Helm the *values-dev.yaml* file from Listing 8.2.

```
version: 1.2.0
```

Listing 8.2 Environment-Specific "values-dev.yaml" File

The values you set in this values file will replace the values from the default values file. The great thing about this merging process is that you can define the values in different ways and bring them into your charts. You can split your values into different files such as the following:

- Company-specific values
- Environment-specific values
- Cluster-specific values

We will go into more detail about the options you have with values and how best to structure them in Section 8.2.2.

> **Note**
>
> The way Helm charts are packaged and provided gives you the opportunity to develop standards for applications in your company. Imagine a SaaS agency that operates the same software for multiple customers. Of course, the configuration has to be slightly different for all customers, but this can be easily adjusted using charts. Instead of maintaining a separate application for each customer, you can work in a standardized way and only address the differences during deployment.

But enough theory for now. You will need Helm on your computer for the following sections. The installation instructions for the Helm CLI for your system can be found at *http://s-prs.co/v596458*.

You can find the appropriate shell completions for your command line at *http://s-prs.co/v596459*.

Concerning the Helm CLI, it works similar to `kubectl`. You can use `helm -h` to display all commands. The completion feature is very helpful in everyday life because nobody needs to know all the calls in detail. You will become familiar with the most important commands in the following sections.

8.1.1 Creating a First Helm Chart

To get started, let's have Helm generate the default chart for us. To do this, you want to run the `helm create humanity-backend` command in your command line, where `humanity-backend` is the name of the chart. Your folder structure should then look as shown in Listing 8.3. Helm uses this command to create all the necessary files and even provides you with a few sample manifests.

```
.
├── Chart.yaml
├── charts
├── templates
│   ├── NOTES.txt
│   ├── _helpers.tpl
│   ├── deployment.yaml
│   ├── hpa.yaml
│   ├── ingress.yaml
│   ├── service.yaml
│   ├── serviceaccount.yaml
│   └── tests
│       └── test-connection.yaml
└── values.yaml
```

Listing 8.3 Folder Structure of Default Helm Chart

You can now get an impression of the structure you have created:

- What does a Helm manifest look like?
- What is different from what you already know?
- Where do you find parallels to what you already know?

We will return to this sample chart again and again in the following sections. It already contains a lot of what you need to know and what Helm enables you to do.

> **Note**
> You should install a Helm plugin for your IDE. I use IntelliJ and have installed the suggested Go template plugin.
>
> This makes your work a little easier, as you can jump directly from the template to the value and generate the manifest in the IDE.

8.1.2 Deploying a Helm Chart via the Command Line Interface

Let's look at how to roll out a chart using the Helm CLI and at the sample humanity-backend Helm chart.

8 Orchestrating Kubernetes Using Helm

If you are in the chart folder, you can roll out the chart via one of two commands:

- `helm install -f values.yaml humanity-backend .`
- `helm upgrade --install -f values.yaml humanity-backend .`

As mentioned previously, in these commands, `humanity-backend` is the name of the release, and the final period (.) references the folder with the Helm chart.

You can use both commands to release the Helm chart in Minikube for the first time. In this case, note that you must add the `--install` option to `helm upgrade`; otherwise, Helm will throw an error as there is no release for upgrading yet. The `--install` option checks if there is already a release, and if not, it will roll out the release like `helm install`. You can use the same `helm upgrade` command when rolling out an update. The `helm install` command would then throw an error because the release has already been rolled out in the cluster.

> **Good to Know**
>
> Typically, I just use the upgrade command in a CI/CD pipeline because then I don't need to distinguish whether a Helm chart has already been released or not.
>
> However, the install command is useful if you want to make sure that there is not yet a release with the corresponding name and do not want to overwrite an old version.

You can use the `-f` option to specify the values file to be used for templating. You can enter as many values files as you like, and the last entry always overwrites the previous one. This allows you to overwrite default values with environment-specific values, for example. An example could look as follows:

`helm upgrade --install -f values.yaml -f values-dev.yaml humanity-backend .`

We'll take a closer look at exactly how to use values in Section 8.2.2.

As you know from the introduction, Helm also has repositories in which the charts are stored. Here, we have deployed the chart directly from the folder. In Figure 8.3, you can see that both methods are possible. You can deploy directly from the repository as well as from the folder. Then you can install the Jenkins chart from the Bitnami repository via the following command, for example:

`helm install jenkins bitnami/jenkins`

Instead of specifying the path to the Helm chart as before, you want to enter `bitnami/jenkins`—that is, `[reponame]/[chartname]`. We'll look at how you can use a repository in Section 8.1.3, while Section 8.1.4 provides more insight into the Jenkins chart itself.

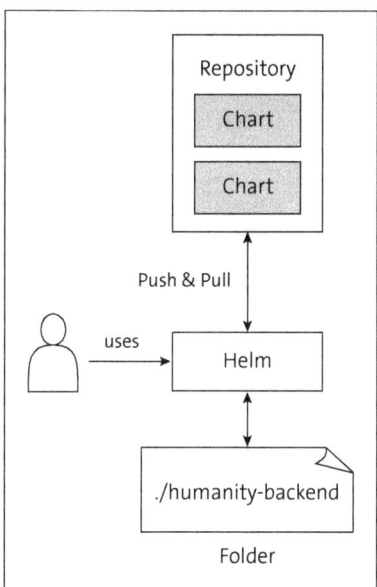

Figure 8.3 Helm Chart in Repositories or as Folder Structure

Good to Know

Whether you store your chart in a Helm repository is a question of your development process. A chart in the repository is ultimately just a packed archive with all files such as templates and values. In Section 8.3.2, you will learn how to package a chart and store it in a repository.

8.1.3 Setting Up and Managing a Helm Repository

One of the advantages of Helm is that you can use numerous open-source charts that are available in public repositories. All you have to do is familiarize your Helm CLI or Lens with the corresponding repositories. Now let's discuss how to integrate new Helm repositories to access these Helm charts.

Let's start with Lens. For this purpose, you need to open Lens and go to the settings. Under the **Kubernetes** menu item, there is a **Helm** area that should look like Figure 8.4. There you can select any known Helm repository via the dropdown menu. Lens uses Artifact Hub as the official site to display the repositories.

Note

You can find Artifact Hub at *https://artifacthub.io/*.

There you can also search for Helm charts and find many open-source applications that you can install directly via Helm.

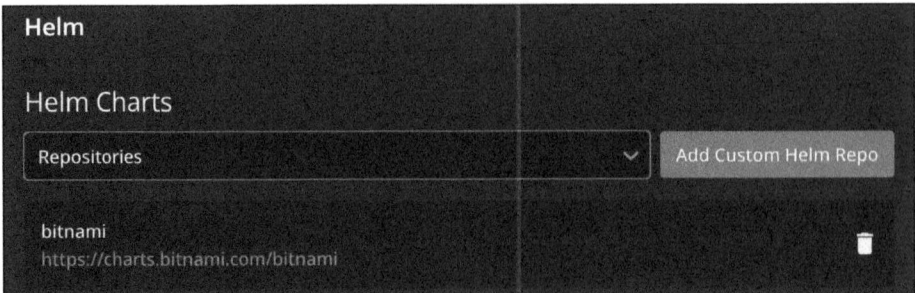

Figure 8.4 Helm Repositories in Lens

As you can see, the Bitnami repository is already selected, and you can add more from the dropdown menu if required.

> [+] **Good to Know**
>
> *Bitnami* is a library of ready-made software packages. You will find the most popular open-source applications from Bitnami already configured and easy to use. Especially when I want to set something up quickly or try out a new tool, I look for Bitnami packages first.

> [!] **Bitnami in Production Environments**
>
> Note that the Helm charts from Bitnami and other providers are not necessarily suitable for your production environments. They are usually designed to be "developer-friendly," so I like to use them for testing. However, there is no guarantee that vulnerabilities, for example, will be patched regularly. In addition, the charts are very large, difficult to understand, and not (only) tailored to your use case. Sometimes it is better to simply write your own chart and follow the KISS principle.
>
> Thus, you should take a closer look if you want to use one of these packages in production and discuss it with your IT department. Helm charts from the internet are very suitable for initial tests on your own playground, but when it gets serious, you should create your own deployments and check exactly what is needed.

If you want to manage Helm repositories in the CLI, you should use the `helm repo` commands. For example, you can use the following command to add the repo for the Harbor tool:

```
helm repo add harbor https://helm.goharbor.io
```

You can use `harbor` to specify the name of the repo in your list and add the URL under which the repo can be reached.

The `helm repo list` command enables you to display your added repos, while `helm repo remove` allows you to remove repos, and `helm repo update` updates your local index of the repository. You need the updated index so that you can also pull Helm charts that have been newly added.

> **Note**
>
> If you have your own Helm repositories in your company, you can also add them via the Helm CLI. Unfortunately, Lens has some difficulties adding it on its own, but once you have added the repo, Lens can also access it.

8.1.4 Deploying a Helm Chart via Lens

Now that you know how to use the Helm repositories, let's look at how you can roll out one of these charts through Lens. We'll use the Jenkins chart from the Bitnami repository for this purpose. To do this, click **Helm · Charts** in your cluster menu in Lens and enter "Jenkins" in the search field. When you click the chart, you will see more information about the chart on the right-hand side, as shown in Figure 8.5—for example:

- The version
- How to install the chart
- How to configure the chart

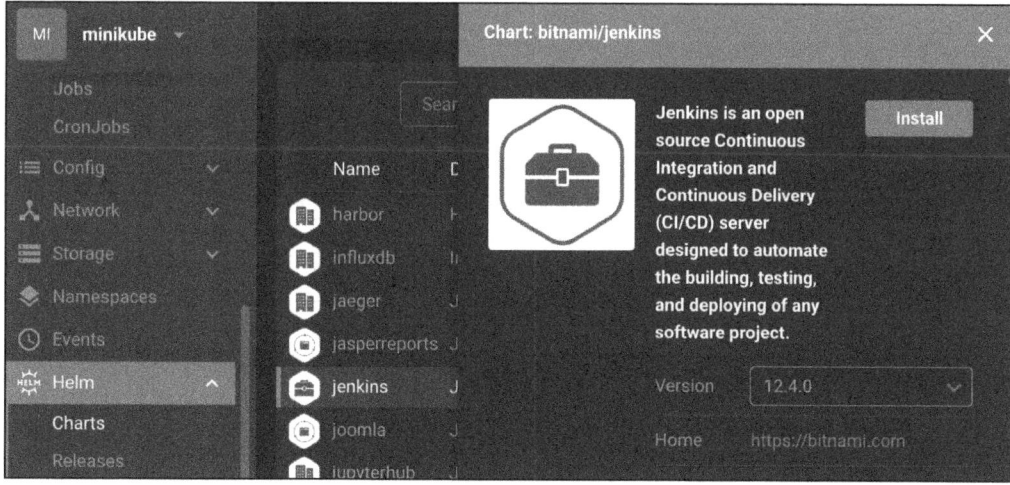

Figure 8.5 Selecting Jenkins Chart in Lens

> **Good to Know**
>
> The difference from the Helm CLI is that you can only use Lens to deploy charts that are stored in a repository. For example, you do not have access to the humanity backend

> chart that is already on your computer. However, we will look at how you can use your own Helm repository in Section 8.3.2.

I use chart version 12.4.0 as this is currently the latest version. If necessary, you can simply use older versions of the software. Now click **Install** in the top-right-hand corner.

The values file of the Helm chart opens at the bottom, which you can use to customize charts (Section 8.2.2). Confirm the settings in the values file by clicking **Install**. That's it. Jenkins will now be rolled out in your cluster.

To be able to access your new Jenkins server, you will of course need more information about it. Helm provides the option of outputting notes that are necessary for the developer. To retrieve these notes, click **Helm · Releases** in Lens and then on your Jenkins release. This will open another window in which you will find the information you need under **Notes**. The notes should look similar to those shown in Listing 8.4.

```
1st Get the Jenkins URL by running:
...
2nd Login with the following credentials
echo Username: user
echo Password: $(kubectl get secret --namespace default jenkins-1698790333
      -o jsonpath="{.data.jenkins-password}" | base64 -d)
```

Listing 8.4 Release Notes of Jenkins Installation

You don't need the URL for now, because you can use simple port forwarding to access Jenkins. However, what you will need is the password that was stored fully automatically as a Kubernetes secret. Bitnami already provides you with the corresponding `kubectl` command. (*Remember:* Your command will look slightly different because your secret has a different name. This means that you cannot copy it directly from the book.) Now check whether you can access the Jenkins GUI.

> [!] **Warning**
> When I use the command to retrieve the secret in my Mac command line, the system appends a percent sign (%) to the password. This could also be the case for you. Pay attention to this and leave it out when copying.

8.1.5 Updating and Deleting Helm Releases

You have already seen the `helm upgrade` command. This command can also be used to import a new version of the chart. There are generally two reasons that you want to install updates: either the chart version has changed because you have made changes to the templates, or you want to import new values.

8.1 Helm: The Kubernetes Package Manager

For both options, simply run the following command and reference the new chart or enter the latest *values.yaml* file:

`helm upgrade --install -f values.yaml humanity-backend` .

In Lens, you can also simply select the release under **Helm • Releases** and then click **Upgrade** as in Figure 8.6 to select a newer version in the window that opens. There, adjust the values.

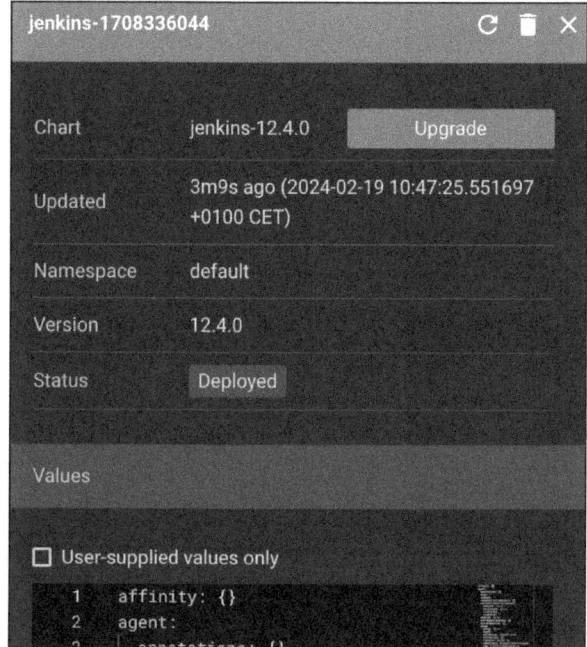

Figure 8.6 Helm Release Upgrade in Lens

You can select the version using the dropdown window as in Figure 8.7.

Figure 8.7 Selecting Upgrade Version

If you only want to change the values, you can do this under **Values**, as shown in Figure 8.6. There you will see the current values. You can adjust them and then click **Save**.

Once you have performed an upgrade, Helm will take care of the rollout. For example, if you have customized a deployment, Helm will make the changes and, if necessary, swap the pods. Helm increments the revision of the release by one with each upgrade. You can see the revision in the release overview.

If you want to delete all Kubernetes resources of a Helm release, this is very simple. Helm remembers which resources belong to a release and can therefore easily update or delete them. You may have already seen this in Section 8.1.2. You can access the releases in Lens under **Helm • Releases** and also delete them there if you click the Helm release and then click **Delete**. If the Helm release is deleted, all Kubernetes resources that were managed by the release are also deleted.

The Helm CLI enables you to view the releases via `helm ls`, and the response should look like Listing 8.5 in a shortened version. Use `helm uninstall jenkins-170` to delete the release.

```
NAME          NAMESPACE   REVISION   STATUS     CHART
jenkins-170   default     1          deployed   jenkins-12.4.0
```

Listing 8.5 Output of "helm ls" Command

8.1.6 Downloading Helm Charts from a Repository

Not only can you install Helm charts directly, but you can also download them. This is very useful, for example, if you want to get a specific inspiration or understand a chart in detail. The command for the download is composed as follows:

`helm pull [reponame]/[chartname]`

For example, if you want to download the Jenkins chart from Bitnami, you can use the `helm pull bitnami/jenkins` command. In Section 8.3, we'll take a look at how such a Helm chart is structured.

> **Note**
>
> If you are unsure about the name of the repo, you can use `helm repo ls` to check how the name is defined.

> **Note**
>
> The `--untar` option makes sure that the chart is unpacked directly.

8.2 Reading and Developing Helm Charts

You have now been introduced to the basic principles of Helm, have used the CLI, and have already deployed your first chart. In this section, we'll dive into the basics of developing Helm charts. If you have looked at the Helm templates from the sample `humanity-backend` chart, you will find a new syntax that differs fundamentally from the previous YAML manifests.

First, we'll look at the templating engine of Helm. You will learn how to use the Go templating engine to create flexible and reusable manifests that you can easily adapt to different environments. You will then learn more about the principle of values in order to fill the templates with values. Then you will be introduced to various template functions, which I have divided into three sections for you. Finally, we'll look at the Helm diff plugin, which helps you to display differences between the current and planned states of your Helm releases.

After reading this section, you will have solid knowledge of how to read and develop Helm charts.

8.2.1 The Templating Engine and the Language of the Charts

You have already written several manifests in the course of the book and deployed them in Kubernetes. If you look at the deployment manifest from the default Helm chart, it looks very strange at first glance. You can find an excerpt of it in Listing 8.6. Even if the syntax seems a little strange at first, the aim of Helm is to create a YAML manifest at the end, as you already know. This is the only way Kubernetes can do anything with the manifest.

```yaml
apiVersion: apps/v1
kind: Deployment
metadata:
  name: {{ include "humanity-backend.fullname" . }}
  labels:
    {{- include "humanity-backend.labels" . | nindent 4 }}
spec:
  {{- if not .Values.autoscaling.enabled }}
  replicas: {{ .Values.replicaCount }}
  {{- end }}
  selector:
    matchLabels:
      {{- include "humanity-backend.selectorLabels" . | nindent 6 }}
  template:
    metadata:
      {{- with .Values.podAnnotations }}
      annotations:
        {{- toYaml . | nindent 8 }}
      {{- end }}
      labels:
        {{- include "humanity-backend.labels" . | nindent 8 }}
        {{- with .Values.podLabels }}
        {{- toYaml . | nindent 8 }}
        {{- end }}
...
```

Listing 8.6 "deployment.yaml" from Default Helm Chart

At the heart of the Helm engine is the templating syntax, which is based on Go templates and is supplemented by some Helm-specific extensions. A major goal is to ensure the reusability of a manifest by means of parameterization. For example, {{ .Values.replicaCount }} refers to a value in *values.yaml* called replicaCount.

Helm uses double curly brackets ({{ }}) as placeholders within the YAML files of Kubernetes to insert dynamic values or execute logic. Everything inside these brackets is evaluated by Helm before the final manifest is passed to Kubernetes. As shown in Figure 8.8, the Helm templating engine will take all templates, import one or more value files, and then process each of the template commands. Helm not only can use values, but also has a wide range of options, which we will now look at in more detail.

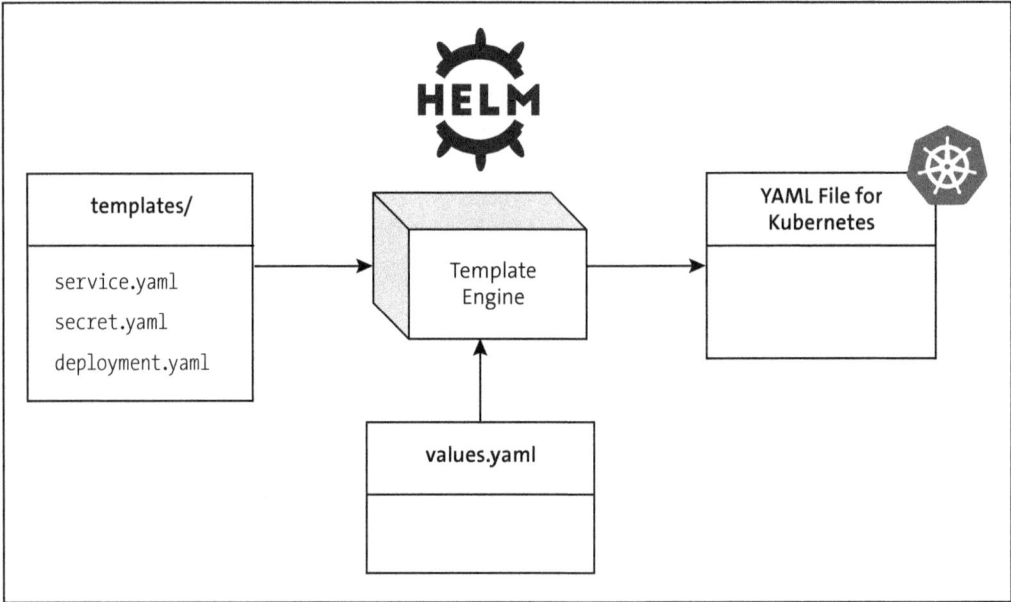

Figure 8.8 Helm Templating Engine

Before we start with the examples, I want to show you how you can carry out templating using Helm. This allows you to track the impact of your changes during the development process. If you are in the folder of the Helm chart, such as in humanity backend, then you can run the helm template . command. Then you will receive the generated manifests as output on the console.

> **Note**
>
> Templating is even easier with plugins for your IDE. I use the Kubernetes plugin for IntelliJ, which also supports Helm.

Built-In Functions

Helm provides a variety of built-in functions that you can use within the templates to manipulate values or evaluate conditions. You connect a function with a | (like the pipe character in Linux) after the value, as shown in Listing 8.7. There I have extended *deployment.yaml* and added the upper and quote functions after the names. This reads the value from the humanity-backend.fullname parameter, then converts everything to uppercase and places it in quotation marks.

```
...
name: {{ include "humanity-backend.fullname" . | upper | quote }}
...
replicas: {{ .Values.replicaCount | default 3 }}
```

Listing 8.7 "deployment.yaml" Extended by Helm Functions

> **Note**
>
> If you look at where exactly some values come from, you will find the *_helpers.tlp* file in the humanity-backend example. This is an advanced way to define and prepare parameters to keep the manifests "clean" and readable.
>
> You can learn more about this at *http://s-prs.co/v596460*.

Try it out for yourself and carry out the templating to see the result. Another example in Listing 8.7 has the default value, which you can define in the template.

One frequently used function is indent or nindent. As an example, take a look at the excerpt from Listing 8.8 of the *deployment.yaml* file. The function inserts a certain number of spaces before the values. This is particularly useful to keep the indentations in generated YAML files correct; doing otherwise will lead to misinterpretations. In our example, four spaces are inserted before each label so that the labels are arranged under the labels object.

```
...
  labels:
    {{- include "humanity-backend.labels" . | nindent 4 }}
...
```

Listing 8.8 Indent Example from "deployment.yaml" File

If you render the template, the result looks like Listing 8.9.

> **Good to Know**
>
> The nindent function differs from indent only in that it also inserts a new line.

```
apiVersion: apps/v1
kind: Deployment
metadata:
  name: release-name-humanity-backend
  labels:
    helm.sh/chart: humanity-backend-0.1.0
    app.kubernetes.io/name: humanity-backend
    app.kubernetes.io/instance: release-name
    app.kubernetes.io/version: "1.16.0"
    app.kubernetes.io/managed-by: Helm
...
```

Listing 8.9 Indent Example Rendered

Another function that you will need often is the conversion to Base64, which is used to encode values in Helm templates. The typical use case is a secret, where the values are expected as a Base64 string.

> **[+] Good to Know**
>
> You can use these functions anywhere in Helm. For example, you could encode the name of the deployment from Listing 8.8 using Base64, but this makes no sense in this case.

In the humanity-backend sample chart, create the new secret file from Listing 8.10 under *templates*. Add the new password: test1234 key-value pair to *values.yaml*.

```
apiVersion: v1
kind: Secret
metadata:
  name: my-secret
type: Opaque
data:
  password: {{ .Values.password | b64enc }}
```

Listing 8.10 Secrets with Base64 Template

After calling helm template ., the generated secret will look as shown in Listing 8.11.

```
apiVersion: v1
kind: Secret
metadata:
  name: my-secret
type: Opaque
```

```
data:
  password: dGVzdDEyMzQ=
```

Listing 8.11 Generated Secret

You have now become familiar with a few of the most common functions and have seen how you can use them. There are many others you can look up if you need them. An overview of this is shown at the following URL: *http://s-prs.co/v596461*.

Built-In Objects

Helm contains so-called built-in objects that you can also reference within the template. For example, these objects contain important information about the chart itself or the current release, which is useful for the configuration and deployment of resources in Kubernetes. The most frequently used built-in objects include the chart and the release.

You can use the values of these objects in the same way as the normal values. In Listing 8.12, you will find an example of the charts object.

```
apiVersion: v1
kind: ConfigMap
metadata:
  name: {{ .Chart.Name }}-config
data:
  chartName: {{ .Chart.Name }}
  chartVersion: {{ .Chart.Version }}
```

Listing 8.12 Sample Chart Object

You can store the information in a ConfigMap there, for example. Alternatively, you can use the chart version within your deployment template. You could use the release as in Listing 8.13, for example, to name your deployment. This is very useful and used often.

> **Note**
> An overview of all built-in objects and their parameters can be found at *http://s-prs.co/v596462*.

```
apiVersion: apps/v1
kind: Deployment
metadata:
  name: {{ .Release.Name }}
```

Listing 8.13 Sample Release Object

8.2.2 Configuring Charts with Values

Let's move on to the Helm values. The values files are at the heart of every chart because this is where the values are stored that turn the template into a real Kubernetes manifest. As mentioned previously, the main reason for using values files in Helm charts is the separation of configuration and code. Instead of writing hard-coded settings directly in the Kubernetes manifests, values allow settings such as image versions, resource limits, and other configurations to be injected dynamically.

The values are also written in YAML. You can apply everything you have learned in Chapter 3, Section 3.2 here. However, depending on the structure, you may need to access the values differently in the templates. Helm distinguishes between a flat structure and a nested structure.

Listing 8.14 shows a *flat structure*. The values are arranged directly under the root element and are ideal for simple configurations. They are easy to understand and easy to change.

```
name: nginx-app
imageName: nginx
imageTag: stable
```

Listing 8.14 Flat "values.yaml" File

The template in Listing 8.15 is also easy to read. It is immediately clear which parameter is being used.

```
spec:
  template:
    spec:
      containers:
        - name: {{ .Values.name }}
          image: "{{ .Values.imageName }}:{{ .Values.imageTag }}"
```

Listing 8.15 Template for Flat "values.yaml"

The *nested structure* in Listing 8.16 is well-suited for more complex charts if you want to group configurations logically. This makes it easier to read the *values.yaml* file, but quickly compromises the readability of the entire template.

```
application:
  name: nginx-app
  image:
    repository: nginx
    tag: stable
```

Listing 8.16 Nested "values.yaml" File

In Listing 8.17, you can already see that the link grows with each nesting level. .Values.imageTag becomes .Values.application.image.tag.

```
spec:
  template:
    spec:
      containers:
      - name: {{ .Values.application.name }}
        image: "{{ .Values.application.image.repository }}:{{ .Values.application.image.tag }}"
```

Listing 8.17 Template for Nested "values.yaml" File

> **Note**
>
> For nested values, Helm recommends that an "existence check" be carried out for each level. You should therefore check that the value is set at all. This also inflates the templates and makes them more difficult to read.
>
> We'll look at the existence check in Section 8.2.3.

Overwriting Values during Rollout

One of the strengths of Helm is the flexibility it offers in configuring deployments by overwriting values in the *values.yaml* file. This allows you to make adjustments to the configuration without having to edit the Helm chart or the default values yourself. This is particularly useful for third-party charts. It gives you two options for setting new values. You can enter the values directly via the command line, or you can enter an additional *values.yaml* file.

By using the --set flag, you can overwrite individual values or nested values. For example, if you want to activate autoscaling in the humanity-backend example, you can do this as follows:

```
helm upgrade --install -f values.yaml
          --set autoscaling.enabled=true humanity-backend .
```

Use this command to overwrite the default value in *values.yaml*, which is set to false.

> **Good to Know**
>
> This variant is a good way of setting secrets such as passwords, especially in CI/CD pipelines. This means that you do not have to check them in as code in your Git repo, but can inject them into the pipeline at runtime.

> **Note**
>
> You can also use `--set` to overwrite a value in a list, but this is not recommended and can be error-prone if the order of the list changes unexpectedly. Let's assume the following:
>
> ```
> containers:
>   - name: nginx
>   - name: database
> ```
>
> In this case, you could adjust the name of the database by using `--set containers[1].name=postgres`.

The second option would be to pass another *values.yaml* file. This is particularly useful if you want to adjust a large number of values and version them via Git. A classic use case is when you have different configurations for each environment. For example, if you want to activate autoscaling in the development environment, simply create a *values-dev.yaml* file as in Listing 8.18.

```
autoscaling:
  enabled: true
```

Listing 8.18 Example of "values-dev.yaml" File

You can then simply extend the command as follows:

```
helm upgrade --install -f values.yaml -f values-dev.yaml \
    humanity-backend .
```

> **Good to Know**
>
> You can extend the list of values files as per your requirements. The file that you specify last in the command has the highest priority and overwrites all values of the previous files and so on.
>
> Other use cases for additional values files include, for example, specific configurations according to regions, AWS accounts, teams, or other criteria. Using different value specifications, you can easily provide the individualized setups of a standardized deployment.

Structure of a Values File

If you develop charts yourself, then a well-structured values file is the be-all and end-all. However, as is so often the case, there is no one true structure. Depending on the size of the chart, the values file will be different or even change in the course of development. An example is shown in Listing 8.19. There we have divided the file into `application settings`, `service configuration`, and `advanced settings`.

```
# Application settings
application:
  name: "nginx-app"
  image:
    repository: "nginx"
    tag: "latest"
  replicaCount: 2
# Service configuration
service:
  type: "LoadBalancer"
  port: 80
# Advanced settings
advanced:
  loggingLevel: "INFO"
```

Listing 8.19 Possible Structure of Values File

You could also define the log level at the root level or generally create a grouping for your application's environment parameters. Also think about other members of your team who may need to use your charts. What could they want to change in the chart without having to touch the templates?

> **Note**
> Define appropriate default values for all configuration options. These values should be selected in such a way that the chart works out of the box, but at the same time can be easily adapted to specific requirements. This may save you one or two configuration files.

If you look at the *values.yaml* file of `humanity-backend`, you will find various related blocks and sometimes individual key-value pairs. Most of them are provided with comments so that it is clear what you are configuring. At first glance, the file is clean and easy to understand.

You can find a beautiful values design there under `resources`. I have inserted the excerpt again in Listing 8.20.

```
resources: {}
...
  # limits:
  #   cpu: 100m
  #   memory: 128Mi
```

```
# requests:
#   cpu: 100m
#   memory: 128Mi
```

Listing 8.20 Resource Configuration in humanity-backend

There is even a commented example of how you can set the configuration. The author of the chart has left it up to you whether you want to include limits and requests. This type of value is perfect if you want to make an entire part of the manifest configurable, which means that you do not have to set each value individually in the template but can swap out the template part to the values file.

In the deployment, exactly what is defined in *values.yaml* is simply inserted at the end. Here is the command from *deployment.yaml*:

```
resources:
  {{- toYaml .Values.resources | nindent 12 }}
```

You can find even more values of this type in *values.yaml*. During development, you should also consider whether you want to swap out entire parts of the template to the values file in order to give the user more flexibility later on.

8.2.3 Conditions in Helm Templates

Conditions in Helm allow you to show or hide parts of a template based on certain values or configurations. A condition checks a value or expression and only executes the template instructions contained therein if the condition is evaluated as true.

Possible application scenarios include the following:

- **Feature flags**
  Activate or deactivate certain components of a chart based on flags in the *values.yaml* file. This is ideal for a step-by-step introduction.

- **Environment-specific configuration**
  Customize configurations depending on the environment in which you are deploying.

- **Dependencies between resources**
  Control the configuration of resources based on the creation of another resource in the same chart. This allows you, for example, to change the replicas of a deployment when you activate a horizontal pod autoscaler.

Let's now take a closer look at the latter example.

In humanity-backend, you will find the *hpa.yaml* file in the templates. The entire object is surrounded by an if statement, as you can see in Listing 8.21. This statement checks whether the `autoscaling.enabled` value is set to `true`, which means that the horizontal pod autoscaler is only generated if you set this value to `true`.

```
{{- if .Values.autoscaling.enabled }}
...
{{- end }}
```

Listing 8.21 Conditions in "hpa.yaml" File

The default values in *values.yaml* can be found in Listing 8.22. Activate the flag there and run the `helm template .` command. Helm should now also display the HPA object.

```
autoscaling:
  enabled: false
  minReplicas: 1
  maxReplicas: 100
  targetCPUUtilizationPercentage: 80
```

Listing 8.22 Autoscaling Values

The *hpa.yaml* file is not the only template in this chart that uses this flag. In *deployment.yaml*, you will find the code from Listing 8.23 because the configuration of the replicas is deactivated in the deployment as soon as an HPA object is created. This makes sense in this case, because you no longer want to define a fixed number of replicas but instead want to use autoscaling.

> **Note**
>
> If an object is not activated by a flag, then the remaining values used under this object are not actually necessary. You should still set default values in *values.yaml* during development so that the user of a chart simply needs to set the flag to `true`.

```
{{- if not .Values.autoscaling.enabled }}
replicas: {{ .Values.replicaCount }}
{{- end }}
```

Listing 8.23 Dependence of Flag for "deployment.yaml"

As in every programming language, you also have the option of defining an `else` for every `if`. An example of this is shown in Listing 8.24. This is very useful if, for example, you want to implement a new API version of the object and use a flag to control which option is active.

```
{{- if .Values.autoscaling.enabled }}
# Manifests for autoscaling
{{- else }}
# Alternative setup if autoscaling is not activated
{{- end }}
```
Listing 8.24 Example of Else Statement

The if statement is also an existence check. In the *hpa.yaml* file, you will find the code from Listing 8.25. This code is inserted into the manifest only if the autoscaling.targetCPUUtilizationPercentage parameter is set.

```
{{- if .Values.autoscaling.targetCPUUtilizationPercentage }}
- type: Resource
  resource:
    name: cpu
    target:
      type: Utilization
      averageUtilization: {{ .Values.autoscaling
          .targetCPUUtilizationPercentage }}
{{- end }}
```
Listing 8.25 Sample Existence Check

Just try it out! Set some different values, think about what the end result should look like, and check it using the helm template . command. Conditions are a powerful tool to make your Helm charts more flexible and reusable.

8.2.4 Other Operations and Control Structures

In addition to the basic if statements, Helm templates provide a variety of advanced operations and control structures that further increase flexibility. In this section, we'll look at some of these advanced techniques and how they can be applied in real life.

With Statement

The need to repeatedly specify long or complex paths to access certain object properties can impair the readability of configuration files and make them more confusing.

The with statement allows you to set the context for the block in which it is used to a specific value. By defining a local context using the with statement, you can refer to the immediate properties of the context object without specifying the full path.

Let's assume you have a deployment and want to structure the values as in Listing 8.26.

8.2 Reading and Developing Helm Charts

```
application:
  name: "nginx-app"
  image:
    repository: "nginx"
    tag: "latest"
  replicaCount: 2
```

Listing 8.26 Values for Deployment

Your deployment will look like Listing 8.27 without the `with` statement. In each parameter, you use `application` repeatedly to refer to the corresponding object in the values file.

```
apiVersion: apps/v1
kind: Deployment
metadata:
  name: {{ .Values.application.name }}
spec:
  replicas: {{ .Values.application.replicaCount }}
  template:
    metadata:
      labels:
        app: {{ .Values.application.name }}
    spec:
      containers:
      - name: {{ .Values.application.name }}
        image: "{{ .Values.application.image.repository }}:{{ .Values.application.image.tag }}"
```

Listing 8.27 "deployment.yaml" File without "with" Statement

If you use `with` to set the context for the deployment to `application`, your *deployment.yaml* file will look like Listing 8.28. As you can see, this makes the lines clearer and avoids unnecessary repetition.

```
{{- with .Values.application }}
apiVersion: apps/v1
kind: Deployment
metadata:
  name: {{ .name }}
spec:
  replicas: {{ .replicaCount }}
  template:
    metadata:
```

```yaml
      labels:
        app: {{ .name }}
    spec:
      containers:
      - name: {{ .name }}
        image: "{{ .image.repository }}:{{ .image.tag }}"
{{- end }}
```

Listing 8.28 "deployment.yaml" Including "with" Statement

Range Statement

When developing and managing Kubernetes applications, developers are often faced with the challenge of dynamically configuring and creating multiple resources. The task becomes particularly complex if the number or configuration of these resources is supposed to be variable and depends directly on the values in *values.yaml*. A typical example is the task of flexibly designing the number of containers within a pod.

You can use the range statement to iterate over lists or maps. Imagine you want to create a pod manifest but make the number of containers configurable via the *values.yaml* file. Listing 8.29 contains an example that allows you to create three containers within the pod, using *values.yaml* from Listing 8.30.

```yaml
apiVersion: v1
kind: Pod
metadata:
  name: multi-container-pod
spec:
  containers:
  {{- range .Values.containers }}
  - name: {{ .name }}
    image: {{ .image }}
  {{- end }}
```

Listing 8.29 Pod Manifest with "range" Function

```yaml
containers:
  - name: web-server
    image: nginx:latest
  - name: app-server
    image: myapp:1.2.3
  - name: helper
    image: helper:latest
```

Listing 8.30 "values.yaml" for "range" Statement

Logical Operators and Comparison Operators

Sometimes a simple if statement is not enough. Helm templates also support the use of logical operators such as and, or, and not to evaluate complex conditions. Helm also supports a range of comparison operators such as eq (equal to), ne (not equal to), lt (less than), le (less than or equal to), gt (greater than), and ge (greater than or equal to) for comparing values.

These operators behave as they do in any other programming language. Listing 8.31 contains a few examples to get a feel for this.

```
{{- if and .Values.enabled .Values.production }}
# enabled and production are both true
{{- if or .Values.beta .Values.preview }}
# beta or preview is true
{{- if not .Values.disabled }}
{{- if eq .Values.environment "production" }}
{{- if lt .Values.replicas 3 }}
```

Listing 8.31 Example of Operators

8.2.5 Helm Diff for Checking Changes

Helm charts can quickly become complex and confusing due to templating. Especially when you are working on charts in a team, it is sometimes difficult to see in advance what the adjustment of a particular value will change. You could then use helm template to generate and check the result, but there is a nice function that simplifies the comparison for you.

The Helm diff plugin easily shows you the differences between two Helm releases. You can use it in various scenarios, for example to

- compare what would change when a release gets updated;
- see the differences between the deployed version and the version of a chart available in the repository; or
- see the effects of changes in values.

> **Good to Know**
>
> For one customer, I outsourced the Helm diff plugin to a separate pipeline step that requires manual approval for certain environments. This helped to identify errors in advance and to carry out a final review of the changes.

To install the plugin, you need to run the following command:

```
helm plugin install https://github.com/databus23/helm-diff
```

For this example, let's again use humanity-backend from Section 8.1.1. It's best to deploy the release again. Now let's activate the horizontal pod autoscaler. Open *values.yaml* and change the value of autoscaling.enabled to true. Now you can use Helm diff to see how the manifests change. To do this, run the following command:

```
helm diff upgrade humanity-backend . --values values.yaml
```

Listing 8.32 shows an excerpt from the output of Helm diff. On your console, the lines that are added should be displayed in green, and those that are removed are marked in red. You can see that the horizontal pod autoscaler object is added and the replicas option is removed from the deployment.

```
...
default, humanity-backend, HorizontalPodAutoscaler (autoscaling) has been added:
-
+ # Source: humanity-backend/templates/hpa.yaml
+ apiVersion: autoscaling/v2
+ kind: HorizontalPodAutoscaler
+ metadata:
+   name: humanity-backend
...
```

Listing 8.32 Helm Diff after Activation of Autoscaling

If you are happy with the changes, you can then release the chart as usual.

> **Note**
>
> The use of Helm diff is perfect for precisely such cases. You change a value and as a result, the templates within several files change as well. Without these tools, such an interaction can easily be overlooked.

Not only can you use Helm diff to check the changes before a release, but you can also compare two revisions of a release with each other. This allows you to check what has changed since the last release during debugging, for example. If you have installed the update with autoscaling and a second revision of the humanity-backend release is available, you can use the following command:

```
helm diff revision humanity-backend 1 2
```

This command compares revision 1 with revision 2.

8.3 Developing Custom Charts

You have now already dealt with Helm and deployed your first Helm charts in Kubernetes. Let's now look at how you can develop a Helm chart for your own application and what you should pay attention to.

As you know, you can also use a Helm chart without a repository and simply check it into the Git repo and roll it out in Kubernetes. This allows you to make use of parameterization, but you will lose a major advantage: the reusability of your chart in your company.

The actual development process for Helm charts is illustrated in Figure 8.9.

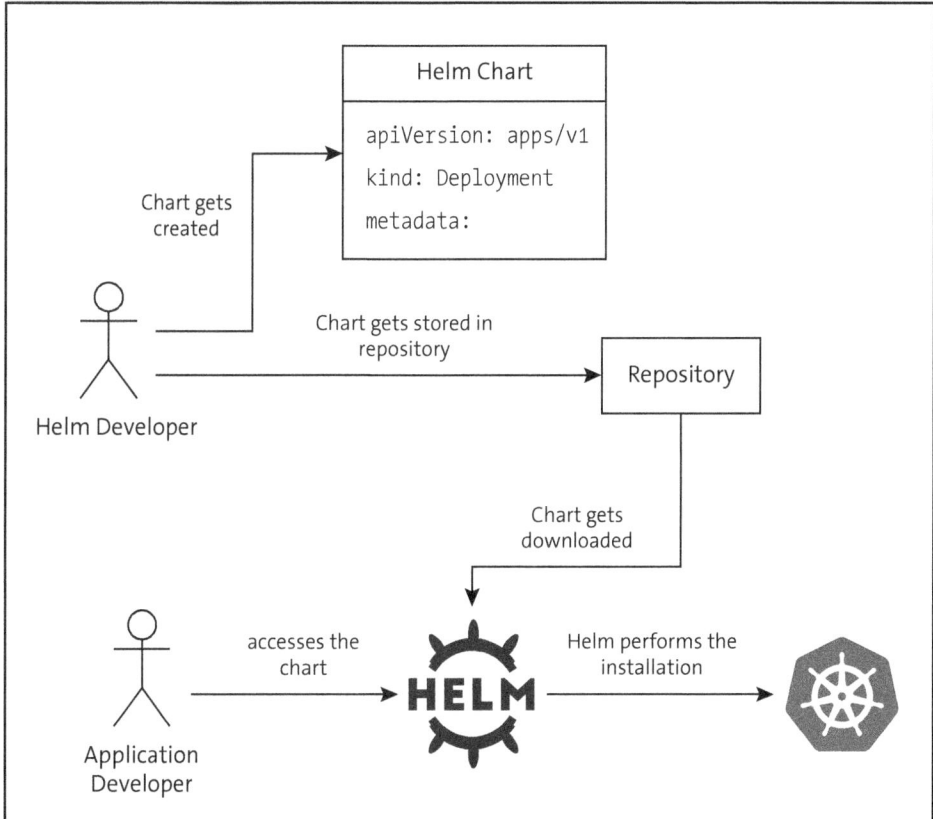

Figure 8.9 Development Process of Helm Charts

Let's assume you're developing a Helm chart for a Postgres database. This database is not only used by yourself; it could also be used in other projects within your company.

Instead of thinking about a perfect Postgres setup for each project, you could think about it once and provide a Helm chart for everyone else. This chart could also be promoted as an inner-sourcing project in which several teams in your company can participate.

If you store the chart in a repository, other developers can easily use it and customize it with values. In this section, we want to take a closer look at the development process, which means you will learn what you need to develop a Helm chart that can be used by others as well. Finally, we will look at the modularity of Helm and how you can write even better charts using dependencies.

8.3.1 The Framework of Your Helm Chart

To develop a chart that other developers also want to use, you need a good framework. Two things are important in this respect:

- A well-maintained *Chart.yaml* file
- Helpful release notes

The *Chart.yaml* file contains all the essential metadata about the chart. If you develop a chart yourself, you should also maintain it. Here are some useful values you should fill in:

- `apiVersion`
  The API version of the chart that Helm uses to interpret the format and functionality of the chart. In the examples here, it is v2 for Helm 3 charts.
- `name`
  The name of the Helm chart. This name must be unique within a Helm repository.
- `version`
  The version of the chart that must follow the semantic versioning schema (SemVer).
- `description`
  A brief description of your chart and its function.
- `keywords`
  A list of keywords associated with the chart. This can help to find the chart in searches.
- `home`
  A URL pointing to the homepage of the project.
- `sources`
  A list of URLs that reference the source code of the software project packaged in the chart.
- `dependencies`
  A list of dependencies on other charts. Here you can specify which other charts are required for this chart to work. We'll take a closer look at the dependencies in Section 8.3.3.
- `maintainers`
  A list of responsible developers so that the user of the chart knows who to contact.

> **Note**
> There are other options for the *Chart.yaml* file. The complete overview can be found at the following address: *https://helm.sh/docs/topics/charts/*.

You have already seen the release notes for other charts such as Jenkins. In the release notes, you should include information such as

- **Notes on user guidance**
  Clear instructions on how the user can interact with the deployed application.
- **Postdeployment steps**
  Inform your users about the necessary steps after installation.
- **Important notes**
  Share information that is relevant to the security, configuration, or usage of the chart.

In the case of Jenkins, this was the output of the default password; in your case, it might be something else.

Release notes are defined in a special file called *NOTES.txt* within the *templates* directory of your Helm chart. The syntax supports the templating engine of Helm, so you can dynamically insert information based on the values of the installation.

Just download the Jenkins chart and open the release notes to get inspired. Listing 8.33 shows an excerpt from the release notes.

```
CHART NAME: {{ .Chart.Name }}
CHART VERSION: {{ .Chart.Version }}
APP VERSION: {{ .Chart.AppVersion }}
** Please be patient while the chart is being deployed **
{{- if .Values.ingress.enabled }}
1st Get the Jenkins URL and associate its hostname to your cluster external IP:
```

Listing 8.33 Excerpt from NOTES.txt File for Bitnami Jenkins

8.3.2 Packaging Charts and Storing Them in the Repository

Once you have developed and tested your own Helm chart, the next step is to prepare it for distribution and use. An essential part of this process is packaging your chart and storing it in a Helm repository. This makes your chart easy to find, versioned, and accessible to other users or teams within your organization. For our example, we use ChartMuseum, a lightweight, easy-to-use tool that serves as a repository for the Helm charts.

A Helm repository has a very simple structure. You can think of it as a collection of packages that contain Helm charts. These repositories allow you to share and publish charts, similar to code in a Git repository. By storing your chart in a repository, you ensure the following:

- **Simple versioning**
  Each version of your chart can be saved in the repository, allowing users to access specific versions.
- **A simple distribution**
  Development teams from your company can easily find and use your charts.
- **A clear dependency management**
  Charts can build on each other. Helm pulls the dependencies from the repositories and takes care of the individual steps.

Installing ChartMuseum

Let's start with the installation of ChartMuseum on Minikube. To install the tool, you must first add the Helm repository of ChartMuseum:

```
helm repo add chartmuseum https://chartmuseum.github.io/charts
helm repo update
```

Then you can install the Helm chart via Lens. To do this, proceed as described in Section 8.1.4. You should be able to find and install the chart as shown in Figure 8.10. It is essential that you set the DISABLE_API parameter to false in the ChartMuseum values. This is the only way you can upload Helm charts later via the API.

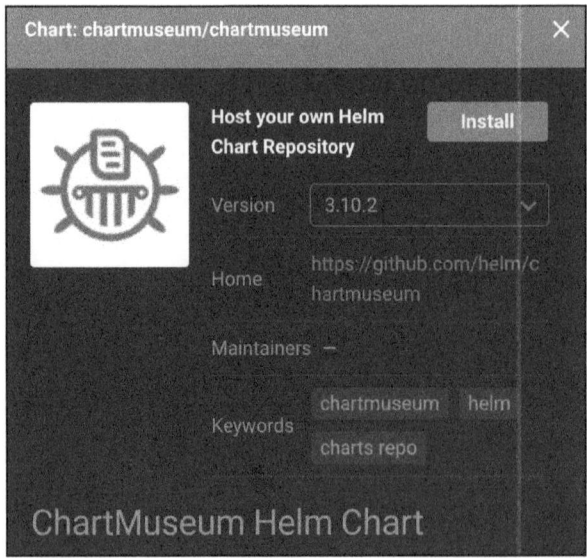

Figure 8.10 Installing ChartMuseum via Lens

8.3 Developing Custom Charts

> **Note**
>
> After adding the tool via `helm repo`, you may need to restart Lens so that it also scans the new Helm repos and displays ChartMuseum.

Once the installation has completed, you'll want to forward a port to the ChartMuseum service via Lens. I have forwarded port 8080 from my computer for this purpose. Now you are ready to file Helm charts in ChartMuseum.

Packaging and Uploading a Helm Chart

In this example, we'll use `humanity-backend` to package the chart immediately and load it into ChartMuseum. For this purpose, you need to go to the chart folder and run the following two commands:

```
helm package .
curl --data-binary "@humanity-backend-0.1.0.tgz" \
    http://localhost:8080/api/charts
```

The first command creates the *humanity-backend-0.1.0.tgz* archive from your Helm chart. The `curl` command uploads the chart to ChartMuseum.

> **Note**
>
> Helm will take all files in the folder with it when `helm package` is executed. Similar to Git, however, Helm takes the *.helmignore* file into account. Here you can add all folders and files Helm is supposed to ignore when creating the package.

Using a Chart from ChartMuseum

Finally, you'll want to check whether the chart really exists in ChartMuseum and whether you can use it. To do this, you first need to add your ChartMuseum as a repository for Helm. Simply use the following command:

```
helm repo add my-chartmuseum http://localhost:8080/
```

Now you should be able to find your chart among the Helm charts in Lens, as shown in Figure 8.11. You have now successfully stored your own Helm chart in a private repository and you can try to install it.

> **Note**
>
> We have created and used ChartMuseum without authentication. However, you should control access to it in your company. Talk to your cluster admins about this as

they may already have another repository solution in use that is connected to an identity management system and therefore provides rights management, because it should be obvious that anyone who can access the Helm charts can carry out far-reaching manipulations. Adequate protection is absolutely essential, aside from test setups and private computers.

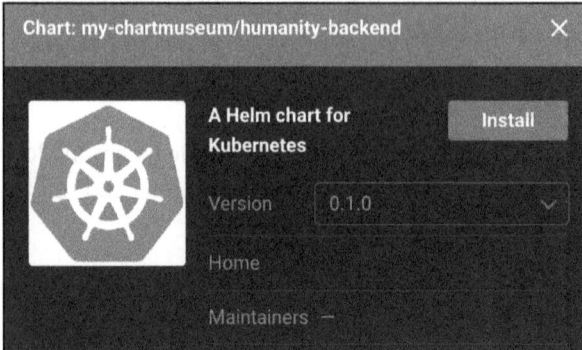

Figure 8.11 "humanity-backend" Chart in ChartMuseum

8.3.3 Managing Dependencies in Helm Charts

Using Helm, you have the option of having your charts build on each other. By defining a dependency on another chart, you can create modular Helm charts. This approach makes it possible to create reusable components that can be used in different projects or under different conditions. It promotes the reuse of code, reduces redundancies, and facilitates the maintenance of complex Kubernetes applications. In this section, we'll look at how Helm handles the management of dependencies between charts.

Figure 8.12 shows what such a dependency can look like. You have stored a versioned chart of your application in a repository. From there, other charts can reference it and use your application in your chart. Use cases for this include the following:

- **Multichart projects**
  These projects have a large chart that you can use to deploy all components of an application. The backend, frontend, and database are completely managed and organized in one chart.
- **Customizing**
  In these cases, you want to customize a chart according to your needs and add additional Kubernetes resources.

We'll take a closer look at both use cases in the course of this section.

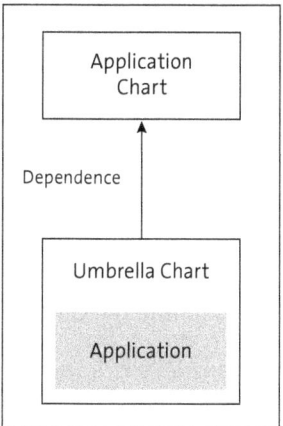

Figure 8.12 Dependencies on Charts

> **Good to Know**
> Charts that use other dependencies are often referred to as *umbrella charts*.

Adding and Updating Dependencies

As an example of a dependency, let's now extend humanity-backend with a Postgres database. We'll use the ready-made chart from Bitnami and enter the dependency in *Chart.yaml*, as shown in Listing 8.34.

```
dependencies:
  - name: postgresql
    version: "14.1.2"
    repository: "https://charts.bitnami.com/bitnami/"
```

Listing 8.34 Dependency in "Chart.yaml" File

Then you need to run the following command:

```
helm dependency update
```

Helm will now carry out several steps. First, it downloads the Helm chart of the Postgres database and stores it in the *charts* path. Helm will then create a file named *Chart.lock*, which looks like the one shown in Listing 8.35.

```
dependencies:
- name: postgresql
  repository: https://charts.bitnami.com/bitnami/
```

```
version: 14.1.2
digest: sha256:9133c60dc762bdd233266d780db912857f04f6033503fc4032ac43be17d18f
generated: "2024-02-21T09:58:03.050996+01:00"
```

Listing 8.35 Chart.lock

This file saves the installed dependency with a hash value that you can use later, such as in the pipeline, to reinstall the dependencies with the same version. Using the `helm dependency build` command, you can install exactly the version that was recorded in the *Chart.lock* file. This means that you do not need to check the subchart into the Git repo, but you can be sure that the wrong version is not installed.

If you want to update the chart now, you have to increase the version of Postgres in *Chart.yaml* file. In this case, I increase the value to version 14.1.3 and run the `helm dependency update` command again. The old Helm chart will be replaced with the new version, and the *Chart.lock* file will be updated.

You should update the dependencies regularly so that you also receive security updates. Unfortunately, every subchart update process is a little time-consuming as you have to look at the changes and check whether anything has changed in the configuration. This can be very time-consuming for large version jumps.

> **Good to Know**
>
> The URL of the repository is the same as the one you find when using `helm repo ls`, because we have entered it there as a local repo. Helm needs the correct URL to the repository here, as the name of the local repo can vary from developer to developer.

Configuring Subcharts

The subcharts are primarily configured using the *values.yaml* file of the main chart. Each subchart has its own *values.yaml* file that defines default values for the subchart. To overwrite or adjust these default values, you need to define values for the subcharts in your own *values.yaml* file in the main chart. To do this, you want to use the name of the subchart as the key.

In this example with the Postgres database, you can add the configuration from Listing 8.36 in *values.yaml* to customize the user name, the password, and the name of the database. The top `postgresql` key references the subchart, while the underlying keys can be found in the *values.yaml* file of the postgres deployment.

```
postgresql:
  global:
    postgresql:
      auth:
        username: "kevinwelter"
```

```
        password: "test1234"
        database: "database"
```

Listing 8.36 Subchart Configuration in "values.yaml"

If you now roll out your chart, you can see in the Postgres pod that the parameters have been adopted, as shown in Figure 8.13.

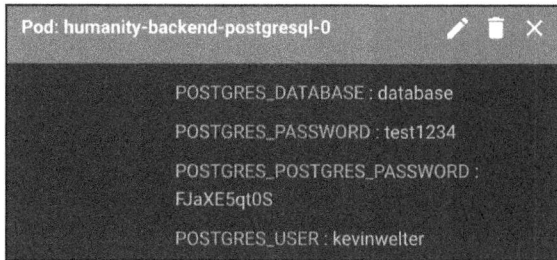

Figure 8.13 Postgres Environment Parameter

> **Note**
> Installing the dependency places the subchart under *charts*. There you can simply look at the *values.yaml* file and determine which configurations you want to adjust.

Multichart Project

For a real-life example of the use of dependencies, let's look at a Helm chart that serves as a wrapper for several subcharts. This allows you to deploy and manage an application consisting of multiple independent components as a whole. The structure could then look like Listing 8.37. In this example, my-application consists of two subcharts—frontend and backend—which are defined as dependencies in the *Chart.yaml* file of my-application. Users can install my-application and Helm takes care of the deployment of the frontend and backend subcharts based on their configurations.

```
my-application/
├── Chart.yaml
├── values.yaml
└── charts/
    ├── frontend/
    │   └── Chart.yaml
    │   ...
    └── backend/
        └── Chart.yaml
        ...
```

Listing 8.37 Sample Structure of Multichart Project

Customizing Charts

In addition to using subcharts and multichart projects, there are situations in which you'll want to customize a Helm chart to your specific needs. This may be the case if you need to integrate additional Kubernetes resources or adapt existing configurations that are not directly covered by the default chart.

For one customer, for example, we enriched the default chart of the Kyverno tool with policies that are to be checked globally for the company. Individual projects then built on this "corporate Kyverno" and implemented additional policy extensions, which were then rolled out in the clusters.

Another example is the extension of an application with ConfigMaps for monitoring purposes. To do this, you can use a default chart as a dependency and create additional ConfigMaps for the Prometheus operator. This means that the default chart can be delivered directly with metrics and alerts.

Listing 8.38 shows an example of what such a setup can look like.

```
my-kyverno/
├── Chart.yaml
├── values.yaml
├── charts/
│   └── kyverno/
│       ├── Chart.yaml
│       └── ... # More files and folders of the Kyverno chart
└── templates/
    ├── policies.yaml # Your individual policies
    └── prometheus-metrics.yaml  # Your ConfigMap for Prometheus
```

Listing 8.38 Folder Structure for Chart Customization

8.4 Conclusion

Now you are familiar with Helm charts and their dependencies and have the tools you need to manage your Kubernetes applications effectively. The ability to define dependencies in your Helm charts helps you to make your applications modular and maintainable. You do not have to reinvent the charts; you can rely on other projects and supplement and adapt your configuration through customizations.

Based on the skills you have now learned, you can approach your Kubernetes applications with a new perspective. When doing so, you should place your focus on modularity, reusability, and easy manageability. Use these options to better structure and manage your applications. Using Helm, you have the control and flexibility to design your deployments exactly as you want them. Happy Helming!

The Author

 **Kevin Welter** is the co-founder and managing director of HumanITy GmbH, which supports large corporations such as Deutsche Bahn, EnBW, and Deutsche Telekom with software development and associated processes. Kevin's goal is to make digitalization simple and attractive for medium-sized businesses.

Index

A

Access control 313
accessMode
   ReadOnlyMany 281
   ReadWriteMany 281
   ReadWriteOnce 281
   ReadWriteOncePod 281
Adapter 99
Admission controller 306
Affinity 124
Alertmanager 343
Ambassador 99, 109
Annotation 97, 118, 136
Antiaffinity 124, 128
API 248
API call flow 51
apiVersion 104
Architecture 45
Atomic 33
Audit 301
Authelia 181
Authentication 300
Authorization 300

B

Backend 32
Batch execution 42
Borg 22
Buildah 101
Busybox 107

C

Capabilities 301
Chaos monkey 29
ChartMuseum 387
cloud-controller-manager 48
Cloud Native Computing Foundation 23
Cluster autoscaler 341
ClusterIP 175
ClusterRole 313
ClusterRoleBinding 313
Common Expression Language (CEL) 254
Compliance 300
Components 45
Config management 42

ConfigMap 97, 152, 249
   environment parameters 158
   Kubernetes API 162
   volume 155
Container 23
containerd 102
Container engines 101
Container Runtime Interface (CRI) 101
Container status
   running 114
   terminated 114
   waiting 114
Container storage interface (CSI) 278, 281
Continuous deployment (CD) 213
Control plane 45
Conway's law 38, 201
CRI-O 102
Cron job 246
Custom resource (CR) 248
Custom resource definition (CRD) .. 235, 248

D

DaemonSet 235, 236
Dashboard 78
Data center 23
Data security 300
default (namespace) 68
Dependents 140
Deployment 97, 138
   creating 142
   rollback 150
   rolling updates 144
DevOps 28
distribution-spec 102
Docker 101
Docker Desktop 54
Dockerfile 104
Downward API 235, 258
Drift detection 224

E

Endpoints controller 49
Ephemeral volume 289
Error handling 22
etcd 47

397

Index

Eviction 134
ExternalName 175, 179

F

Falco 236
Feature flags 378
Field selectors 121
Flask 330
Folder structure 204
Frontend 32
fzf 78

G

Git
    branch 202
    branching 207
    commit 202
    Kubernetes manifests 203
    merge 202
    repository 202
    tag 202
Git flow 208
GitHub flow 209
GitLab flow 210
GitOps 223
Google 22
Governance 300
Graceful shutdown 115
Grace period 116, 117
Grafana 343, 353
gRPC 101

H

Headless 175
Health check 326
Helm
    chart 358
    ChartMuseum 387
    dependencies 378
    environment-specific configuration 378
    feature flags 378
    Lens 85
    release 358
    repository 358
    templating 225
    values.yaml 359
High availability 43
History 21
Homebrew 54

Horizontal pod autoscaler (HPA) 335
Horizontal scaling 42
Hub and spoke 52

I

Idempotence 191
Image 61
image-spec 102
Ingress 97, 171, 180
Init container 110

J

Job 235, 239
JSON object 136

K

kube-apiserver 47
Kubeconfig file 65
kube-controller-manager 48
kubectl
    alias 70
    autocompletion 64
    commands 68
    configuration 65
    installing 61
    Linux 62
    macOS 62
    Minikube 61
    secrets 167
    versions 62
    Windows 63
kubectl api-resources 76
kubectl apply 73
kubectl create 71
kubectl delete 73
kubectl describe 74
kubectl exec 75
kubectl get 70
kubectl logs 74
kubectl port-forward 75
kubectl replace 72
kubectx 77
Kubelet 50
kube-node-lease 68
kubens 77
Kube proxy 50
kube-public 68
Kubernetes
    advantages 42

Index

Kubernetes (Cont.)
   dashboard 61
   disadvantages 44
   features 41
   promise 37
Kubernetes Test Tool (KUTTL) 215
kube-scheduler 47
kube-system 68
Kustomize 225
Kyverno 308

L

Label 97, 118
Least privileged 302
Lens
   adding clusters 88
   cluster metrics 83
   custom resource definitions 88
   custom resources 88
   Helm 85
   licensing terms 81
   pod action bar 87
   port forwarding 83
   resources 86
   terminal 87
libcontainer 102
Lift and shift 321
Limit range 312, 325
Liveness probe 326, 327
LoadBalancer 175
Load balancing 41
Log collector 107
Logging 301

M

Masters 45
maxSurge 147
maxUnavailable 147
Minikube
   container registry 59
   controlling 58
   dashboard 78
   kubectl 61
   launching 58
   Linux 55
   macOS 54
   Windows 57
Mitigations 299
Monitoring 342
Monorepo 211

N

Namespace 67, 77
Naming convention 207
Network address translation (NAT) 173
Network file system (NFS) 279
Never Outgrow 39
Node affinity 124
Node controller 49
NodePort 175, 178
NodeSelector 122
NoOps 39

O

Open Container Initiative (OCI) 101
OpenLens 81
Operators 255
   architecture 256
   PostgreSQL 257
Overcommitment 323
Owners 140

P

Pause container 100
Persistence 32
Persistent volume 273
   csi 278
   fc 278
   hostPath 278
   iscsi 278
   local 278
   nfs 278
   storage types 278
Persistent volume claim 273
Pets and cattle 28
Pi cluster 89
   hardware 90
   installation 92
   Kubeconfig file 93
   SSH 92
   Wi-Fi 91
Pipeline
   architecture 218
   Kubernetes 213
   linting 214
Planet Scale 37
Pod 96, 98, 104
   communication 173
Pod affinity 128
Podman 101

Pod management policy
- *OrderedReady* 269
- *Parallel* 269

Pod phases
- *Failed* 113
- *Pending* 113
- *Running* 113
- *Succeeded* 113
- *Unknown* 113

Pod priority 235, 261
Pod resources 322
Pod security admission 304
Pod security policy 304
Pod shell 88
Policies 308
Policy management 300
Port forwarding 75, 83
Preemption 235, 261
PriorityClass 261
Privileges 301
Projected volume 291, 295
Prometheus 343

Q

Queue worker 239
Quorum 47

R

RabbitMQ 242
Raft 47
Raspberry Pi 89
Readiness probe 326, 327
Reconciliation loop 190
ReplicaSet 97, 138, 140
Replication controller 49
Resource management 22, 301
Resource quota 311, 325
Restart policy 114
Retention policy 272
Role 313, 315, 317
RoleBinding 313, 317
Rollback 41, 150
Rolling updates 144
Rollout 41
runC 102
Run K8s Anywhere 40
runtime-spec 102

S

Scalability 43
Secret management 42
Secrets 97, 152
- *container registry* 168
- *environment parameters* 165
- *kubectl* 167
- *volume* 166

Security context 302
sed 225
Selectors 119
Self-healing 42, 326
Separation of concerns 31
Service 97
- *communication* 174
- *end point* 174
- *load balancing* 174
- *service discovery* 174

Service account 315
Service account controller 49
Service discovery 41
Sidecar 99, 106, 109
Single point of failure 26
Single source of truth 223
Software-defined storage 265
SonarQube 299
Startup probe 326, 327
Stateful 29
StatefulSet 266
- *OnDelete* 270
- *RollingUpdate* 270

Stateless 29
Storage 278
Storage orchestration 42
Subjects 315
systemd 238

T

Taint 133
- *NoExecute* 134
- *NoSchedule* 134
- *PreferNoSchedule* 134

Templating
- *Helm* 225
- *Kustomize* 225

Tolerations 133

U

UID 301

V

Versioning 263
Version management 200
Vertical pod autoscaler (VPA) 339
Virtual machine 24
volumeBindingMode 284
Volume snapshot 291
Vulnerability management 300

W

Winget 57
Workers 45, 49

Y

YAML
   alias 196
   anchor 196
   comments 199
   data types 194
   indentations 194
   linting 199
   single-line 197
   syntax 192
   weaknesses 197
yq 225

- Get hands-on practice with Docker, from setup to orchestration
- Work with Dockerfiles, Docker Compose, GitLab, and Docker Hub
- Learn about project migration, security, Kubernetes, and more

Bernd Öggl, Michael Kofler

Docker

Practical Guide for Developers and DevOps Teams

Learn the ins and outs of containerization in Docker with this practical guide! Begin by installing and setting up the platform. Then master the basics: get to know important terminology, understand how to run containers, and set up port redirecting and communication. You'll learn to create custom images, work with commands, and use key containerization tools. Gain essential skills by following exercises that cover common tasks from packaging new applications and modernizing existing applications to handling security and operations.

491 pages, pub. 01/2023
E-Book: $44.99 | **Print:** $49.95 | **Bundle:** $59.99

www.rheinwerk-computing.com/5650

- Get hands-on practice with Git
- Understand branches, commands, commits, workflows, and more
- Learn to use GitHub, GitLab, and alternative Git platforms

Bernd Öggl, Michael Kofler

Git

Project Management for Developers and DevOps Teams

Get started with Git—today! Walk through installation and explore the variety of development environments available. Understand the concepts that underpin Git's workflows, from branching to commits, and see how to use major platforms, like GitHub. Learn the ins and outs of working with Git for day-to-day development. Get your versioning under control!

407 pages, pub. 10/2022
E-Book: $44.99 | **Print:** $49.95 | **Bundle:** $59.99

www.rheinwerk-computing.com/5555

- Your complete, cross-distribution, professional guide to Linux, for beginners and advanced users
- Get detailed instructions for installation, configuration, and administration, on both desktop and server
- Set up security, virtualization, and more

Michael Kofler

Linux

The Comprehensive Guide

Beginner or expert, professional or hobbyist, this is the Linux guide you need! Install Linux and walk through the basics: working in the terminal, handling files and directories, using Bash, and more. Then get into the nitty-gritty details of configuring your system and server, from compiling kernel modules to using tools like Apache, Postfix, and Samba. With information on backups, firewalls, virtualization, and more, you'll learn everything there is to know about Linux!

1178 pages, pub. 05/2024
E-Book: $54.99 | **Print:** $59.95 | **Bundle:** $69.99

www.rheinwerk-computing.com/5779

- Learn to work with scripting languages such as Bash, PowerShell, and Python
- Get to know your scripting toolbox: cmdlets, regular expressions, filters, pipes, and REST APIs
- Automate key tasks, including backups, database updates, image processing, and web scraping

Michael Kofler

Scripting

Automation with Bash, PowerShell, and Python

Developers and admins, it's time to simplify your workday. With this practical guide, use scripting to solve tedious IT problems with less effort and fewer lines of code! Learn about popular scripting languages: Bash, PowerShell, and Python. Master important techniques such as working with Linux, cmdlets, regular expressions, JSON, SSH, Git, and more. Use scripts to automate different scenarios, from backups and image processing to virtual machine management. Discover what's possible with only 10 lines of code!

470 pages, pub. 02/2024
E-Book: $44.99 | **Print:** $49.95 | **Bundle:** $59.99

www.rheinwerk-computing.com/5851

- The complete Python 3 handbook
- Learn basic Python principles and work with functions, methods, data types, and more
- Walk through GUIs, network programming, debugging, optimization, and other advanced topics

Johannes Ernesti, Peter Kaiser

Python 3

The Comprehensive Guide

Ready to master Python? Learn to write effective code, whether you're a beginner or a professional programmer. Review core Python concepts, including functions, modularization, and object orientation and walk through the available data types. Then dive into more advanced topics, such as using Django and working with GUIs. With plenty of code examples throughout, this hands-on reference guide has everything you need to become proficient in Python!

1036 pages, pub. 09/2022
E-Book: $54.99 | **Print:** $59.95 | **Bundle:** $69.99

www.rheinwerk-computing.com/5566

- Your all-in-one guide to JavaScript
- Work with objects, reference types, events, forms, and web APIs
- Build server-side applications, mobile applications, desktop applications, and more

Philip Ackermann

JavaScript

The Comprehensive Guide

Begin your JavaScript journey with this comprehensive, hands-on guide. You'll learn everything there is to know about professional JavaScript programming, from core language concepts to essential client-side tasks. Build dynamic web applications with step-by-step instructions and expand your knowledge by exploring server-side development and mobile development. Work with advanced language features, write clean and efficient code, and much more!

982 pages, pub. 08/2022
E-Book: $54.99 | **Print:** $59.95 | **Bundle:** $69.99

www.rheinwerk-computing.com/5554